Insights on Resiliency and Urban Development

Ungku Norani Sonet
Taylor's University, Malaysia

Tamil Salvi Mari
Taylor's University, Malaysia

A volume in the Advances in Civil and Industrial Engineering (ACIE) Book Series

IGI Global
PUBLISHER of TIMELY KNOWLEDGE

Published in the United States of America by
IGI Global
Engineering Science Reference (an imprint of IGI Global)
701 E. Chocolate Avenue
Hershey PA, USA 17033
Tel: 717-533-8845
Fax: 717-533-8661
E-mail: cust@igi-global.com
Web site: http://www.igi-global.com

Library of Congress Cataloging-in-Publication Data

Names: Sonet, Ungku Norani, 1987- editor. | Mari, M. Tamilsalvi, 1969- editor.
Title: Inisghts on resiliency and urban development / edited by Ungku Sonet, M. Tamilsalvi Mari.
Description: Hershey,PA : Engineering Science Reference, [2024] | Includes bibliographical references and index. | Summary: "By offering a comprehensive examination of the intricacies that lie in building architecture and urban planning and design that not only withstand climate change, economic fluxes, and increased population but also flourish amidst these changes, this book fosters understanding of sustainable urban development"-- Provided by publisher.
Identifiers: LCCN 2023055018 (print) | LCCN 2023055019 (ebook) | ISBN 9798369316504 (hardcover) | ISBN 9798369316511 (ebook)
Subjects: LCSH: Sustainable urban development. | City planning--Environmental aspects. | Sustainable design.
Classification: LCC HT241 .H348 2024 (print) | LCC HT241 (ebook) | DDC 307.1/416--dc23/eng/20231229
LC record available at https://lccn.loc.gov/2023055018
LC ebook record available at https://lccn.loc.gov/2023055019

British Cataloguing in Publication Data
A Cataloguing in Publication record for this book is available from the British Library.

Advances in Civil and Industrial Engineering (ACIE) Book Series

Ioan Constantin Dima
University Valahia of Târgovişte, Romania

ISSN:2326-6139
EISSN:2326-6155

Mission

Private and public sector infrastructures begin to age, or require change in the face of developing technologies, the fields of civil and industrial engineering have become increasingly important as a method to mitigate and manage these changes. As governments and the public at large begin to grapple with climate change and growing populations, civil engineering has become more interdisciplinary and the need for publications that discuss the rapid changes and advancements in the field have become more in-demand. Additionally, private corporations and companies are facing similar changes and challenges, with the pressure for new and innovative methods being placed on those involved in industrial engineering.

The **Advances in Civil and Industrial Engineering (ACIE) Book Series** aims to present research and methodology that will provide solutions and discussions to meet such needs. The latest methodologies, applications, tools, and analysis will be published through the books included in **ACIE** in order to keep the available research in civil and industrial engineering as current and timely as possible.

Coverage

- Construction Engineering
- Earthquake engineering
- Hydraulic Engineering
- Materials Management
- Operations Research

IGI Global is currently accepting manuscripts for publication within this series. To submit a proposal for a volume in this series, please contact our Acquisition Editors at Acquisitions@igi-global.com or visit: http://www.igi-global.com/publish/.

The Advances in Civil and Industrial Engineering (ACIE) Book Series (ISSN 2326-6139) is published by IGI Global, 701 E. Chocolate Avenue, Hershey, PA 17033-1240, USA, www.igi-global.com. This series is composed of titles available for purchase individually; each title is edited to be contextually exclusive from any other title within the series. For pricing and ordering information please visit http://www.igi-global.com/book-series/advances-civil-industrial-engineering/73673. Postmaster: Send all address changes to above address.

Titles in this Series

For a list of additional titles in this series, please visit: www.igi-global.com/book-series

Emerging Engineering Technologies and Industrial Applications
Younes El Kacimi (Ibn Tofail University, Morocco) and Khaoula Alaoui (Ibn Tofail University, Morocco)
Engineering Science Reference • copyright 2024 • 426pp • H/C (ISBN: 9798369313350) • US $300.00 (our price)

Emerging Materials, Technologies, and Solutions for Energy Harvesting
Shilpa Mehta (Auckland University of Technology, New Zealand) Arij Naser Abougreen (University of Tripoli, Libya) and Shashi Kant Gupta (Eudoxia Research University, USA)
Engineering Science Reference • copyright 2024 • 368pp • H/C (ISBN: 9798369320037) • US $315.00 (our price)

Innovations in Machine Learning and IoT for Water Management
Abhishek Kumar (Chandigarh University, India) Arun Lal Srivastav (Chitkara University, India) Ashutosh Kumar Dubey (University of Castilla-La Mancha, Spain) Vishal Dutt (AVN Innovations Pvt. Ltd., India) and Narayan Vyas (AVN Innovations Pvt. Ltd., India)
Engineering Science Reference • copyright 2024 • 312pp • H/C (ISBN: 9798369311943) • US $300.00 (our price)

Intelligent Engineering Applications and Applied Sciences for Sustainability
Brojo Kishore Mishra (NIST Institute of Science and Technology (Autonomous), India & Saveetha College of Liberal Arts and Sciences, SIMATS, Chennai, India)
Engineering Science Reference • copyright 2023 • 542pp • H/C (ISBN: 9798369300442) • US $270.00 (our price)

Global Science's Cooperation Opportunities, Challenges, and Good Practices
Mohamed Moussaoui (National School of Applied Sciences of Tangier (ENSAT), Abdelmalek Essaadi University, Morocco)
Engineering Science Reference • copyright 2023 • 357pp • H/C (ISBN: 9781668478745) • US $245.00 (our price)

Handbook of Research on Inclusive and Innovative Architecture and the Built Environment
Ng Foong Peng (Taylor's University, Malaysia) and Ungku Norani Sonet (Taylor's University, Malaysia)
Engineering Science Reference • copyright 2023 • 526pp • H/C (ISBN: 9781668482537) • US $315.00 (our price)

IGI Global
PUBLISHER of TIMELY KNOWLEDGE

701 East Chocolate Avenue, Hershey, PA 17033, USA
Tel: 717-533-8845 x100 • Fax: 717-533-8661
E-Mail: cust@igi-global.com • www.igi-global.com

Table of Contents

Detailed Table of Contents

The demand for well-planned, sustainable, and resilient city infrastructure cannot be timelier, for the urban share of Globe's population continues to rise. Cities are complex systems with intricate dynamics, necessitating a thorough understanding of how different elements such as sewerage systems, culture centres, transport systems, energy consumption, and sewage management affect the success of an urban centre. Simulation tools offer a drawing board kind of set-up to urban planners where they experiment and compare various scenarios to select the best. The design of sustainable urban infrastructure and its resilience have also been discussed here. Urban planners may use simulation tools to model and evaluate various possible vulnerabilities and corresponding response options. Further, the chapter discusses technologies such as spatial databases, AI and machine learning models, and building information modeling in urban simulation environments, and also an international case study of urban growth simulation in Portland, Oregon, USA.

Heritage architecture education is important to develop contextual understanding and cultural consciousness among architecture students. Conventional teaching approaches are inadequate to cultivate students' cultural knowledge of specific places and community contexts. This chapter addresses this gap by evaluating how problem-based learning (PBL) as a teaching approach in a measured drawing module contributes to cultural consciousness among architecture students. It adopts quantitative survey research design to evaluate students' cultural consciousness based on knowledge of place identity, understanding of community social contexts, and recognition of community cultural contexts. Two student cohorts completing the module participated as respondents, providing insights into PBL's impact. The findings substantiate the adoption of PBL as an effective pedagogy to enhance students' cultural consciousness, by developing contextual and cultural sensitivity in design cognition. It is also useful in guiding the design of teaching approaches for culturally related subjects in other disciplines.

Historic buildings (HB) are invaluable embodiments of culture and identity. Retrofitting HB has become a strategic imperative in Malaysia, given their substantial presence of building stock, moreover these buildings grappled with contemporary challenges. Energy retrofit allows adaptation and reuse while prioritizing environmental sustainability. However, such notions are still in their infancy stage, hindered by the gap between understanding heritage preservation techniques and energy efficiency consideration. The lack of inclusive guidance is an obstacle, restricting widespread application of retrofit intervention in HB conservation. This chapter aims to identify current limitations of Malaysia's Conservation Guidelines (HBCG) through a comparative analysis with similar international guidelines. Findings reveal lack of interpretations for contemporary green issues, unlike the latter which allows broader interpretation. The paper suggests additional guidance for practitioners in navigating environmental challenges, emphasizing intangible considerations such as heritage values and identity.

This abstract presents a comprehensive literature review on the critical factors driving China's successful adoption of industrialized building systems (IBS). Aiming at improving project efficiency and quality, this paper drawing insights from global experiences by utilizing the sysmatic literature review (SLR) method based on 78 IBS-related papers from Scopus database. The study emphasizes the significance of cognitive incentives, regulations, education, technology, construction management, the performance of IBS, culture, and tradition in the IBS implementation process. Ultimately, this chapter provides guidance for effective IBS implementation, distilled from regulatory, workforce, technological, collaborative, and cultural dimensions.

Caribbean islands are some of the most vulnerable countries in the world to experience extreme flooding resulting from tropical cyclones, storm surges, and excessive rainfall. Persistent changes in the global climate have made countries like Trinidad and Tobago hot spots of climate change. As a result, the occurrence of extreme flooding events is likely to persist and is made worse by the barriers to adaptation and mitigation, such as having limited knowledge of climate change and how to plan for its impact at the individual level, as well as limited financial resources and technical know-how. Using a secondary research methodology, this chapter seeks to investigate the role that artificial intelligence (AI) plays in urban planning and design in responding to extreme flooding events in Trinidad and Tobago. This chapter discusses not only the historical occurrence of flooding events, and its causes, but also how the possible use and integration of AI in the planning and design of urban spaces in Trinidad and Tobago can assist in reducing the ill effects of such events.

Museum cultural and creative products (MCCP) are seen as carriers of museum cultural resources and are important for the sustainable development of cultural heritage, Nevertheless, the cultural aspect is often overlooked within the realm of sustainable product design. Hence, a primary objective of this research endeavour was to address this gap by developing a comprehensive framework that clarifies the interconnectedness between CS and MCCP. Furthermore, an additional objective of this study was to put forth a conceptual framework and specific indicators for the assessment of the cultural sustainability of MCCP. The indicator sets were designed through the literature and analytic hierarchy process. This process has yielded a tool that will undergo additional testing and practical implementation. The findings of this study will enhance comprehension of cultural sustainability within the context of MCCP and provide guidance for future MCCP design aligned with the principles of sustainable development.

Despite notable technological progress, the construction industry has slowly embraced information and communication technologies (ICT). Cloud computing, a tool potentially enhancing information management productivity, operates through remote servers linked via the Internet. This allows for data storage and application accessibility from any device without requiring prior installation. This research addresses this technological gap by exploring how quantity surveying firms perceive the integration of cloud computing. The study aims to evaluate these firms' inclination to adopt cloud computing and identify the obstacles impeding its implementation. A quantitative approach and questionnaires were adopted to uncover quantity surveying firm's readiness for technological advancement. This study provides valuable insights into how these firms can reshape their operational and communicative practices in the context of IR 4.0, preparing them for the essential technological shifts that will shape future industry standards.

COVID-19, a new respiratory disease identified in December 2019, is one of the most recent risks to human health around the world. The purpose of this research is to examine the effects of COVID-19 on the ecological footprints of five cities across India: Delhi, Bhopal, Bareilly, Betul, and Razole, using the

Key findings revealed diverse needs and provisions influenced by social factors and living infrastructure. Ultimately, accommodating the changing behaviors and characteristics of the elderly was crucial for developing infrastructure that promoted positive and healthy aging.

Preface

The concept of resiliency in urban development has never been more crucial than it is today. With the multifaceted challenges posed by climate change, economic fluctuations, and rapid population growth, the need for cities to not only endure but also prosper has become paramount. *Insights on Resiliency and Urban Development* brings together an eclectic mix of perspectives and expertise to address these pressing concerns. As editors, we have had the privilege of curating a collection of research and practical insights that we believe will serve as an invaluable resource for a diverse audience of professionals, researchers, and policymakers.

This book is a testament to the collaborative efforts of government leaders, architects, urban planners, quantity surveyors, landscape architects, sustainability advocates, and engaged citizens. Together, they have contributed to a comprehensive guide that outlines innovative methods for cultivating resilient urban 'sanctuaries.' Our contributors delve into the intricacies of architecture, urban design, urban planning, sustainability, and community development, offering a vision of modern cities as spaces that foster growth, resilience, and community cohesion.

Divided into three thematic sections, the book provides a holistic view of sustainable urban development:

Sustainable Urban Design and Planning: This section explores the principles of liveable cities, sustainable urban design, sustainable architecture, urban landscapes, tropical urban design, and urban housing and communities. It highlights how thoughtful design can enhance urban living and create spaces that are both functional and inspiring.

Conservation, Education, and Awareness: Here, we examine the roles of education, leadership, urban heritage, cultural preservation, future digital technologies, and resilient design for climate action and disaster preparedness. This section underscores the importance of knowledge dissemination and cultural awareness in fostering resilient urban environments.

Sustainable Construction and Infrastructure: Focusing on the technical aspects of urban resilience, this section covers construction and project management, green building practices, asset and facilities management, and the integration of Building Information Modeling (BIM) and digital modeling. These discussions provide practical strategies for creating robust and sustainable infrastructure.

Through vibrant illustrations, compelling case studies, and practical strategies, this book offers a rich resource for anyone dedicated to shaping sustainable, inclusive, and forward-thinking cities. It provides the theoretical groundwork supported by real-world examples, illustrating how cities can transform into robust, sustainable sanctuaries. Additionally, the book emphasizes the importance of community involvement in urban development, facilitating more inclusive, resilient, and sustainable communities.

Our target audience spans a wide range of disciplines, including academics, architecture, urban design, urban planning, landscape architecture, quantity surveying, construction management, business management, psychology, computer science, engineering, and environmental science. By offering insights into the management of expertise, knowledge, and information across various types of work communities and environments, we aim to support executives and professionals concerned with organizational development.

We hope that this book will foster a deeper understanding of sustainable urban development and encourage cross-disciplinary collaboration. By broadening the scope and impact of this critical research area, we aspire to contribute to the creation of resilient urban communities that can thrive amidst the inevitable changes of our world.

Chapter 1: Urban Planning and Design Simulation to Develop Sustainable and Resilient Cities

Vikram Singh, Ch. Devi Lal University, Sirsa, India
Sanyogita Singh, Panjab University, India

In an era where urban populations are rapidly increasing, the demand for meticulously planned, sustainable, and resilient city infrastructure has never been more urgent. This chapter delves into the complex dynamics of urban systems, encompassing elements such as sewerage systems, cultural centers, transport networks, energy consumption, and sewage management. It emphasizes the use of simulation tools that allow urban planners to experiment with and compare various scenarios to identify optimal solutions. The chapter further discusses the integration of technologies like spatial databases, artificial intelligence (AI), machine learning models, and Building Information Modeling (BIM) in urban simulation environments. A case study on urban growth simulation in Portland, Oregon, USA, is included to illustrate practical applications of these concepts.

Chapter 2: Cultural Consciousness in Heritage Architecture Education With a Problem-Based Learning Approach

Jing Hao Koh, School of Architecture, Building and Design, Malaysia
Khairool Aizat Ahmad Jamal, School of Architecture, Building and Design, Malaysia

This chapter addresses the critical need for cultural consciousness in heritage architecture education. Traditional teaching methods often fall short in imparting a deep understanding of cultural and contextual nuances to architecture students. The chapter evaluates the effectiveness of Problem-Based Learning (PBL) in a measured drawing module to enhance students' cultural consciousness. Using a quantitative survey research design, it examines the impact of PBL on students' knowledge of place identity and their understanding of community and cultural contexts. The findings advocate for PBL as a superior pedagogical approach to cultivate cultural sensitivity in design cognition, offering valuable insights for teaching culturally related subjects across various disciplines.

Chapter 3: A Review and Comparative Study of Retrofit for the Energy Efficiency of Malaysian Heritage Buildings

Khairool Aizat Ahmad Jamal, Taylor's University, Malaysia
Siti Norlizaiha Harun, Universiti Teknologi MARA, Malaysia
Shahrul Yani Said, Universiti Teknologi MARA, Malaysia

This chapter explores the intersection of heritage conservation and energy efficiency in Malaysia's historic buildings. It highlights the strategic importance of retrofitting heritage buildings to address contemporary environmental challenges while preserving cultural identity. The chapter identifies gaps in Malaysia's Conservation Guidelines by comparing them with international standards, revealing a lack of provisions for modern green issues. It suggests enhanced guidelines to help practitioners balance heritage values with environmental sustainability, providing a roadmap for broader and more effective retrofit interventions in heritage building conservation.

Chapter 4: A Review of Critical Factors of Successfully Implementing IBS in China Based on Scopus Database

Xiaonuo Kang, Taylor's University, Malaysia
Kenn Jhun Kam, Taylor's University, Malaysia

Focusing on the adoption of Industrialized Building Systems (IBS) in China, this chapter conducts a comprehensive literature review using the Scopus database. It identifies critical factors that have driven successful IBS implementation, such as cognitive incentives, regulatory frameworks, education, technology, construction management, performance metrics, and cultural traditions. By synthesizing insights from 78 IBS-related papers, the chapter offers a detailed guide for effective IBS adoption, addressing regulatory, workforce, technological, collaborative, and cultural dimensions to enhance project efficiency and quality.

Chapter 5: The Role of AI in Urban Design and Planning in Responding to Extreme Flooding events in Trinidad and Tobago

Roshnie Doon, Global Labor Organization (GLO), Essen, Germany

Trinidad and Tobago are particularly vulnerable to extreme flooding due to tropical cyclones, storm surges, and excessive rainfall, exacerbated by climate change. This chapter investigates the role of Artificial Intelligence (AI) in urban planning and design to mitigate the impacts of these flooding events. Using a secondary research methodology, it examines historical flooding occurrences and explores how AI can be integrated into urban planning to improve resilience. The chapter provides insights into how

AI technologies can aid in planning and designing urban spaces to reduce the adverse effects of extreme flooding in Trinidad and Tobago.

Chapter 6: Achieving Cultural Sustainability through Museum Cultural and Creative Products (MCCP)

Hui Liu, Taylor's University, China
Siti Norzaini Zainal Abidin, The Design School at Taylor's, Malaysia
Filzani Illia Ibrahim, The Design School at Taylor's, Malaysia
Cheng An, The Design School at Taylor's, Malaysia
Wang Lu, Wenzhou University, China

This chapter explores the role of Museum Cultural and Creative Products (MCCP) in achieving cultural sustainability. It addresses the often-overlooked cultural aspects within sustainable product design and proposes a comprehensive framework to connect cultural sustainability with MCCP. The research develops specific indicators for assessing MCCP's cultural sustainability, using literature reviews and the analytic hierarchy process. The resulting tool aims to enhance understanding and guide future MCCP design in alignment with sustainable development principles.

Chapter 7: Cloud Computing Adoption at Quantity Surveying Firms: Are We Ready?

Norhazren Izatie Mohd, Universiti Teknologi Malaysia, Malaysia
Nur Nilam Sari Md Yusoff, Universiti Teknologi Malaysia, Malaysia
Hamizah Liyana Tajul Ariffin, Universiti Teknologi Malaysia, Malaysia
Ahmad Faiz Azizi Ahmad Fauzi, Universiti Teknologi Malaysia, Malaysia

Despite advancements in technology, the construction industry has been slow to adopt Information and Communication Technologies (ICT), including cloud computing. This chapter examines the readiness of Quantity Surveying firms to integrate cloud computing into their operations. Through a quantitative approach using questionnaires, the study assesses perceptions and identifies barriers to adoption. The findings provide insights into how these firms can leverage cloud computing to improve information management, operational efficiency, and readiness for future technological shifts in the context of Industry 4.0.

Chapter 8: Comparative Analysis of Ecological Footprint in Five Cities of India: A Pandemic Analysis

Neha Pranav Kolhe, Maulana Azad National Institute of Technology, India
Surabhi Mehrotra, Maulana Azad National Institute of Technology, Bhopal, India
Aditi Parihar, Maulana Azad National Institute of Technology, Bhopal, India

The COVID-19 pandemic has had significant impacts on urban ecological footprints. This chapter conducts a comparative analysis of the ecological footprints of five Indian cities: Delhi, Bhopal, Bareilly, Betul, and Razole. Using the Global Footprint Network's calculator tool, it examines six major factors—food, shelter, transportation, commodities, and services—through primary questionnaires. The study employs statistical tests like T-tests and ANOVA to analyze differences in ecological footprints, revealing significant variations among cities, particularly in larger urban centers like Delhi. The findings provide insights into how the pandemic has reshaped ecological footprints and suggest strategies for sustainable urban development.

Chapter 9: Consequences of COVID-19 Pandemic on Declining Cities, Urban Vacant Land Uses, Green Infrastructure, and Their Impact on Climate Change Hazards

José Vargas-Hernandez, Budapest Centre for Long-term Sustainability, Mexico
Omar Vargas-González, Tecnológico Nacional de México, Campus Cd. Guzmán, Mexico
Omar Guirette-Barbosa, Universidad Politécnica de Zacatecas, Mexico
Selene Castañeda-Burciaga, Universidad Politécnica de Zacatecas, Mexico

This chapter examines the implications of the COVID-19 pandemic on declining cities, urban vacant lands, and green infrastructure. It explores how urban vacant land, often viewed negatively, can be repurposed to develop green infrastructure that mitigates climate change hazards and promotes economic and socio-ecological development. Using an analytical-descriptive method based on theoretical and empirical literature reviews, the chapter concludes that strategic policies and incentives can transform vacant lands into valuable green spaces, contributing to urban resilience and sustainability.

Chapter 10: Integration of Building Information Modelling (BIM) into the Construction Project Scheduling

Lam Tatt Soon, Taylor's University, Malaysia
Jasmine Yi Xuan Liew, Taylor's University, Malaysia
Tze Shawn Lim, Taylor's University, Malaysia

Kenn Jhun Kam, Taylor's University, Malaysia
Boon Tik Leong, Taylor's University, Malaysia

Building Information Modelling (BIM) offers significant potential to enhance construction management by integrating time or schedule-related information into project workflows. This chapter investigates the application of 4D BIM in construction management, focusing on its ability to streamline project scheduling and improve efficiency. Using a case study of a residential building in Malaysia, the research develops a 4D BIM scheduling process and demonstrates its practical implementation. The chapter serves as a guide for developers and contractors, illustrating how 4D BIM can optimize construction sequences and project outcomes.

Chapter 11: Educational Dynamics in Urban Study: Unraveling Student Preferences and Expectations in BQS Study

Kenn Jhun Kam, Taylor's University, Malaysia
Wei Vern Tan, Taylor's University, Malaysia
Tze Shwan Lim, Taylor's University, Malaysia
Lam Tatt Soon, Taylor's University, Malaysia
Chin Ai Ling Shirley, Taylor's University, Malaysia

Understanding the factors that influence students' decisions when choosing a university for tertiary education is crucial for maintaining competitive program quality. This chapter explores the determinants influencing students' enrollment in the Bachelor of Quantity Surveying (BQS) degree program at Taylor's University. Using a questionnaire survey, the study examines factors such as employability, academic quality, campus environment, university reputation, geographic location, and influences from family and peers. The findings provide insights into students' preferences and experiences, offering practical recommendations for improving enrollment rates and enhancing program appeal.

Chapter 12: Research Methodology for Cost Modelling Development on Smart Living Features Implementation in Malaysian Housing

Koh Chieng, Universiti Teknologi Malaysia, Malaysia
Fara Mustapa, Universiti Teknologi Malaysia, Malaysia

Smart Living housing in Malaysia, characterized by home automation systems, offers enhanced comfort, security, and energy efficiency. This chapter proposes a robust research methodology for developing a cost model for Smart Living features in Malaysian housing. By addressing the complexities of cost modeling, including data collection, analysis, and interpretation, the chapter provides a comprehensive framework to guide future research in this area. The methodology is designed to help stakeholders, including developers, policymakers, and researchers, in making informed decisions regarding the adoption and implementation of smart living technologies in housing projects.

Chapter 13: Exploring Smart Home Technologies for Energy Efficiency and Sustainable Living in Malaysia

Ahmad Zulhakim Abdullah, Universiti Teknologi Malaysia, Malaysia
Farah Izwani Ismail, Universiti Teknologi Malaysia, Malaysia

Smart home technologies have the potential to revolutionize residential energy efficiency and promote sustainable living practices. This chapter delves into the latest advancements in smart home systems and their applications in Malaysia. It examines various smart technologies such as automated lighting, smart thermostats, energy monitoring systems, and their impact on reducing energy consumption. Through case studies and data analysis, the chapter highlights how these technologies can be integrated into Malaysian homes to foster sustainable living. It also discusses challenges and recommendations for widespread adoption, offering insights for policymakers, developers, and homeowners.

Chapter 14: Examining the Impact of Green Roofs on Urban Heat Island Effect in Dense Urban Areas

Muhammad Imran, University of Malaya, Malaysia
Nur Ain Ismail, University of Malaya, Malaysia

Urban heat islands (UHI) are a growing concern in densely populated cities, contributing to higher temperatures and increased energy demand. This chapter investigates the effectiveness of green roofs in mitigating the UHI effect in dense urban areas. By analyzing temperature data from various case studies and simulations, it demonstrates how green roofs can reduce surface and ambient temperatures. The chapter also explores the additional benefits of green roofs, such as improved air quality, stormwater management, and enhanced biodiversity. Recommendations for implementing green roof initiatives in urban planning are provided, with a focus on policy implications and community engagement.

Chapter 15: Enhancing Disaster Resilience in Urban Communities through Community-Based Approaches

Anita Sharma, Indian Institute of Technology, India
Rajesh Kumar, Indian Institute of Technology, India

Disaster resilience in urban communities requires proactive strategies that involve local populations. This chapter focuses on community-based approaches to enhance resilience against natural disasters such as floods, earthquakes, and cyclones. It presents various models and frameworks that engage community members in disaster preparedness, response, and recovery efforts. Through case studies from different regions, the chapter illustrates successful community-driven initiatives and their outcomes. It also discusses the role of local governments, NGOs, and international organizations in supporting these efforts. Practical recommendations for fostering community resilience through education, training, and resource allocation are provided.

In closing, the urgency of developing resilient urban environments has never been more apparent. Our world faces unprecedented challenges from climate change, rapid urbanization, economic uncertainties, and social transformations. *Insights on Resiliency and Urban Development* addresses these challenges head-on by compiling a diverse array of research, practical strategies, and innovative solutions aimed at fostering sustainable, resilient, and vibrant cities.

The collective expertise and dedication of our contributors—from government leaders and urban planners to architects, sustainability advocates, and engaged citizens—shine through in this comprehensive guide. Together, they provide a multifaceted approach to urban resilience, emphasizing the critical interplay between sustainable design, community engagement, cultural awareness, and advanced technologies.

This book is more than just a reference; it is a call to action for all stakeholders involved in urban development. We hope that the insights and strategies presented here will inspire professionals across disciplines to collaborate and innovate, driving forward the creation of urban environments that are not only sustainable but also adaptable and inclusive.

Our goal is to provide a resource that supports the ongoing efforts to build cities capable of withstanding and thriving amidst the myriad challenges of the 21st century. By fostering a deeper understanding of the principles and practices of urban resilience, we aim to contribute to the global conversation on sustainable urban development and inspire concrete actions that lead to lasting, positive change.

As editors, we extend our deepest gratitude to all contributors and readers who are committed to the vision of resilient, sustainable, and inclusive urban futures. We believe that through shared knowledge, innovative thinking, and collaborative efforts, we can create urban sanctuaries that offer not just survival but prosperity for generations to come.

Ungku Norani Sonet
University College Dublin, Ireland

TamilSalvi Mari
Taylor's University, Malaysia

Chapter 1
Urban Planning and Design Simulation to Develop Sustainable and Resilient Cities

Vikram Singh
https://orcid.org/0000-0001-5757-9111
Ch. Devi Lal University, Sirsa, India

Sanyogita Singh
Panjab University, India

ABSTRACT

The demand for well-planned, sustainable, and resilient city infrastructure cannot be timelier, for the urban share of Globe's population continues to rise. Cities are complex systems with intricate dynamics, necessitating a thorough understanding of how different elements such as sewerage systems, culture centres, transport systems, energy consumption, and sewage management affect the success of an urban centre. Simulation tools offer a drawing board kind of set-up to urban planners where they experiment and compare various scenarios to select the best. The design of sustainable urban infrastructure and its resilience have also been discussed here. Urban planners may use simulation tools to model and evaluate various possible vulnerabilities and corresponding response options. Further, the chapter discusses technologies such as spatial databases, AI and machine learning models, and building information modeling in urban simulation environments, and also an international case study of urban growth simulation in Portland, Oregon, USA.

INTRODUCTION

The importance of urban planning and design in determining the future of cities has never been greater as the globe continues to experience unprecedented urbanisation (UN-Habitat, 2016). Given that urban regions are home to more than half of the world's population (United Nations, 2018), it is critical to embrace cutting-edge tactics that can result in the growth of resilient and sustainable cities. The crucial role that urban planning and design simulations play in accomplishing these goals while addressing the particular difficulties that urban regions face in the twenty-first century is explored in this chapter.

DOI: 10.4018/979-8-3693-1650-4.ch001

Sustainable and Resilient Cities

Urbanisation presents opportunities as well as difficulties. Cities serve as centres for innovation, economic development, and cross-cultural interaction. On the other side, rapid urban growth puts a burden on resources, exacerbates environmental problems, and exposes people to a range of dangers, such as socioeconomic inequality and climate change-related catastrophes (Seto et al., 2014; Angel et al., 2011). Cities need to work towards sustainability and resilience to navigate this complicated urban environment.

According to Alberti et al. (2003), sustainability in this context refers to the prudent use of resources, diminished environmental effect, and enhanced quality of life for locals. It includes factors like social fairness, energy conservation, environmentally friendly transportation, and efficient land usage (Echenique et al., 2012). The capacity of a city to endure, adapt to, and recover from shocks and pressures while sustaining key functions is referred to as resilience (Grimm et al., 2008; Berke et al., 2015). To achieve sustainability and resilience, one must take meticulous preparation and foresight.

The Role of Simulation in Urban Planning and Design

The dynamic and interrelated structure of contemporary urban systems may be difficult to handle using traditional urban planning methods, which frequently rely on static models and historical data (Batty, 2013). More adaptable and data-driven methods are also required due to the uncertainties surrounding future trends, particularly the effects of climate change and population expansion (Batty et al., 2012). Simulations of urban planning and design are useful in this situation. Fig. 1 depicts the general modeling and simulation process in abstract form.

Figure 1. A general simulation model

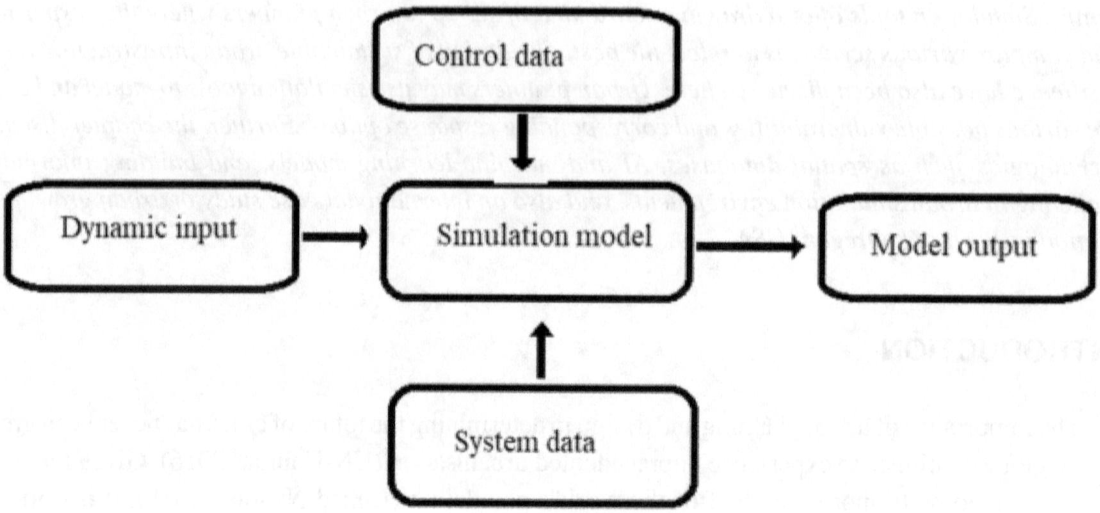

Figure 1. A general simulation model

Urban planning and design simulations make use of sophisticated computational modelling, data analytics, and visualisation technologies to provide interactive, information-rich representations of urban landscapes (Wegener & Fürst, 2011; Waddell, 2002). Planners, designers, and decision-makers can use these simulations to examine a variety of situations and their possible effects on sustainability and resilience. Cities can improve their capacity to absorb shocks and pressures while simultaneously achieving sustainability goals by simulating various development strategies and evaluating their effects (Geertman & Stillwell, 2009).

The goal of this chapter is to clearly explain to the readers why urban planning and design simulations are essential tools in the effort to build sustainable and resilient cities, as well as how they can help to create a more prosperous urban future. Traditional urban planning techniques, which usually rely on static models and historical data, may find it challenging to deal with the dynamic and interconnected nature of contemporary urban systems (Batty, 2013). There is also a need for more flexible and data-driven approaches given the uncertainty surrounding future trends, particularly the impacts of climate change and population expansion (Batty et al., 2012). In this case, simulations of urban planning and design are helpful.

In order to create dynamic, information-rich representations of urban environments, urban planning and design simulations combine advanced computer modelling, data analytics, and visualisation technologies (Wegener & Fürst, 2011; Waddell, 2002). These simulations can be used by planners, designers, and decision-makers to assess a range of scenarios and their potential impacts on sustainability and resilience. By simulating different growth strategies and analysing their effects, cities can increase their ability to withstand stresses and shocks while also reaching sustainability goals (Geertman & Stillwell, 2009).

The goal of this chapter is to clearly explain to the readers why urban planning and design simulations are essential tools in the effort to build sustainable and resilient cities, as well as how they can help to create a more prosperous urban future.

REVIEW OF EXISTING WORK

With the help of various modelling methodologies referred to as "urban simulation," urban planners and designers can more effectively comprehend, assess, and optimise urban systems. The scope, scale, and approach of these simulations differ, but they all offer valuable insights into various aspects of urban evolution. In this section, use cases of urban simulation, namely, transportation system simulations are indispensable tools in urban planning and design, enabling cities to create efficient, sustainable, and resilient transportation networks. These simulations address traffic flow, public transit planning, multimodal integration, sustainability, real-time management, environmental impact, disaster resilience, and data-driven decision-making, all of which are critical for achieving the goals of urban sustainability and resilience.

The basic objective of land use and zoning simulations is to simulate the use, development, and zoning of land parcels within a city. These simulations can be used by urban planners to assess the effects of land use constraints, identify potential redevelopment sites, and optimise how land is distributed among various users (Wegener & Fürst, 2011). Simulations of land cover changes track alterations in urban landscapes, including deforestation, urban sprawl, and green space expansion. These models assess the impact on local ecosystems, biodiversity, and the provision of ecosystem services (Seto et al., 2012). By preserving and enhancing green spaces, cities can improve environmental sustainability and enhance residents' quality of life.

Transport simulations show how people and things move throughout cities. They examine the movement of people within cities, taking into account issues like traffic flow, congestion, public transportation, and the results of transportation restrictions. These simulations are necessary to build efficient transport networks and get rid of gridlock (Batty, 2013). Transportation system simulations also assess the environmental impact of urban mobility. They model emissions, fuel consumption, and air quality, allowing planners to identify areas with high pollution levels and prioritize improvements in sustainable transportation modes (Seto et al., 2012). Simulations also play a role in sustainable transportation and energy planning. They evaluate the environmental impact of transportation modes, renewable energy integration, and the reduction of greenhouse gas emissions (Waddell, 2002).

Transportation system simulations are crucial for assessing the resilience of transportation networks to natural disasters and emergencies. They help identify vulnerable transportation infrastructure and guide investment in disaster preparedness and recovery measures. Simulation tools are used to assess a city's vulnerability to climate-related disasters, such as floods, hurricanes, and wildfires. These models help identify high-risk areas, evaluate evacuation plans, and design resilient infrastructure (Berke et al., 2015).

Environmental sustainability is a central concern in urban planning and design. Simulation tools have emerged as vital instruments for assessing the potential impacts of urban development on climate, ecosystems, and natural resources. The major objectives of environmental simulations are the assessment and mitigation of the environmental effects of urban expansion. They model urban ecological footprints, energy use, and water and air quality. Environmental simulations are essential for achieving sustainability goals and minimising the ecological impact of urban areas, according to Echenique et al. (2012).

Environmental simulations examine the sustainable management of natural resources such as water, energy, and land. By modeling resource consumption and demand, planners can optimize resource allocation, promote energy efficiency, and reduce waste (Echenique et al., 2012). Environmental simulations assist in planning urban green infrastructure, including parks, wetlands, and green roofs. These features enhance urban resilience by providing flood control, temperature regulation, and habitat conservation (Geertman & Stillwell, 2009). Eco-footprint simulations calculate the ecological footprint of a city, measuring its demand for natural resources and its carbon emissions. By comparing this footprint to the city's ecological capacity, planners can gauge its sustainability and identify areas for improvement (Batty et al., 2012).

Environmental simulations assess air quality and pollution levels resulting from urban activities. They model emissions from transportation, industrial processes, and energy consumption, allowing planners to identify areas with poor air quality and develop strategies for improving it (Alberti et al., 2003). Environmental simulations are crucial for assessing public health impacts related to environmental factors. They help predict health outcomes related to air quality, heat stress, and exposure to pollutants, enabling proactive measures to protect public health (Patz et al., 2015).

How vulnerable a city is to climate-related perils like heatwaves, hurricanes, and flooding is determined through simulations of climate change and resilience. These models help urban planners and decision-makers create plans to improve a city's resilience, adapt to shifting climatic conditions, and reduce the risks associated with catastrophic weather occurrences (Patz et al., 2015). Climate simulation models help predict how urbanization may alter local climate patterns. These models consider factors such as urban heat island effects, changes in precipitation patterns, and temperature variations (Patz et al., 2015). By simulating these impacts, planners can make informed decisions about mitigating heat stress, managing water resources, and reducing the urban heat island effect.

The main subjects of social and demographic simulations include urban population dynamics, migration patterns, and social interactions. By helping planners comprehend the evolving needs and preferences of urban residents, they enable decision-making in the fields of housing, healthcare, education, and social services (Waddell, 2002). Economic and real estate simulations are used to study the economic dynamics of metropolitan areas, including property values, market trends, and employment opportunities. Decision-makers can utilise these simulations to make well-informed decisions on investments, economic strategies, and land use (Alberti et al., 2003).

The adaptable method of agent-based modelling (ABM) simulates how certain agents interact with other agents (representing residents, businesses, or other entities) in a city. ABM is used to study emerging behaviours such as market behaviour, social dynamics, and traffic patterns. It offers a bottom-up perspective on urban systems, claim Batty et al. (2012). In scenario planning and simulations, visualisation tools are used to create scenarios and communicate the likely outcomes of urban development to stakeholders. By including members of the public, lawmakers, and urban planners in the decision-making process, these simulations promote transparency (Geertman & Stillwell, 2009). These urban simulations can be used in tandem and are frequently integrated to provide a comprehensive understanding of urban systems. The appropriateness of a particular simulation model depends on the specific objectives of urban planning and design, the accessibility of data, and the complexity of the urban challenges being addressed.

Simulations are essential for planning and optimizing public transportation systems, including bus, subway, and light rail networks. Transit simulation models consider factors such as passenger demand, routes, schedules, and fleet management. They assist in designing efficient public transportation networks that reduce congestion, emissions, and reliance on private vehicles (Waddell, 2002). Many cities are adopting multimodal transportation systems that combine various modes of travel, such as walking, cycling, public transit, and ride-sharing services. Simulation models for multimodal transport integration help planners assess the feasibility and effectiveness of these systems, ensuring seamless connectivity and accessibility for residents (Alberti et al., 2003).

Urban sustainability goals often include reducing car dependency and promoting sustainable modes of transportation. Simulation models evaluate the impacts of initiatives like bike lanes, pedestrian-friendly infrastructure, and electric vehicle charging stations. These assessments guide decisions on sustainable mobility investments (Echenique et al., 2012). Simulation-based traffic management systems enable cities to respond to real-time traffic conditions. Adaptive traffic signal control, dynamic route guidance, and incident management simulations help alleviate congestion and minimize travel delays, contributing to urban resilience by ensuring the efficient functioning of transportation networks during disruptions (Patz et al., 2015).

In the pursuit of sustainable and resilient urban planning and design, one of the critical aspects to consider is the city's ability to respond effectively to disasters and emergencies. Emergency response and disaster management simulation play a pivotal role in assessing and enhancing a city's preparedness and resilience (Smith, 2019). The utilization of emergency response and disaster management simulations in urban planning and design serves several vital purposes such as: assessing urban resilience (FEMA, 2017), training urban personnel (Smith, 2019), and testing urban plans (FEMA, 2017).

The integration of emergency response and disaster management simulations into urban planning and design has several far-reaching benefits such as: enhanced urban coordination (Smith, 2019), improved urban decision-making (FEMA, 2017), and optimized urban resource allocation (Smith, 2019). Ultimately, the integration of emergency response simulations into urban planning and design safeguards the safety and well-being of urban communities by equipping the city with the necessary skills and experience to protect and assist residents during real urban crises (FEMA, 2017). Emergency response and disaster management simulation is a vital component of urban planning and design for sustainable and resilient cities. By employing simulations, urban planners and authorities can assess, train, and enhance their urban emergency response strategies, ultimately ensuring the safety and security of urban communities in the face of disasters and emergencies.

In urban planning, the decisions made regarding city design and development have far-reaching consequences that affect the quality of life, environmental sustainability, and resilience of urban areas. Urban planners often find themselves at a crossroads, tasked with selecting the most suitable design alternatives from a range of possibilities. This section explores the critical process of choosing between alternate designs and highlights the pivotal role that simulation plays in aiding this decision-making process. The selection of design alternatives is a pivotal phase in urban planning and design for several key reasons: sustainability integration (Batty & Marshall, 2019), resilience enhancement (Colding & Barthel, 2019), and community well-being (Gehl, 2018).

Use of virtual geographic environments (VGEs) as a unique platform for urban modelling and simulation in order to address the relevant issues has been reported in Lin, et al. (2022). Authors explore the history, definition, and unique characteristics of VGEs in this proposal, setting the stage for their use in modelling and simulating urban systems. In this research, authors present an extensive VGE system designed to handle the intricacies of city settings. two case studies to illustrate the effectiveness of VGEs in urban modelling and simulation at various scales: air pollution dispersion simulation in the Pearl River Delta (PRD) urban agglomeration and crowd evacuation simulation during a fire emergency. Both case studies use VGEs for collaborative modelling.

In order to overcome the issues arising in meticulous planning and design of resilient urban settlements, Zeng, (2022) uses a virtual reality application platform. Goal of the article is to give users an access to a more detailed and immersive environment for production or instructional reasons, which will promote the advancement and application of science and technology. Authors in Peng, et al. (2023) have created a typology to summarise and visualise the four stages of AI in urban planning, from planning support to plan-making: AI-assisted, AI-augmented, AI-automated, and AI-autonomized planning. This is based on our scoping review of the body of research on the application of AI in urban planning. Based on a scoping literature review, this paper aims to address these concerns in the context of AI-enabled planning and, in doing so, develop a typology of AI in urban planning that will classify and delineate the evolution of AI in urban planning, from AI-assisted and AI-augmented planning to AI-automated and ultimately AI-autonomized planning. This article makes the case that urban planners' duties will change from handling planning mechanisms like data analysis to solving urban issues. Their emphasis will change from making planning to doing urban research, which looks at how diverse stakeholders interact with the built environment and regulations. This change in emphasis will provide in-depth understanding of how cities operate, what their citizens require, and how policies affect fairness while leaving the mechanics of planning to AI systems.

The research reported in Liu, et al. (2020) suggests an ideal urban growth scenario that balances urban development with environmental conservation in terms of the distribution of urban sprawl and the rate of urban expansion. Authors contend that the results of our ESP-based urban expansion simulation offer fresh insight into forecasting urban sprawl and may help planners and legislators make decisions that will promote greener urban development. Additionally, a paradigm for future research to model ecologically friendly urban expansion was proposed in this work. Even with this study's creative design, there are two areas that warrant more research. First, the impact of human activity on ecological services may vary depending on the stage of urban development. Therefore, additional sociodemographic characteristics, such as GDP, population, and income, may be taken into account while building ESPs and planning for urban expansion. Second, both natural and human elements should be spatiotemporally examined in order to capture the driving factors and produce a better urban growth simulation, given the apparent complexity and uncertainty of urban growth in the setting of fast urbanisation.

Salem, et al. (2021) offers a thorough analysis of Delhi's urban expansion dynamics, including its driving forces, spatial patterns, trends, and intensity, setting it apart from the numerous other studies that have examined the topic. The findings indicate that Delhi has had a sharp increase in population between 1989 and 2020, which is at odds with sustainable urban growth. Every geographical unit in Delhi had its urban expansion intensity determined, with the East South East (ESE) direction showing the strongest tendency and South South West coming in second (SSW). The LRM was employed to identify and enhance our comprehension of the physical forces responsible for this urban growth as well as to identify the most likely sites and trends of urban expansion in Delhi. According to the LRM's results, Delhi's urban expansion was mostly influenced by its closeness to major thoroughfares and urban hubs. In addition, a simulation of Delhi's future urban growth was carried out using the LRM's findings. The model's ability to forecast urban expansion was evaluated using the ROC curve. Lastly, the simulation's correctness was evaluated using the relative operating characteristic approach. The validity of the model was demonstrated by the strong ROC curve score of 0.8646. Nonetheless, the study suggests that future investigations for assessing and modelling urban growth in Delhi should make use of contemporary artificial intelligence-based models.

He, et al. (2022) have built a model to simulate changes in land usage in Dalian City, China's Jinpu New Area. According to the findings, the model validation stage's overall accuracy and Kappa coefficient were 83.57 percent and 0.768, respectively. The above results demonstrate that the proposed model complies with existing maritime development and protection regulations and is capable of capturing the nuances of coastal reclamation construction in the area. This study shows that the simulation model can potentially enhance the accuracy and efficiency of urban design.

Singh, et al. (2021) find that participatory modelling is becoming more and more used across all domains during our research, which emphasises the importance of clarifying the field. The application of participatory modelling is universal, hence this study's scope was notably broad. As a result, the focus was narrowed down to comprehend the specific field of urban logistics and manufacturing. The research presented in this article illustrates the point of crossing between the domains, namely simulation-based participatory modelling. This paper reports a comprehensive literature evaluation of the last ten years' worth of research on urban and production logistics utilising simulation-based participatory modelling. Looking at this body of work authors have noted how increased processing power modelling affected participatory modelling. According to the report, while traditional qualitative research methods like interviews continue to be widely used in the field, a new trend in high-performance computing research is beginning to emerge.

Zhang, at al. (2021) examines Jinan's urban growth patterns from 2000 to 2015 using time series DMSP/OLS NTL and NDVI data, and classified these patterns into six categories. The first kind was the stable, low-density urban type with increasing vegetation (LDSU-VI). It was mainly distributed in southern Jinan, mostly mountains and forests. The second type was the urban expansion and low density stable vegetation type (UE-LDSV) mainly located in the suburban areas of Jinan. The third type was the low density stable urban and high density stable vegetation type (LDSU-HDSV) and is mainly in the south of Jinan. The fourth type was high density stable urban and low density stable vegetation type (HDSU-LDSV) mainly found in the center of Jinan. Urban expansion and high-density stable vegetation (UE-HDSV), primarily in Jinan's suburbs, constituted the fifth kind. Urban expansion and vegetation degeneration (UE-VD) was the sixth kind, primarily found in Jinan's surrounding suburbs. Based on a review of urban development trends, conservative urban development concepts, and environmental preservation, three urban growth simulation scenarios were constructed. From 2015 to 2030, the urban simulations showed a tendency of moderate urban expansion. The pace of urban growth was highest in the historical scenario. Because of their varying degrees of constraint, scenarios 2 and 3 showed slower trajectories in urban expansion from 2015 to 2030. From the standpoint of conservative urban development and environmental preservation, they present the prospects for urban growth.

Urban growth and planning models and methodologies, including geographic information systems, cellular automata, genetic algorithms, and regression models have been studied in Hassan & Elhassan (2020). The primary goal of this work is to compile the 70 scientific publications that address urban growth in order to conduct a review and determine the most crucial goal, elements, methods, and outcomes for the most effective strategy to investigate urban expansion. Each article must primarily address urban growth modelling and methodologies, utilising a wide range of data and factors, according to the selection criteria. Because there is a lack of optimal or exemplary ways for applying analytical tools to analyse urban expansion, this work attempts to fill the vacuum left by the lack of the best methods for studying urban growth. Thus, the goal of this study is to facilitate researchers' ability to combine appropriate methodologies to obtain results that are consistent with their hypothesis. The findings state that in order to ensure that simulation or prediction models can provide accurate and realistic methods, two or more methodologies and models must be combined. Additionally, this evaluation found that the most common components for basic data analysis were the usage of land use and land cover maps, satellite pictures, and remote sensing.

Findings of Yeomans & Kozlova (2023) encourage urban policymakers to investigate the full range of model consequences while conducting comprehensive sensitivity studies of the most important system variables using the systematic SimDec approach. Computational system modelling has been shown to be essential for the development of effective urban policies because it often yields a more comprehensive understanding of the expected effects of suggested solutions and can offer a wider range of insights into the underlying causes of many potential problems. It is demonstrated how the novel SimDec analytical method may be effectively applied to visualising a sensitivity analysis of an urban systems model and to the analysis of urban planning. Based on the results, authors argue that SimDec should be used generally in sensitivity analyses of urban systems because of its: (i) remarkable ability to uncover hidden dependencies when assessing intricate computer models; (ii) ease of use in conveying a multitude of potentially hidden perspectives to decision-makers; (iii) overall superior visualizable analytic proficiencies; and (iv) significantly reduced computational complexities.

MATHEMATICAL MODELING OF URBAN GROWTH

Urbanisation is a complex phenomenon that is influenced by several factors, including population growth, economic growth, restrictions on land use, and infrastructure spending. Simulations of urban expansion are crucial for controlling and understanding it. They employ computational models and data analysis techniques to forecast how cities might develop over time and assist urban planners and politicians in making educated choices. This section discusses urban growth simulations and their significance in the context of sustainability and resilience.

Population Growth Modeling

Mathematical models are frequently used in urban growth simulation to forecast and comprehend how cities grow and change over time. The Logistic Growth Model, which is based on the idea of population growth, is one often employed model. A population's (or, in the case of urban expansion, a city's area or infrastructure) change over time is described by the logistic growth model, a differential equation. The mathematical model of population growth is given in equation (i) below:

$$\frac{dP}{dt} = r\,P\,(1 - \frac{P}{K}) \text{ (i)}$$

Where:

- $\frac{dP}{dt}$ represents the rate of change of the population (or urban area) with respect to time.
- P is the current population (or urban area) at time t.
- r is the intrinsic growth rate, which represents the maximum growth rate when the population is small.
- K is the carrying capacity, which represents the maximum population.

Simulations of urban growth frequently begin with projections of population growth and demographic shifts. Models of population dynamics, such as the conventional logistic growth model (Verhulst, 1838), adapted for urban contexts, shed light on the anticipated long-term rise of urban populations. These models take into account migration, mortality, and birth rates. (Seto et al., 2012).

Land Use Modeling

Examples of land use change models that are widely used to simulate how land parcels inside of a city will be allotted for different uses include cellular automata (CA) and Markov chain models (Wegener & Fürst, 2011). For instance, CA divides the metropolitan region into cells and specifies guidelines for how land use changes occur based on factors like proximity to existing infrastructure or land use regulations.

The allocation of land in a city for various uses, such as residential, commercial, industrial, and recreational, can be understood and simulated using mathematical models of urban land use. The Urban Land Usage Allocation Model, which may be written mathematically as follows, is one popular model for urban land usage.

Let:

- L represents the total land area within a city.
- L_i represents the land area allocated to land use category i, where i ranges over all relevant categories (e.g., residential, commercial, industrial, etc.).
- P_i represents the population or demand associated with land use category i.
- C_i represents the cost or profitability associated with land use category i.

- U_i represents a utility function that quantifies the desirability or attractiveness of land use category i to residents or businesses.

The goal of the model is to find the optimal allocation of land area L_i to each land use category i while maximizing some objective function, typically representing the overall welfare, utility, or economic value of the city. This land allocation can be formulated as an optimization problem (ii) below:

Maximize:

$\Sigma_i(U_i * P_i * L_i)$ (ii)

Subject to equations/inequalities (iii) and (iv) below:

$\Sigma_i(L_i) \leq L$ (total land area constraint) (iii)

$L_i \geq 0$ (non-negativity constraint) (iv)

By dividing up the available land into diverse uses in a way that maximises overall welfare, the objective function seeks to maximise the overall utility or value of the city. The total land area restriction makes sure that the overall amount of land allotted to categories doesn't go beyond the total amount of land that is actually available in the city. Each category's allotted land area is guaranteed to be non-negative by the non-negativity condition.

Depending on the setting, the availability of data, and the objectives of the urban planning or research project, the precise utility functions U_i and limitations may change. To improve the models' realism and accuracy as representations of actual urban land, researchers and urban planners frequently calibrate and validate them using historical land-use data and observed patterns of urban development.

Environmental Impact Assessment Modeling

Assessment of the environmental effects of urban expansion is also aided by simulations. They can simulate the changes in land use, carbon emissions, and water use brought on by urban growth. This knowledge is crucial for creating green cities with small ecological footprints (Alberti et al., 2003).

A mathematical model for environmental simulation is a powerful tool for studying and predicting the behavior of natural systems and ecosystems. One widely used model for simulating environmental processes is the Lotka-Volterra Predator-Prey Model, which describes the dynamics between two interacting species, such as predators and prey, and can be extended to simulate broader environmental interactions. The basic Lotka-Volterra predator-prey model is represented mathematically by a system of differential equations:

Let:

- $P(t)$ represent the population of prey species at time t,
- $H(t)H(t)$ represent the population of predator species at time t,
- r represent the intrinsic growth rate of the prey population,
- a represent the predation rate (the rate at which predators consume prey),

- b represent the reproduction rate of predators per prey consumed, and
- d represent the death rate of predators in the absence of prey.

The mathematical model for environmental impact assessment is represented by a couple of equations (v) and (vi):

$$\frac{dP}{dt} = rP - aPH \text{ (v)}$$

$$\frac{dH}{dt} = -dH - bPH \text{ (vi)}$$

The first equation describes how the prey population changes over time. Prey reproduce at the rate bPH and are consumed by predators at a rate aPH, where a represents the predation rate and H represents the predator population. The second equation describes how the predator population changes over time. Predators experience natural death at a rate d and gain energy from consuming prey at a rate bPH, where b represents the reproduction rate of predators per prey consumed.

The Lotka-Volterra model is a simplified representation of predator-prey interactions and does not capture all the complexities of real ecosystems. However, it serves as a foundational model that can be extended and adapted to simulate more complex environmental interactions, including competition, symbiosis, and food web dynamics. To simulate environmental systems, additional factors and differential equations can be incorporated to represent variables like resource availability, habitat quality, environmental disturbances, and other relevant factors. These extended models can provide valuable insights into the dynamics of ecosystems and the potential impacts of various environmental changes and management strategies.

Traffic Flow Modeling

Simulation models for traffic flow, such as microscopic and macroscopic traffic models, help urban planners and transportation engineers understand how vehicles move through the city's road network (Batty, 2013). Microscopic models focus on individual vehicle behavior, while macroscopic models examine traffic flow at a broader level. These simulations can predict congestion patterns and inform road design and traffic management strategies (Batty et al., 2012). One commonly used mathematical model for transportation simulation is the Traffic Flow Model. Traffic flow models describe the movement of vehicles on road networks and are essential for analyzing traffic patterns, congestion, and optimizing transportation systems. One classical traffic flow model is the Lighthill-Whitham-Richards (LWR) Model.

The LWR model is a macroscopic traffic flow model that considers traffic density as a function of space and time. It represents the relationship between traffic density ρ and traffic velocity v within a continuous medium. The fundamental equation (eq. vii) of the LWR model is:

$$\frac{\partial \rho}{\partial t} + \frac{\partial(\rho v)}{\partial x} = 0 \text{ (vii)}$$

In this equation:

- ρ represents the traffic density (the number of vehicles per unit length).

- t represents time.
- x represents the spatial coordinate.
- v represents the traffic velocity.

The equation essentially states that the change in traffic density with respect to time plus the change in the product of traffic density and velocity with respect to space is equal to zero, indicating the conservation of vehicles within the road segment.

Key characteristics of the LWR model:

- Conservation Law: The model is based on the principle of conservation of vehicles within a road segment.
- Traffic Flow: It describes the flow of traffic as a function of density and velocity.
- Wave Propagation: The model can simulate the propagation of traffic waves and shockwaves, which represent the formation and dissipation of congestion.
- Jam Density: The model can define a critical traffic density, known as the jam density, at which traffic comes to a standstill.
- Fundamental Diagram: The relationship between traffic density and traffic velocity can be plotted to create a fundamental diagram, showing the phases of free flow, congestion, and maximum flow.

Extensions and variations of the LWR model, such as the use of flux functions and boundary conditions, can make it more realistic and suitable for specific transportation scenarios. It is often used in traffic management, congestion prediction, and transportation system optimization. Keep in mind that this is a simplified macroscopic model, and real traffic systems are more complex, involving factors like individual driver behavior, traffic signals, road geometry, and more. Therefore, more detailed models like microscopic traffic simulations (e.g., using agent-based models) are also used to capture finer-grained interactions among vehicles.

CHALLENGES AND PITFALLS

While simulation-based decision support systems offer invaluable assistance in urban planning and design, they come with their own set of challenges and potential pitfalls that need to be navigated carefully.

Data Accuracy and Availability

Challenge: One of the primary challenges in simulation-based decision-making is ensuring the accuracy and availability of data. Simulations heavily rely on data inputs, including demographic, environmental, and infrastructure data. If this data is outdated, incomplete, or inaccurate, it can lead to flawed results and unreliable decision-making (Batty & Marshall, 2019).

Pitfall: Planners might be tempted to use available data even if it is of poor quality due to budget constraints or time pressures. This can result in inaccurate simulations and misguided decisions.

Model Complexity

Challenge: Urban systems are inherently complex, with numerous interconnected variables. Creating simulation models that accurately capture this complexity can be challenging. As models become more detailed, they also become more computationally intensive, potentially leading to longer simulation times and increased resource requirements (Batty & Marshall, 2019).

Pitfall: Overly complex models can be difficult to understand and interpret. Planners may struggle to communicate the results effectively to stakeholders, hindering decision-making processes.

Uncertainty and Assumptions

Challenge: All simulations involve a level of uncertainty. Urban environments are dynamic and subject to change due to unforeseen events. Models often rely on assumptions about future conditions, and if these assumptions are incorrect, it can lead to misinformed decisions (Colding & Barthel, 2019).

Pitfall: Planners may make critical decisions based on simulations without fully acknowledging the uncertainties and assumptions inherent in the models. This can result in a false sense of confidence in the chosen design alternative.

Stakeholder Engagement

Challenge: Involving a diverse range of stakeholders in the decision-making process is essential for creating sustainable and resilient cities. However, effective stakeholder engagement can be challenging and time-consuming. Balancing the needs and preferences of various groups can be complex (Gehl, 2018).

Pitfall: Inadequate stakeholder engagement can lead to community dissatisfaction, resistance to proposed designs, and delays in project implementation. Additionally, conflicting stakeholder interests can make it difficult to reach a consensus on the best design alternative.

Ethical Considerations

Challenge: As simulations become more advanced, ethical considerations come into play. Decisions made based on simulation outcomes can have profound social, economic, and environmental impacts. Ethical questions regarding data privacy, equity, and unintended consequences must be addressed (Gehl, 2018).

Pitfall: Neglecting ethical considerations can lead to decisions that disproportionately affect vulnerable communities or compromise data privacy. It can also erode public trust in urban planning processes.

Resource Constraints

Challenge: Developing and running simulation models can be resource-intensive. Cities with limited budgets or technical expertise may face challenges in implementing simulation-based decision support systems effectively.

Pitfall: Attempting to use simulation tools without adequate resources can result in incomplete or poorly executed simulations, leading to unreliable outcomes and suboptimal design choices.

Interdisciplinary Collaboration

Challenge: Urban planning requires collaboration between various disciplines, including urban design, engineering, environmental science, and sociology. Effective communication and collaboration between experts from these fields can be challenging, as they may use different terminology and approaches.

Pitfall: Miscommunication or lack of collaboration between disciplines can result in incomplete or biased simulations. It's essential to ensure that experts from diverse backgrounds work together effectively.

PORTLAND, OREGON: URBAN GROWTH SIMULATION

Portland, Oregon, often referred to as the "City of Roses," has emerged as a prominent case study for urban growth simulation due to its dedication to sustainability and urban planning (Smith, 2019). With a rapidly growing population and a strong emphasis on preserving its natural environment, Portland presents a unique opportunity for urban planners and researchers seeking to develop sustainable and resilient cities.

Urban growth simulations in Portland have involved running multiple scenarios to evaluate different development strategies, namely, (i) Business-as-usual - a baseline scenario that extrapolates current trends in population growth, land use, and infrastructure development (Smith & Johnson, 2018), (ii) Smart growth - a scenario emphasizing compact and sustainable development, often with an emphasis on public transportation, mixed land uses, and green infrastructure (Gehl, 2010), (iii) Environmental conservation - z scenario that prioritizes the preservation of natural areas and the reduction of urban sprawl (Portland Parks & Recreation, 2019), and (iv) Climate resilience – to model the effects of climate change on urban growth and evaluating strategies for adapting to increased temperatures, sea-level rise, and extreme weather events (Oregon Climate Change Research Institute, 2020).

Data Sources

Urban growth simulations in Portland rely on a diverse range of data sources. Census data from the U.S. Census Bureau provides critical insights into population trends, household composition, and migration patterns (U.S. Census Bureau, 2020). Geographic Information System (GIS) data, including land use maps, transportation networks, and environmental features, offer a spatial context for urban growth modeling (City of Portland, 2021). Additionally, economic data, such as information about local economies, employment opportunities, and economic development plans, are essential for understanding growth drivers (Portland Development Commission, 2020). Environmental data, encompassing natural resources, climate data, and environmental regulations, play a pivotal role in sustainable urban planning (Oregon Department of Environmental Quality, 2019).

Modeling Techniques

Portland urban growth simulation has used at least three models for simulating various processes of the urban system, namely, Agent-based Modeling (ABM) for simulating interactions between different "agents" of urban systems, Cellular Automata (CA) for simulating urban expansion, an System Dynamics (SD) for modeling the long-term impact of urban growth policies.

Agent-based Modeling: Agent-based modeling is a widely employed technique for simulating urban growth in Portland. It involves creating individual "agents" that represent households, businesses, or other entities. These agents interact with each other and their environment, making decisions about land use, housing, and transportation based on predefined rules and preferences (Benenson & Torrens, 2004). ABM allows for the exploration of various scenarios and the assessment of their impacts on urban growth (Railsback & Grimm, 2019).

Cellular Automata: Cellular automata models have been extensively used to predict urban expansion in Portland. These models divide the study area into cells, each with specific rules for land use changes. Cellular automata models simulate the evolution of land use patterns over time, taking into account factors such as proximity to transportation nodes, land zoning regulations, and population density (White & Engelen, 1993).

System Dynamics: System dynamics models have played a crucial role in assessing the long-term impacts of urban growth policies in Portland. These models consider feedback loops and time delays, making them suitable for examining the sustainability of urban development strategies (Sterman, 2000). System dynamics models have been used to evaluate the consequences of alternative transportation plans on urban sprawl and traffic congestion (Dabbour & Sliuzas, 2013).

Findings

The findings from urban growth simulations in Portland have influenced the following policy decisions: (1) simulations help evaluate the impact of new transit lines or road networks on urban sprawl and accessibility (City of Portland, 2017), (2) planners can use simulation results to assess the effectiveness of zoning regulations in guiding urban growth toward sustainable patterns (Portland Bureau of Planning and Sustainability, 2016), and (3) simulation outcomes guide decisions regarding the protection of green spaces, parks, and natural habitats within the urban area (The Trust for Public Land, 2019).

CONCLUSION

In conclusion, climate and environmental impact simulations are essential tools for urban planners and designers working towards sustainable and resilient cities. These simulations encompass climate modeling, air quality assessment, land cover analysis, resource management, disaster resilience planning, green infrastructure design, eco-footprint analysis, sustainable transportation and energy modeling, and public health assessments. By utilizing these tools, cities can make informed decisions to mitigate environmental risks, reduce ecological footprints, and create urban environments that are both sustainable and resilient. The process of choosing between alternate designs in urban planning is a critical juncture in shaping the future of cities. With the growing emphasis on sustainability and resilience, simulation-based decision support has emerged as an indispensable tool. It empowers urban planners and designers to make informed choices that align with the principles of sustainability, resilience, and community well-being, ultimately contributing to the development of more livable and vibrant cities.

While simulation-based decision support systems offer invaluable assistance in urban planning and design, they come with their own set of challenges and potential pitfalls that need to be navigated carefully. The amount of effort and the degree of human expertise required in the mathematical modeling of urban processes count among the prominent challenges in urban simulation. Simulation experiments are computation-intensive and might not yield the expected results even after investing a substantial time and finances. Moreover, the results of the simulation experiments, too, are highly cryptic and require expensive human resources. Urban growth simulation in Portland, Oregon, serves as a valuable tool for policymakers, urban planners, and researchers striving to create sustainable and resilient cities. By leveraging data-driven models and scenario analysis, Portland can continue its commitment to urban development that balances growth with environmental conservation and quality of life.

REFERENCES

Alberti, M., Booth, D., Hill, K., Coburn, B., Avolio, C., Coe, S., & Spirandelli, D. (2003). The impact of urban patterns on aquatic ecosystems: An empirical analysis in Puget lowland sub-basins. *Landscape and Urban Planning*, 64(3), 175–198.

Angel, S., Parent, J., Civco, D. L., Blei, A., & Potere, D. (2011). The dimensions of global urban expansion: Estimates and projections for all countries, 2000-2050. *Progress in Planning*, 75(2), 53–107. 10.1016/j.progress.2011.04.001

Batty, M. (2013). *The new science of cities*. MIT Press. 10.7551/mitpress/9399.001.0001

Batty, M., Besussi, E., & Chin, N. (2012). Traffic evolution: Exploring the historical morphology of London's street network. *Environment and Planning. B, Planning & Design*, 39(5), 838–860.

Batty, M., & Marshall, S. (2019). *Cities, Planning and Simulation*. Routledge.

Benenson, I., & Torrens, P. M. (2004). *Geosimulation: Automata-based modeling of urban phenomena*. John Wiley & Sons. 10.1002/0470020997

Berke, P. R., Cooper, J., Aminto, M., Reams, M., & Lee, J. S. (2015). Local mitigation planning to reduce social vulnerability: Southeast Florida's regional climate change compact. *Mitigation and Adaptation Strategies for Global Change*, 20(6), 991–1013.

City of Portland. (2021). GIS data. https://www.portland.gov/bps/maps-and-gis/gis-data

Colding, J., & Barthel, S. (2019). Urban Greening. In Newton, R. W. (Ed.), *Water and Urban Development Paradigms: Towards an Integration of Engineering, Design and Management Approaches* (pp. 77–98). Springer.

Dabbour, E., & Sliuzas, R. V. (2013). Integrating system dynamics and cellular automata for urban simulation: Prospects and challenges. In *Proceedings of the 7th International Conference on Geographic Information Science (GIScience 2012)* (pp. 63-77). IEEE.

Echenique, M. H., Mingardo, G., & Jara-Diaz, S. R. (2012). On the income elasticity of car travel demand. *PLoS One*, 7(9), e43515.

Federal Emergency Management Agency (FEMA). (2017). *Homeland Security Exercise and Evaluation Program (HSEEP) overview*. FEMA. https://www.fema.gov/sites/default/files/2020-07/2020_hseep_overview.pdf

Geertman, S., & Stillwell, J. (2009). Planning support systems: An inventory of current practice. *Computers, Environment and Urban Systems*, 33(3), 189–199.

Gehl, J. (2010). *Cities for people*. Island Press.

Grimm, N. B., Faeth, S. H., Golubiewski, N. E., Redman, C. L., Wu, J., Bai, X., & Briggs, J. M. (2008). Global change and the ecology of cities. *Science*, 319(5864), 756–760. 10.1126/science.115019518258902

Hassan, M. I., & Elhassan, S. M. M. (2020). Modelling of urban growth and planning: A critical review. *Journal of Building Construction and Planning Research*, 8(4), 245–262. 10.4236/jbcpr.2020.84016

He, F., Yang, J., Zhang, Y., Sun, D., Wang, L., Xiao, X., & Xia, J. (2022). Offshore Island Connection Line: A new perspective of coastal urban development boundary simulation and multi-scenario prediction. *GIScience & Remote Sensing*, 59(1), 801–821. 10.1080/15481603.2022.2071056

Lin, H., Bingli, X., Yuting, C., Wenhang, L., Lan, Y., & Jie, H. (2022). VGEs as a New Platform for Urban Modeling and Simulation. *Sustainability (Basel)*, 14(13), 7980. 10.3390/su14137980

Liu, X., Wei, M., & Zeng, J. (2020). Simulating Urban Growth Scenarios Based on Ecological Security Pattern: A Case Study in Quanzhou, China. *International Journal of Environmental Research and Public Health*, 17(19), 7282. 10.3390/ijerph1719728233028037

Oregon Climate Change Research Institute. (2020). *Oregon's Fourth Climate Assessment Report*. OCCRI. https://www.occri.net/

Oregon Department of Environmental Quality. (2019). *Data and reports*. Oregon Department of Environmental Quality. https://www.oregon.gov/deq/pages/data-and-reports.aspx

Patz, J. A., Gibbs, H. K., Foley, J. A., Rogers, J. V., & Smith, K. R. (2015). Climate change and global health: Quantifying a growing ethical crisis. *EcoHealth*, 12(4), 447–455.

Peng, Z.-R., Lu, K.-F., Liu, Y., & Zhai, W. (2023). The Pathway of Urban Planning AI: From Planning Support to Plan-Making. *Journal of Planning Education and Research*, 0(0), 0739456X231180568. Advance online publication. 10.1177/0739456X231180568

Portland Bureau of Planning and Sustainability. (2016). *Comprehensive Plan*. Portland Bureau of Planning and Sustainability. https://www.portland.gov/bps/comprehensive-plan

Portland Development Commission. (2020). *Economic development*. Portland Development Commission. https://www.pdc.us/economic-development/

Portland Parks & Recreation. (2019). *Parks, gardens & natural areas*. Portland Parks & Recreation. https://www.portland.gov/parks

Railsback, S. F., & Grimm, V. (2019). *Agent-based and individual-based modeling: A practical introduction*. Princeton University Press.

Salem, M., Bose, A., Bashir, B., Basak, D., Roy, S., Chowdhury, I. R., Alsalman, A., & Tsurusaki, N. (2021). Urban Expansion Simulation Based on Various Driving Factors Using a Logistic Regression Model: Delhi as a Case Study. *Sustainability (Basel)*, 13(19), 10805. 10.3390/su131910805

Seto, K. C., Güneralp, B., & Hutyra, L. R. (2012). Global forecasts of urban expansion to 2030 and direct impacts on biodiversity and carbon pools. *Proceedings of the National Academy of Sciences of the United States of America*, 109(40), 16083–16088. 10.1073/pnas.121165810922988086

Singh, A., Baalsrud Hauge, J., & Wiktorsson, M. (2021). Simulation-Based Participatory Modelling in Urban and Production Logistics: A Review on Advances and Trends. *Sustainability (Basel)*, 14(17), 17. 10.3390/su14010017

Smith, J. A. (2019). *Disaster Response and Recovery*. Wiley.

Smith, J. D. (2019). *Sustainable urban development in the age of climate change: Lessons from Portland.* Routledge.

Smith, J. D., & Johnson, K. (2018). Urban growth modeling of Portland, Oregon: Incorporating land use policy scenarios into cellular automata modeling. *Landscape and Urban Planning*, 175, 103–114.

Sterman, J. D. (2000). *Business dynamics: Systems thinking and modeling for a complex world.* Irwin/McGraw-Hill.

The Trust for Public Land. (2019). *Our work in Oregon.* TPL. https://www.tpl.org/our-work/oregon

UN-Habitat. (2016). *New Urban Agenda.* UN. https://.org/new-urban-agenda

United Nations. (2018). *World urbanization prospects: The 2018 revision.* UN. https://population.un.org/wup/

U.S. Census Bureau. (2020). *Census data.* US Census Bureau. https://data.census.gov/

Verhulst, P. F. (1838). Notice sur la loi que la population suit dans son accroissement. *Correspondance Mathématique et Physique*, 10, 113–121.

Waddell, P. (2002). UrbanSim: Modeling urban development for land use, transportation, and environmental planning. *Journal of the American Planning Association*, 68(3), 297–314. 10.1080/01944360208976274

Wegener, M., & Fürst, F. (2011). Land-use transport interaction: State of the art. *SSRN* 1887602.

Yeomans, J. S., & Kozlova, M. (2023). Extending system dynamics modeling using simulation decomposition to improve the urban planning process. *Frontiers in Sustainable Cities*, 5, 1129316. 10.3389/frsc.2023.1129316

Zeng, L. (2022). *Research on Dynamic Simulation Technology of Urban 3D Landscape Based on Virtual Reality APP Platform.* In 2022 2nd International Conference on Networking, Communications and Information Technology (NetCIT), Manchester, United Kingdom. 10.1109/NetCIT57419.2022.00107

Zhang, Y., Zhao, L., Zhao, H., & Gao, X. (2021). Urban development trend analysis and spatial simulation based on time series remote sensing data: A case study of Jinan, China. *PLoS One*, 16(10), e0257776. 10.1371/journal.pone.025777634618811

Chapter 2
Cultural Consciousness in Heritage Architecture Education With a Problem-Based Learning Approach

Jing Hao Koh
https://orcid.org/0000-0003-3831-8025
School of Architecture, Building, and Design, Malaysia

Khairool Aizat Ahmad Jamal
https://orcid.org/0000-0003-4674-581X
School of Architecture, Building, and Design, Malaysia

ABSTRACT

Heritage architecture education is important to develop contextual understanding and cultural consciousness among architecture students. Conventional teaching approaches are inadequate to cultivate students' cultural knowledge of specific places and community contexts. This chapter addresses this gap by evaluating how problem-based learning (PBL) as a teaching approach in a measured drawing module contributes to cultural consciousness among architecture students. It adopts quantitative survey research design to evaluate students' cultural consciousness based on knowledge of place identity, understanding of community social contexts, and recognition of community cultural contexts. Two student cohorts completing the module participated as respondents, providing insights into PBL's impact. The findings substantiate the adoption of PBL as an effective pedagogy to enhance students' cultural consciousness, by developing contextual and cultural sensitivity in design cognition. It is also useful in guiding the design of teaching approaches for culturally related subjects in other disciplines.

INTRODUCTION

Architecture plays a significant role in reflecting the cultural identity and history of a place and its community. This encompasses various aspects of building design which should consider local environmental factors and meet the needs of the people while respecting the values and traditions of the com-

DOI: 10.4018/979-8-3693-1650-4.ch002

munity. The International Union of Architects (Union Internationale des Architectes) acknowledges this significance and envisions a diverse and inclusive architectural education that recognises human, social, cultural, urban, architectural, and environmental aspects as outlined in the UNESCO-UIA Charter for Architectural Education (UIA, 2023). It is therefore crucial for architectural students to learn about the significance of each place and features of local socio-cultural importance. However, the current approach to architectural education tends to prioritise universal and global values, often neglecting local and regional significance (Gunewardene, 2016). Olweny (2020) suggests a shift of focus on local contexts when developing architectural education curriculum. Embaby (2014) argues that both the universal and local values of place and community should be comprehensively integrated, rather than being treated as separate entities. Recognising this, the UNESCO-UIA Charter for Architectural Education emphasises the role of architectural heritage education in developing knowledge of sustainability, resilience, social contexts and sense of place in building design. This would nurture design thinking for creative approaches in addressing the transformation of cultural process (UIA, 2023, p3).

Although the significance of cultural heritage has been recognised in the architectural education framework (Clarke et al., 2020; Djabarouti & O'Flaherty, 2019; Jadresin Milic & Nikezic, 2018; Lapadula & Quiroga, 2012), related subjects such as architectural history receive little attention from educators, because they are separated from the architectural design domain (Li, 2018). Pasha (2020) reports that the role of culture in architectural education is over-simplified and reduced. Furthermore, current approaches to teaching architectural history remain limited in achieving the desired outcomes. Li (2016) asserts that architectural history courses have been marginalised compared to architectural design courses in academic institutions. Inadequate teaching time hinders the effective delivery of the course. In addition, an over-reliance on lectures as the primary teaching method leads to passive learning among students, reducing their engagement with the subject. Embaby (2014) argues that conventional teaching of architectural history often neglects its practical applications, overlooking the crucial link between cultural heritage and architectural education. This highlights the need for educators to recognise the importance of incorporating cultural heritage as an integral part of a holistic architectural curriculum, without discriminating against other aspects.

ADDRESSING THE NEED FOR A PROBLEM-BASED LEARNING APPROACH

As scholars generally agree that local values should be given equal respect in architecture education, it is considered important to adopt a place-based approach in understanding the contexts that constitute these values. Ng (2013) observes that placing emphasis on contexts enables students to cultivate sensitivity towards a place while also expanding design creativity. Similarly, Nikezić & Marković (2015) have highlighted the value of place-based learning in enhancing students' awareness of spatial-cultural contexts, thereby influencing their environmental literacy and developing a sense of responsibility. However, conventional teaching approaches towards heritage architecture education often adopt a generic learning scope that lacks real-world contexts involving specific places, communities, and cultural systems. This may not sufficiently engage students in developing a deeper cultural understanding of unique places and community contexts.

In response to this, Smith & Sobel (2004) suggest a learning process that emphasises the connections between place and society, allowing real issues affecting specific localities and communities to be addressed. Consequently, an effective educational approach in heritage architecture should integrate a problem-solving framework, incorporating Problem-Based Learning (PBL) to develop students' ability to address real issues related to place and society.

According to Bigelow (2004), PBL is an instructional method that uses problems to help students develop problem-solving skills and acquire basic knowledge. Guerra & Kolmos (2011) define PBL as an inquiry process where students work on real and complex problems that guide their learning. Banerjee & Graaf (2010) describe PBL as a shift from traditional teaching to a student-centred approach that integrates different subjects and skills through thematic curriculum design. Alhassan et al., (2020) describe PBL as a student-centred approach focusing on interactive and identity-directed learning while incorporating background knowledge and skills. In PBL, the role of the instructor changes from that of a knowledge provider to that of a facilitator of the learning process (Guerra & Kolmos, 2011), allowing students to acquire knowledge through problem-solving. Place-based learning is rooted in PBL (Alhassan et al., 2020, p.74). As demonstrated in studies by Curto et al. (2021) and Valentín-Gamazo et al. (2021), place-based learning in heritage architecture can benefit from PBL to achieve the intended learning outcomes.

While PBL has been widely adopted across various disciplines, its application in heritage architecture education has been relatively limited. Consequently, there is an urgent need for research into the integration of PBL into the curriculum and methodology of heritage architecture education. This paper aims to bridge the gap identified in conventional heritage architecture education pedagogy, by examining how the PBL instructional approach used in a measured drawing module facilitates the development of students' cultural consciousness.

THE CULTURAL CONSCIOUSNESS FRAMEWORK IN ARCHITECTURE

In a broad sense, culture refers to the beliefs, traditions, way of thinking, religion, and lifestyle of a social group. This encompasses the cognitive processes that define the values and behaviour of a group (Hessam & Sotoue 2016). Culture is viewed as a holistic concept collectively shared by a society, including both tangible and intangible aspects, and has the ability to change and develop over time (Pasha, 2020, p.113). The relationship between culture and architecture has been extensively studied. Architecture serves as a tangible manifestation of a society's collective consciousness, expressing and communicating its ideas, values, and beliefs (Hendrix, 2010). This has a profound impact on architectural representation. Architecture involves the interdisciplinary understanding of culture, including the study of material culture and spatial cultures, and focuses on the transformative process of creating and cultivating human living spaces within the environment (Troiani et al., 2013). According to Rapoport (2005), culture functions as a "design for living" that guides the way of life and provides a framework for understanding the patterns and structures of a society. It also serves to distinguish and highlight the differences between social groups. The impact of culture on architecture is strongly evident, with values, practices, activities strongly influencing the design of contemporary human inhabitation (Al Husban et al., 2021).

Recognising the importance of cultural diversity is essential, as stated in the Universal Declaration on Cultural Diversity by UNESCO. This declaration emphasises that cultural diversity should be seen as a shared heritage and valued for the benefit of present and future generations (UNESCO, 2002). One must be culturally conscious to appreciate cultural diversity. Cultural consciousness refers to the process of developing an understanding and awareness of different cultures, which can lead to a deeper knowledge and appreciation of other individuals and contexts (Páez & Albert, 2012). This also fosters the development of respect and tolerance towards cultural diversity (McKeen, 2019). As architecture is closely linked to human societies, cultural consciousness offers a perspective to understand architecture based on cultural parameters, which will foster appreciation towards the embodied values held by both historical and contemporary societies. Many studies have emphasised the importance of preserving cultural heritage with cultural consciousness, which plays a vital role in achieving cultural sustainability (De Merode et al., 2004; McMINN & Polo, 2005; Jokilehto, 2007; Postalcı & Atay, 2018). Therefore, incorporating a cultural consciousness framework into architectural study is crucial. Rapoport's framework on the cultural dimensions of architecture makes a significant contribution in this regard.

Rapoport's framework incorporates cultural and social variables into the built environment. Social variables include family and kinship structure, social networks, roles, statutes, social institutions (Figure 1). Cultural variables involve specific values derived from a particular worldview, expressed through ideas, images, and meanings. These values shape people's lifestyles and result in a set of defined activity systems. The activity system then influences the design of the built environment, including the organisation of space, time, meaning, and communication, as well as the features of the cultural landscape. Culture in architecture can be seen as the physical manifestation of a society's values (beliefs, traditions, religion) in the environment (buildings and landscape), accommodating the specific activity systems expressed through social variables (family and kinship structure, social networks, roles, statutes, social institutions, and so on).

From this framework, the physical manifestation of the environment indicates the place, where one can experience the landscape and see the buildings. Social variables and society values constitute a place's social contexts. Values and the activity system suggests the cultural contexts associated with a place. By categorising these elements, we can derive the cultural consciousness attainment parameters, as shown in Figure 2. A person's cultural consciousness of a place can be measured by three indicators, namely knowledge of place identity, social contexts of the community understanding and recognition of cultural contexts of the community.

Place identity knowledge acquisition involves the perception of meanings associated with a spatial setting, determined by the physical environment and human values (Chen & Xu, 2017). It also encompasses a sense of distinctiveness that sets a place apart from others, while maintaining a continuity of self-conceptualization over time (Morel-EdnieBrown, 2012. Factors such as history, daily activities, shared memories, physical features of the site, architectural styles, historical events, and traditional businesses contribute to the cognitive understanding of a place's identity.

Context refers to any components which may characterize the situation of an entity, relevant to the interaction between a user and a situation (Dey, 2001). Social contexts therefore include social variables that generate meanings specific to the users in relation with their interaction with groups of people. Understanding a community's social contexts involves understanding the kinship, family units, organisational structure, and social network related to the users.

A community's cultural contexts recognition refers to the values and activity systems of the community. Values may be indoctrinated through faiths and beliefs, which have significant effects on lifestyle through religion, customs, art, food, clothing, language and other factors as the ideal images or schemata that define the activity system. Examples of such activities include prayers, rituals, consecration of ancestors and gathering of community members.

Figure 1. Rapoport's framework of cultural dimension in architecture

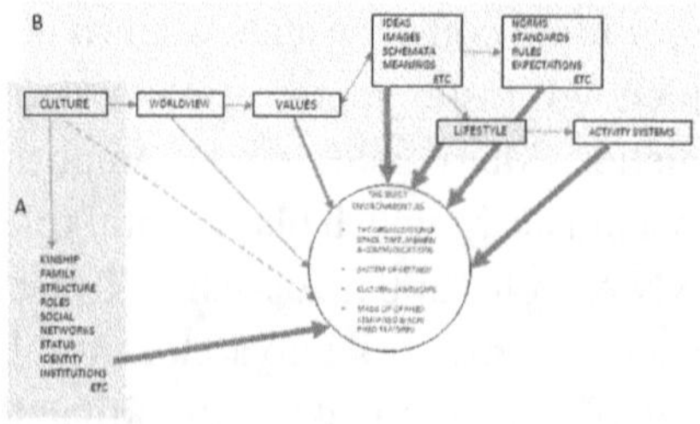

Rapoport (2005)

Figure 2. Derivation of the cultural consciousness framework from Rapoport's framework of cultural dimensions in architecture

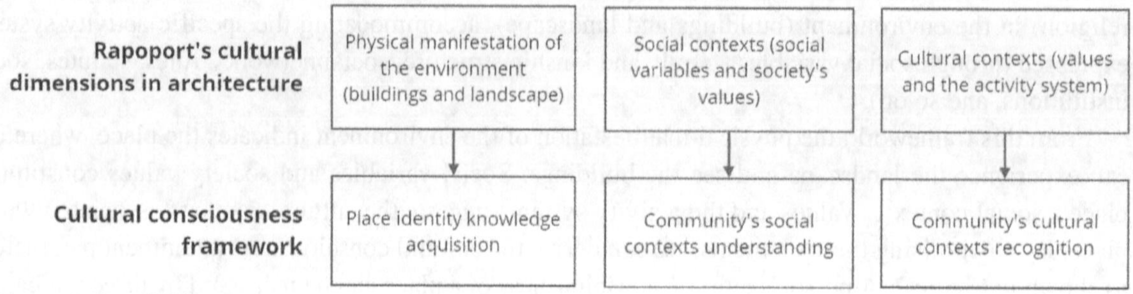

Source: Authors (2023)

MEASURED DRAWING AS PROBLEM-BASED LEARNING INSTRUCTIONAL APPROACH IN HERITAGE ARCHITECTURE EDUCATION

Measured drawing is a method adopted in heritage documentation, a part of architecture heritage conservation process. Its importance lies in providing valuable insight into historical significance, documenting conditions, interpretation, and guiding conservation efforts (Andrews et al., 2010). Warden & Woodcock (2005) emphasise the significance of history and pedagogical values offered by historic architecture documentation. This enhances students' observational skills to develop three-dimensional understanding of the building, and allows them to accurately describe the building and appreciate its craftmanship.

Measured drawing is taught in many institutions as a learning approach in heritage architecture education. Akboy-İlk (2017) defines measured drawing as a formalised documentation product that records existing conditions of the building's architectural setting after it was constructed, enabling students to learn about the architectural contexts and expression of cultural values by analysing and interpreting the local environment, social and cultural settings. This involves fieldwork study within specific localities and communities, which leads students to address real problems defined under the framework of PBL.

The instructional approach of measured drawing is designed to address three defined problems: place identity, social contexts, and cultural contexts. It involves documenting the physical and non-physical attributes of identified heritage buildings. The working process begins by identifying a place or community with heritage and historical significance. Fieldwork is carried out by the students in a group. Site analysis is conducted to understand the tangible and intangible characteristics of the place and community. The documentation of buildings involves physical measurements and historical data collection through interviews, photography and archive materials review. Findings of the documentation are presented in the form of architectural drawings produced using Building Information Modelling (BIM) software, a scale model and a report. Acquired knowledge is disseminated and shared through exhibitions and sharing sessions organised with the local communities.

This process aligns with the PBL process involving five stages as proposed by Schechter (2011), shown in Table 1. The first and most essential stage involves problems identification. This shall introduce students to the complex, real-world problem within as specific context acting as learning stimulus. At this stage, students will identify, reframe and analyse the problems to generate hypothesis and derive learning goals. This requires students to define the real-world problems in heritage architecture learning: place identity, social contexts, and cultural contexts of the selected place and community. In Stage 2, students will engage in the process of problem-solving by actively seeking information. Using measured drawing, students will identify both tangible and intangible attributes of a place and create a plan for conducting fieldwork. Once on-site, they will select appropriate techniques for measuring, choose reliable resources for data collection from individuals or institutions, and utilise efficient methods for producing drawings. In Stage 3, students are required to generate and test the solutions, supported by data and existing research, using on-site measurement, interviews, photographic documentation, and the study of archive materials. The solution also incorporates collaborative cloud-based 3D modelling using BIM software and a physical scaled model creation. In Stage 4 students are required to review and appraise the findings, which involves reviewing the progress of 3D modelling and coordinating it using the BIM 360 cloud service application. Report findings are reviewed using the Google Drive application. In Stage 5, students engage in reflection on their learning experience and share the outcomes, through exhibition of physical scaled models, drawings, and reports. Additionally, they also participate in seminars and knowledge-sharing sessions to disseminate their acquired knowledge and experiences.

Table 1. Specifications of tasks involved in measured drawing module's PBL instructional approach

PBL Stages	Description	Specification of Tasks in Measured Drawing Module
1. Identify Problems	Define the problems	• Define place identity • Define social contexts • Define cultural contexts
2. Develop Solution Strategies	Identify what are needed to solve the problems	• Identify place contexts • Fieldwork planning • Determine appropriate measuring techniques • Select relevant individual/ institution and reliable resources for data collection • Identify efficient 3D modeling methods
3. Execute Solution Strategies	Make and test the best solution	• On-site measurement • Interviews, photography documentation and study of archive materials • Collaborative cloud-based 3D modeling using BIM software • Physical scaled model creation
4. Finding Review & Evaluation of Results	Appraise and review the findings	• 3D modeling progress monitoring and coordination review using BIM 360 cloud service application • Cloud-based report findings review using Google Drive application
5. Outcomes Sharing	Reflect on learning experience and share the outcomes	• Exhibition of physical scaled model, drawings and report • Seminar and sharing session

RESEARCH METHODS

This research investigates the cultural consciousness attainment of students who experienced the measured drawing module as part of the heritage architecture education. Quantitative survey research allows examination of specific aspects such as demographic features, behavioural displays and attitudes of an identified population. An online survey research design was adopted to measure architecture students' cultural consciousness acquisition of place identity knowledge, community's social contexts understanding and community's cultural contexts recognition through the PBL instructional approach adopted in the measured drawing module.

It seeks to answer the following research questions: (1) Will the PBL instructional approach of measured drawing module enable students' acquisition of place identity knowledge? (2) Will the PBL instructional approach of measured drawing module significantly affect students' understanding of the community's social contexts? (3) Will the PBL instructional approach of measured drawing module significantly affect students' recognition of the community's cultural contexts? The research questions lead to formulating three hypotheses as follows:

Hypothesis One H1: The PBL instructional approach adopted in the measured drawing module will have significant effects on students' acquisition of place identity knowledge.

Hypothesis Two H2: It is expected that the PBL instructional approach adopted in the measured drawing module will have substantial effects on students' understanding of the community's social contexts.

Hypothesis Three H3: It is likely that the PBL instructional approach adopted in the measured drawing module will significantly affect students' recognition of the community's cultural contexts.

The subject information in Table 2 shows a sample of 40 students who have completed the measured drawing module. Cohort X comprises 47.5% of the total respondents, who were tasked with documenting heritage buildings in Klang, Malacca, Kedah, and George Town, Penang. The heritage buildings encompass traditional Chinese shophouses, a traditional Chinese temple, a traditional Malay house, Malay palace, and church. The remaining 52.5% of respondents belonged to cohort Y and are focused on documenting a traditional Chinese temple, a church, and a colonial bungalow in Johor Bahru, Johor. There were 10.5% more female respondents than male respondents.

*Table 2. Personal characteristics * group crosstabulation*

Personal Characteristics		Total	Percentage (%)
Cohort	X	19	47.5
	Y	21	52.5
Total		40	100
Gender	Female	21	52.5
	Male	19	47.5
Total		40	100

An online questionnaire invitation was posted in the university's official Learning Management System (LMS) and social media to ensure wider accessibility. Email and social media invitations containing the survey link were also delivered to the target participants. Responses were directly recorded on a database file on the website server. As illustrated in Table 3, the online questionnaire consists of four sections to assess the following variables:

- Questions 1 & 2: respondents' personal characteristics (cohort and gender);
- Questions 3(a)-3(e): respondents' ratings on place identity knowledge acquisition;
- Questions 4(a)-4(e): respondents' ratings on community's social contexts understanding;
- Questions 5(a)-5(e): respondents' ratings on community's cultural contexts recognition

Responses were measured using Likert scaling method with 5 ratings (1 - strongly disagree to 5 - strongly agree). Respondents' personal characteristics (cohort and gender) were kept constant to ensure that they do not affect their evaluation. They were used to test the correlations between the respondents' personal characteristics with their ratings. Ratings on the place identity knowledge acquisition, community's social contexts understanding, and community's cultural contexts recognition categories would indicate the extent to which the PBL instructional approach adopted in the measured drawing module significantly affects students.

Table 3. Online questionnaire questions

Place Identity Knowledge Acquisition	Community Social Contexts Understanding	Community Cultural Contexts Recognition
3(a) I can identify the characteristics of the physical fabric of the place I have documented.	**4(a)** I am able to interpret the relationship between spatial layout of the building with the social structure of the users associated with it.	**5(a)** I can describe the cultural meanings of the ornamentation features of the building.
3(b) I can describe the distinctive building facades and architectural styles of the building I have documented.	**4(b)** I can describe the social status of the community related to the building I have documented.	**5(b)** I am able to gain understanding of the values and beliefs of the users or community related to the building.
3(c) I am able to tell about the place's history from my documentation work.	**4(c)** I am able to gain understanding of the organizational structure of the community related to the building I have documented.	**5(c)** I am able to gather knowledge of the communal activities i.e. prayers, rituals, consecration of ancestors, gathering of community members and etc.
3(d) I am able to gain knowledge about the place attributes by documenting identified traditional businesses and trades related to the community.	**4(d)** I can relate the social network of the community with the building I have documented.	**5(d)** I can identify distinctive cultural displays of the users or community associated with the building i.e. customs, art, food, clothing, language, ornamentations & etc.
3(e) I can differentiate the features of the place I documented from another place that I have been to.	**4(e)** I am conscious that community social contexts had significantly shaped the built environment of the place I have documented.	**5(e)** I can appreciate the cultural system of the community related to the building I documented.

Greene and D'Oliveira (1982) suggest that non-parametric statistical tests are appropriate for experimental data that are measured at ordinal level or nominal level. The data in this study are based on response-rating scales (ordinal data). Furthermore, non-parametric statistical tests are used when it's not possible to obtain an assumption of normal distribution (Sedgwick, 2015), particularly for the ordinal data in this study. Lucas (1991) also recommends the use of non-parametric statistical tests for small sample size. Hence, non-parametric tests are appropriate to analyse the data.

The analysis firstly examined the independence of respondents' personal characteristics using two-dimensional chi-square test using SPSS software. If the p value at the significance level of 0.05 ($p < 0.05$) is more than 0.05, respondents' ratings are not influenced by differences in their personal characteristics. To test hypotheses H1, H2 and H3, one-dimensional chi-square test was calculated. If the p value at the significance level of 0.05 ($p < 0.05$) is less than 0.05, the null hypothesis is rejected, indicating that the PBL instructional approach adopted in the measured drawing module has significant effects on students' cultural consciousness attainment. Figure 3 illustrates the research framework.

Figure 3. Research framework

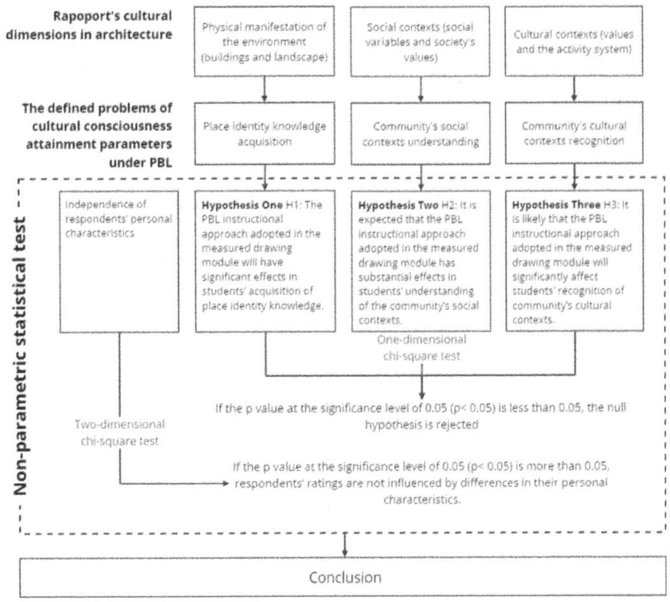

FINDINGS

Respondents' Characteristics Independence

Respondents' personal characteristics independence was examined using two-dimensional chi-square test. The results in Figure 4 show that overall p value for all tested variables were more than 0.05 ($p>0.05$), indicating that respondents' ratings were not influenced by differences in their personal characteristics. The personal characteristics factors were kept constant and did not significantly affect the evaluation.

Figure 4. Results of the two-dimensional chi-square test to test respondents' characteristics (cohort and gender) independence

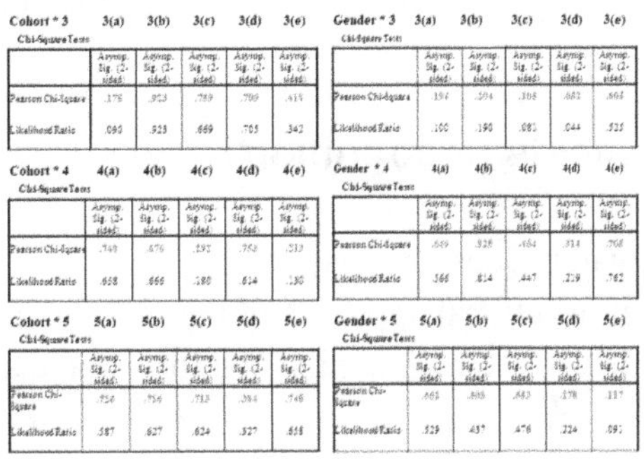

Place Identity Knowledge Acquisition

The null hypothesis for Research Question 1 is:

H_0: it is predicted the PBL instructional approach of the measured drawing module will have no significant effects on students' acquisition of place identity knowledge.

Table 4. Chi-square test statistics for place identity knowledge acquisition

Question	3(a)	3(b)	3(c)	3(d)	3(e)
Chi-Square(a)	26.600[a]	7.850[b]	18.000[a]	6.650[b]	23.600[a]
df	3	2	3	2	3
Asymp. Sig.	**.000**	**.020**	**.000**	**.036**	**.000**

a. 0 cells (0.0%) have expected frequencies less than 5. The minimum expected cell frequency is 10.0.
b. 0 cells (0.0%) have expected frequencies less than 5. The minimum expected cell frequency is 13.3.

From Table 4, p value for all questions 3(a)-3(e) in the place identity knowledge acquisition section were less than 0.05 ($p<0.05$). H_0 was rejected, indicating that the PBL instructional approach of the measured drawing module has significant effects on students' acquisition of place identity knowledge.

Community Social Contexts Understanding

The null hypothesis for Research Question 2 is:

H_0: It is predicted the PBL instructional approach of the measured drawing module will have no substantial effects on students' understanding of the community's social contexts.

Table 5. Chi-square test statistics for community's social contexts understanding

Question	4(a)	4(b)	4(c)	4(d)	4(e)
Chi-Square(a)	21.200[a]	10.400[a]	20.600[a]	44.750[b]	13.800[a]
df	3	3	3	4	3
Asymp. Sig.	**.000**	**.015**	**.000**	**.000**	**.003**

c. 0 cells (0.0%) have expected frequencies less than 5. The minimum expected cell frequency is 10.0.
d. 0 cells (0.0%) have expected frequencies less than 5. The minimum expected cell frequency is 8.0.

From Table 5, the p-values for all questions 4(a)-4(e) in the community's social contexts understanding section were less than 0.05 ($p<0.05$). H_0 was rejected, indicating that the PBL instructional approach of the measured drawing module has significant effects on students' understanding of the community's social contexts.

Community Cultural Contexts Recognition

The null hypothesis for Research Question 3 is:

H_0: It is predicted the PBL instructional approach of the measured drawing module will not significantly affect students' recognition of the community's cultural contexts.

Table 6. Chi-square test statistics for community's cultural contexts recognition

Question	5(a)	5(b)	5(c)	5(d)	5(e)
Chi-Square(a)	54.750[a]	30.800[b]	35.600[b]	22.200[b]	21.200[b]
df	4	3	3	3	3
Asymp. Sig.	**.000**	**.000**	**.000**	**.000**	**.000**

e. 0 cells (0.0%) have expected frequencies less than 5. The minimum expected cell frequency is 8.0.
f. 0 cells (0.0%) have expected frequencies less than 5. The minimum expected cell frequency is 10.0.

From Table 6, p value for all questions 5(a)-5(e) in the community's cultural contexts recognition section were less than 0.05 ($p<0.05$). H_0 was rejected, indicating that the PBL instructional approach adopted in the measured drawing module would significantly affect students' recognition of the community's cultural contexts.

DISCUSSIONS

Cultivating Cultural Consciousness Through PBL-Designed Learning Experiences

PBL serves as a pivotal tool in identifying and framing specific issues within real-world contexts. This is particularly significant in developing an understanding of the interplay between a place's characteristics and the formation of place identity. This establishes the problem that guides students' learning in heritage architecture, by exploring how the distinctive features of a place contribute to shaping its unique identity. Students are directed by a clear framework to document, analyze and interpret the characteristics of place identity, which ultimately shapes their cognitive understanding. Without a clear framework initiated by a problem, students may not be able to develop a depth of inquiry, and such learning would remain superficial.

As active participation engages and motivates students in their learning (Bergmark & Westman, 2018; Sharoff, 2019), PBL incorporates students' active learning in developing and executing solution strategies to address the established problem. Architectural drawings provide insights into the spatial configuration, building technology, and symbolic representations that exert profound influences on culture and society in terms of architectural production. However, a good execution of data collection is essential, involving physical measurements of buildings and gathering of research materials for drawings production and historical analysis. Students need to formulate the most effective ways to collect data, as reflected in the fieldwork planning where specific tasks are allocated to each group member. A clear work schedule outlines milestones and provides a roadmap to the successful completion of tasks within the stipulated timeframe.

Another important aspect of PBL is the development of collaboration in the learning process (Ariyanto & Muslim, 2019; Hendarwati et al., 2021; Murray-Harvey et al., 2013; Saldo & Walag, 2020). The cloud service provides an excellent collaborative platform for modeling work production and facilitating progress reviews. This platform allows the instructor to view the 3D models in real-time on the cloud and provide comments with direct annotations. Furthermore, the integration of Google Documents into the workflow enables students to make real-time amendments and facilitates the simultaneous viewing of the report by all team members. These settings not only increase overall productivity, but also contribute to a more efficient findings review and evaluation process.

Social engagement as a part of PBL process enhances students' learning experience (Arnold, 2022; Chung, 2019; Leidig & Oakes, 2021). Participating in exhibitions and seminars offers opportunities to further engage with real communities, thereby enriching students' learning experiences by connecting with real-world contexts. These interactions extend beyond traditional classroom settings and allow students to view their projects from a more holistic perspective. As they showcase their findings, students not only reflect on their individual learning experiences but also actively share the knowledge they have acquired with the public.

Figure 5. The PBL-designed learning experience in cultivating students' cultural consciousness, from the left, the physical measurement, the cloud-based model review, and an outcome sharing session with the local community

Source: Authors' (2023)

Acquiring Place Identity Knowledge

The findings from the study reveal that the PBL instructional approach adopted in the measured drawing module has positively contributed towards students learning. A large percentage of students, as indicated by the "Agree" and "Strongly agree" ratings in the survey, recognised the specific attainment of cultural consciousness elements in all three cultural dimensions of architecture.

The data presented in Figure 6 illustrate the extent to which students acquired knowledge of place identity through their learning experience. Indicators include identifying the physical fabric characteristics; describing the distinctive building façade and architectural styles; telling the place's history; understanding the place attributes through traditional business and trades; and distinguishing the features of the place.

Figure 6. Respondents' ratings for place identity knowledge inquiry

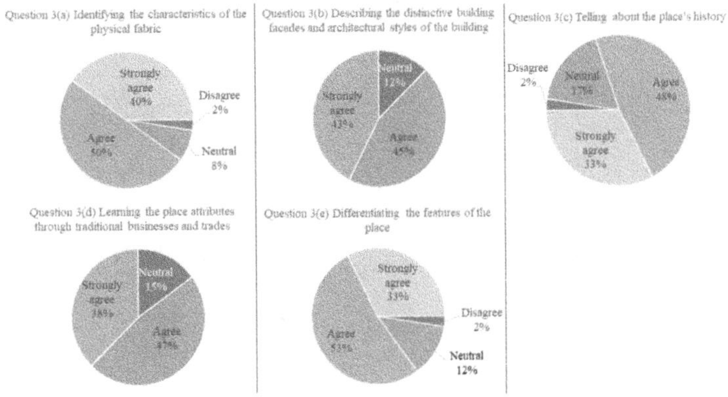

Most of the respondents agreed that they were able to acquire place identity knowledge from their learning experiences. As shown in Figure 6, 90% of them were able to identify the characteristics of the physical fabric of the place, such as the urban layout, road networks, buildings, public spaces, and so on. 87.5% of the students were able to describe the distinctive building facades and architectural styles of the heritage buildings they documented. This knowledge is further enhanced with the understanding of the place's history, as claimed by 85% of the respondents. 85% of the students agreed that they managed to learn about the place attributes through traditional businesses and trades. 85% of the students agreed that their learning experience enabled them to distinguish the place features they studied from another place that they have been to. It is evident that students were able to develop their cognitive understanding of a distinctive characteristics that form the place identity. This reflects the significance of place-specific architectural education as emphasised by Smith & Sobel (2004) and Nikezić & Marković (2015).

Understanding of Community Social Contexts

Figure 7 illustrates how students demonstrate understanding of community social contexts guided by the PBL approach in their measured drawing project. This encompassed interpreting the relationship between spatial layout and social structure, describing the social status of the community, understanding the community's organizational structure, establishing connections with the social network, and recognizing the influence of community social contexts on shaping the characteristics of a place.

Figure 7. Respondents' ratings for community social contexts understanding

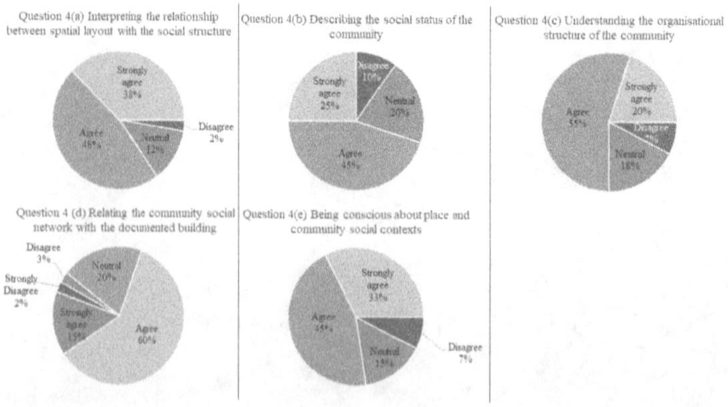

A significant number of students, at 85%, demonstrated the ability to interpret the relationship between the spatial layout of a building and the social structure of the users connected to it. This highlights their awareness of the interplay between architecture and societal dynamics. A substantial 70% of students were able to provide descriptions of the social status of the community associated with the documented heritage buildings, while 75% of the respondents managed to gain an understanding of the organizational structure within the community linked to the documented building. A further 75% of students were able to draw connections between the social network of the community and the buildings they documented. 77.5% of students recognized the pivotal role of community social contexts in significantly shaping the built environment of the place documented. These findings reflect the students' ability to develop comprehension of the broader social context related to heritage architecture and community spaces. They were also able to demonstrate analytical skills in identifying the influence of societal forces on the architectural development of a particular community. These findings highlight the significance of social-architectural knowledge and how PBL contributes to a deeper understanding of this relationship.

Recognising the Community Cultural System

Figure 8 demonstrates the students' ability to identify the cultural system within the community. This involved describing the cultural meaning of ornamentation features of the building; grasping the values and beliefs associated with the building; familiarizing themselves with communal activities; identifying distinctive cultural displays; and gaining an appreciation for the community's cultural system.

Figure 8. Respondents' ratings for community cultural contexts recognition

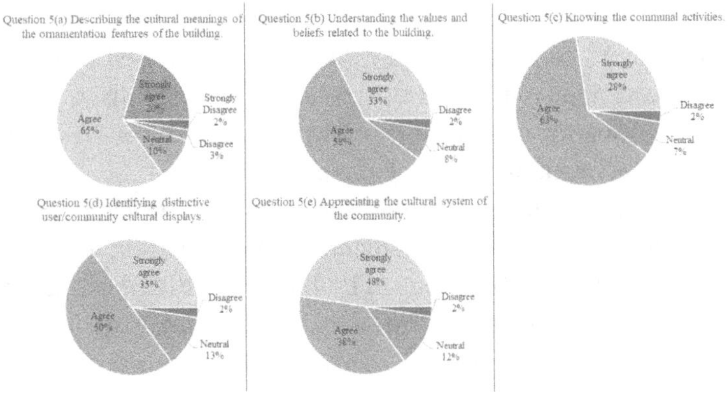

The questionnaire findings show that through their involvement in heritage building documentation, 85% of the respondents were able to describe the cultural meanings underlying the ornamentation features of the buildings. Furthermore, a vast majority of students, at 90%, understood the values and beliefs held by the users and communities associated with the buildings they documented. In addition, an equivalent 90% of the students effectively gathered knowledge concerning the communal activities, encompassing prayers, rituals, ancestral consecrations, community gatherings, and more. 85% of the respondents showcased their ability to identify distinctive cultural displays of the users or communities associated with the buildings. Finally, a substantial 85% of the students demonstrated appreciation towards the cultural system embedded within the communities related to the buildings they documented. These findings indicate students' profound understanding of cultural influences in shaping architecture. Through heritage architecture documentation, they learned about the aesthetic and symbolic dimensions of architectural design and the empathetic approach to cultural perspectives. The students' comprehensive insight into the cultural practices within the communities is evidently portrayed. This understanding reflects not only their cognitive learning in heritage architecture, but also their respect for cultural diversity of the various communities they encountered during their study. Recognizing the cultural activities within a community strengthens learners' knowledge of a place (Koh & Ahamad, 2023). This reinforces the role of PBL in enhancing understanding of culture and architecture.

FUTURE RESEARCH DIRECTIONS

Further investigation on how PBL affects students' cultural consciousness could take into account their personal experiences. A qualitative approach in a subsequent study could also further explore the scope of experience to identify emerging themes that highlight PBL's contribution in their learning. Additionally, investigating the long-term effects of PBL on students' ability to integrate cultural sensitivity into their design studio projects should be considered for future research.

CONCLUSION

In addition to addressing technical aspects, architectural design challenges often encompass complex dynamics related to place, community, and culture. As emphasised by the International Union of Architects, architectural education must inculcate problem-solving skills to address these multifaceted aspects (UIA, 2023, p.5). Adopting a problem-based learning framework is crucial in training students to address issues that intersect with these critical dimensions.

Jadresin Milic et al. (2022) highlight that the advancement of teaching and learning architectural history depends on heritage, conservation, community engagement, digital technology, and expertise in measuring and drawing buildings. According to Philokyprou (2011), heritage architecture education equips students with foundational knowledge and analytical skills for assessing vernacular buildings and historical settlements across various dimensions. This cultivates cultural consciousness by recognising the diversity of cultures, particularly in societies with diverse cultural characteristics. Therefore, learning from the real-world contexts through heritage architecture documentation appears to be significant.

As Akboy-İlk (2007) highlights, measured drawing allows the establishment of deep relationship between the inquirer with the architectural fabric throughout the process of documentation, analysis, observation, measuring and interpretation. Such an enquiry process, supported by a clearly defined PBL framework, enables students to elucidate the meanings of architecture within a specific place and community.

Findings from the research substantiates the adoption of PBL as an effective pedagogy to improve students' cultural consciousness attainment. Positive evaluation outcomes of this research serve as an important indicator of students' ability to develop contextual and cultural sensitivity. With the acquired learning experiences involving place identity knowledge, community social contexts, and their corresponding cultural system, students become able to develop their design cognition consciousness towards place contexts which is critical in preserving its cultural sustainability. The PBL framework is also useful in guiding the design of teaching approaches for cultural related subjects in other disciplines.

ACKNOWLEDGMENTS

This research received no specific grant from any funding agency in the public, commercial, or not-for-profit sectors. The authors would like to thank all the students who were involved in this research.

REFERENCES

Akboy-İlk, S. (2017). Crafting the architectural measured drawings. *The Plant Journal*, 2(1), 39–61.

Al Husban, S. A. M., Al Husban, A. A. S., & Al Betawi, Y. (2021). The Impact of the Cultural Beliefs on Forming and Designing Spatial Organizations, Spaces Hierarchy, and Privacy of Detached Houses and Apartments in Jordan. *Space and Culture*, 24(1), 66–82. 10.1177/1206331218791934

Alhassan, N. U., Aede, H. M., Hassan, A. M., & Idris, A. M. (2020). *Effects of place-based and activity-based approaches in technical education, interest and retention.*

Andrews, D., Bedford, J., Blake, B., Cromwell, T., & Lea, R. (2010). *Measured and Drawn: Techniques and practice for the metric survey of historic buildings | Historic England* (J. Bedford & H. Papworth, Eds.). English Heritage.

Ariyanto, S. R., & Muslim, S. (2019). Collaborative Problem-Based Learning Models Implementation in Vocational High Schools. *1st Vocational Education International Conference (VEIC 2019)*, (pp. 238–245). IEEE. 10.2991/assehr.k.191217.039

Arnold, M. (2022). Problem-based Learning and Community Engagement: A Service-Learning Project with Social Pedagogues. In Fahr, U., Alessandra, K., Angenent, H., & Eßer-Lüghausen, A. (Eds.), *Hochschullehre erforschen* (pp. 199–214). Springer Fachmedien Wiesbaden., 10.1007/978-3-658-34185-5_11

Banerjee, H. K., & Graaff, E. D. (1996). Problem-based learning in architecture: Problems of integration of technical disciplines. *European Journal of Engineering Education*, 21(2), 185–195. 10.1080/03043799608923402

Bergmark, U., & Westman, S. (2018). Student participation within teacher education: Emphasising democratic values, engagement and learning for a future profession. *Higher Education Research & Development*, 37(7), 1352–1365. 10.1080/07294360.2018.1484708

Bigelow, J. D. (2004). Using problem-based learning to develop skills in solving unstructured problems. *Journal of Management Education*, 28(5), 591–609. 10.1177/1052562903257310

Chen, S., Wang, S., & Xu, H. (2017). Influence of Place Identity on Residents' Attitudes to Dark Tourism. *Journal of China Tourism Research*, 13(4), 338–356. 10.1080/19388160.2017.1401023

Chung, E. Y. (2019). Facilitating learning of community-based rehabilitation through problem-based learning in higher education. *BMC Medical Education*, 19(1), 433. 10.1186/s12909-019-1868-431752842

Clarke, N., Kuipers, M., & Stroux, S. (2020). Embedding built heritage values in architectural design education. *International Journal of Technology and Design Education*, 30(5), 867–883. 10.1007/s10798-019-09534-4

Curto, R., Barreca, A., Coscia, C., Ferrando, D. G., Fregonara, E., & Rolando, D. (2021). The Active Role of Students, Teachers, and Stakeholders in Managing Economic and Cultural Value, Urban and Built Heritage. *The Interdisciplinary Journal of Problem-Based Learning*, 15(1), 1. 10.14434/ijpbl.v15i1.29626

De Merode, E., Smeets, R., & Westrik, C. (2004). Linking universal and local values: Managing a sustainable future for world heritage. *World Heritage Paper*, (13).

Dey, A. K. (2001). Understanding and using context. *Personal and Ubiquitous Computing*, 5(1), 4–7. 10.1007/s007790170019

Djabarouti, J., & O'Flaherty, C. (2019). Experiential learning with building craft in the architectural design studio: A pilot study exploring its implications for built heritage in the UK. *Thinking Skills and Creativity*, 32, 102–113. 10.1016/j.tsc.2019.05.003

Embaby, M. E. (2014). Heritage conservation and architectural education: "An educational methodology for design studios". *HBRC Journal*, 10(3), 339–350. 10.1016/j.hbrcj.2013.12.007

Greene, J. H., & D'Oliveira, M. (1982). *Learning to use statistical tests in psychology: A student's guide*. Open University Press.

Guerra, A., & Kolmos, A. (2011). Comparing problem based learning models: suggestions for their implementation. *PBL across the disciplines: Research into best practice*, 3-16.

Gunewardene, N. (2016). Documentation of vernacular heritage in architectural education: "Measured Drawing of Devala Buildings" with Level 1 students of architecture, University of Moratuwa, Sri Lanka. *The International Journal of Environmental Studies*, 73(4), 502–511. 10.1080/00207233.2016.1178983

Hendarwati, E., Nurlaela, L., Bachri, B., & Sa'ida, N. (2021). Collaborative problem based learning integrated with online learning. [iJET]. *International Journal of Emerging Technologies in Learning*, 16(13), 29–39. 10.3991/ijet.v16i13.24159

Hendrix, J. S. (2012). Architecture as the Psyche of a Culture. In Emmons, P., Lomholt, J., & Hendrix, J. S. (Eds.), *The cultural role of architecture: Contemporary and historical perspectives*. Routledge.

Hessam, G. K., & Sotoue, E.-D. (2016, January 01). Evaluation of the relationship between culture and traditional architecture and its effects on design quality improvement. *International Journal of Applied Engineering Research: IJAER*, 11(3), 2120–2123.

Jadresin Milic, R., McPherson, P., McConchie, G., Reutlinger, T., & Singh, S. (2022). Architectural history and sustainable architectural heritage education: Digitalisation of heritage in New Zealand. *Sustainability (Basel)*, 14(24), 16432. 10.3390/su142416432

Jadresin Milic, R., & Nikezic, A. (2018). Communicating Heritage Through Intertwining Theory and Studio Based Course in Architectural Education. In Amoruso, G. (Ed.), *Putting Tradition into Practice: Heritage, Place and Design* (pp. 626–635). Springer International Publishing. 10.1007/978-3-319-57937-5_65

Jokilehto, J. (2007). An international perspective to conservation education. *Built Environment*, 33(3), 275–286. 10.2148/benv.33.3.275

Koh, J. H., & Ahamad, N. A. (2023). A Phenomenological Study of Learner's Virtual Place Experience. In *Handbook of Research on Inclusive and Innovative Architecture and the Built Environment* (pp. 368–386). IGI Global. 10.4018/978-1-6684-8253-7.ch020

Lapadula, M. I., & Quiroga, C. (2012). Heritage as a pedagogical resource and platform for exploration in architectural design education. *Journal of Architecture (London)*, 17(4), 591–607. 10.1080/13602365.2012.709028

Leidig, P., & Oakes, W. (2021). Model for project-based community engagement. *International Journal for Service Learning in Engineering. Humanitarian Engineering and Social Entrepreneurship*, 16(2), 1–13.

Li, R. M. (2016). *Jianzhu lishi kecheng jiaoxuezhong de kunhuo yu duice: Yi Qingdao Binhai xueyuan jianzhulishi kecheng weili* [Problems of Architectural History Teaching and their Solutions: A Case of the Qingdao Binhai College Architectural History Course]. *Jiancai yu zhuangshi [Construction Materials & Decoration], 1*, 176–177.

Li, W. (2018). A Research on Undergraduate Architecture Teaching Approach Based on Integration of Architectural Design and Architectural History Teaching. *Creative Education*, 9(12), 12. 10.4236/ce.2018.912135

Lucas, H. C., Jr. (1989). Methodological issues in information systems survey research. The Information Systems Research Challenge: Survey Research Methods, New York.

McKeen, H. (2019). Culturally Relevant Preparedness Using STEAM Integrative Teaching in the Classroom. In Keengwe, J., & Onchwari, G. (Eds.), *Handbook of Research on Assessment Practices and Pedagogical Models for Immigrant Students* (pp. 127–138). IGI Global. 10.4018/978-1-5225-9348-5.ch008

McMINN. J., & Polo, M. (2005, September). Sustainable Architecture as a Cultural Project. In *The 2005 World Sustainable Building Conference, Tokyo* (pp. 27-29).

Morel-EdnieBrown, F. (2012). Wither Genius Loci?: The City, Urban Fabric and Identity in Perth, Western Australia. *The Role of Place Identity in the Perception, Understanding, and Design of Built Environments*, 209.

Murray-Harvey, R., Pourshafie, T., & Reyes, W. S. (2013). What Teacher Education Students Learn about Collaboration from Problem-Based Learning. *Journal of Problem Based Learning in Higher Education*, 1(1), 114–134.

Ng, F. P. (2013). Values of Learning Through 'Place-making" in the Design Studio. *Archnet-IJAR: International Journal of Architectural Research*, 7(1), 86–98.

Nikezić, A., & Marković, D. (2015). Place-based education in the architectural design studio: Agrarian landscape as a resource for sustainable urban lifestyle. *Sustainability (Basel)*, 7(7), 9711–9733. 10.3390/su7079711

Olweny, M. R. O. (2020). Architectural education in sub-Saharan Africa: An investigation into pedagogical positions and knowledge frameworks. *Journal of Architecture (London)*, 25(6), 717–735. 10.1080/13602365.2020.1800794

Páez, M., & Albert, L. (2012). Cultural consciousness. In Banks, J. A. (Ed.), *Encyclopedia of diversity in education* (Vol. 1, pp. 510–510). SAGE Publications, Inc., 10.4135/9781452218533.n160

Pasha, Y. N. (2020). Presence Of Culture In Architectural Education: A Case Study Of METU School of Architecture. *METU Journal of the Faculty of Architecture*, 37(1), 95–116. 10.4305/METU.JFA.2020.1.7

Philokyprou, M. (2011). Teaching conservation and vernacular architecture. *Journal of Architectural Conservation*, 17(2), 7–24. 10.1080/13556207.2011.10785086

Postalcı, I. E., & Atay, G. F. (2018). Rethinking on Cultural Sustainability in Architecture: A Reading on Projects of Behruz Çinici.

Rapoport, A. (2005). Culture, architecture, and design: Locke Science Pub. *Co., Chicago, US.*

Saldo, I. J. P., & Walag, A. M. P. (2020). Utilizing problem-based and project-based learning in developing students' communication and collaboration skills in physics. *American Journal of Educational Research*, 8(5), 232–237.

Schechter, C. (2011). Switching cognitive gears. *Journal of Educational Administration*, 49(2), 143–165. 10.1108/09578231111116707

Sedgwick, P. (2015). A comparison of parametric and non-parametric statistical tests. *BMJ (Clinical Research Ed.)*, 350(apr17 1), 350. 10.1136/bmj.h205325888112

Sharoff, L. (2019). Creative and innovative online teaching strategies: Facilitation for active participation. *The Journal of Educators Online*, 16(2), n2. 10.9743/JEO.2019.16.2.9

Smith, G. A., & Sobel, D. (2014). *Place-and community-based education in schools*. Routledge. 10.4324/9780203858530

Troiani, I., Ewing, S., & Periton, D. (2013). *Architecture and Culture: Architecture's Disciplinarity*. Architecture and Culture., 10.2752/175145213X13760412749917

UIA. (2023). *UNESCO-UIA Charter for Architectural Education*. Union Internationale des Architectes.

UNESCO. (2002). *UNESCO universal declaration on cultural diversity: A vision, a conceptual platform, a pool of ideas for implementation, a new paradigm*. UNESCO.

Valentín-Gamazo, D. V., Durango, N. T., Garcia-Herrero, F., Blanco, M. I. M., & de las Moras, M. C. R. (2021). Project-Based Learning with historic buildings: Immersion in real training environments for the Degree in Technical Architecture. *Proceedings of the International Conference of Innovation, Learning and Cooperation 2021, 3129*. IEEE.

Warden, R., & Woodcock, D. (2005). Historic documentation: A model of project based learning for architectural education. *Landscape and Urban Planning*, 73(2), 110–119. 10.1016/j.landurbplan.2004.11.003

KEY TERMS AND DEFINITIONS

Cultural Consciousness: Understanding and recognition of diverse cultures, resulting in a deeper knowledge and appreciation of other individuals and contexts.

Cultural Context: Values and activity systems of the community.

Measured Drawing: A standardised method of documenting the existing architectural conditions of a building.

Place Identity: The perceived meanings derived from the spatial setting that determine the values associated with the human relationship with the place.

Problem-Based Learning: An instructional method that utilizes problems to help students develop problem-solving abilities and acquire fundamental knowledge.

Social Context: The settings depicted by social variables and society values of a particular place.

Chapter 3
A Review and Comparative Study of Retrofit for the Energy Efficiency of Malaysian Heritage Buildings

Khairool Aizat Ahmad Jamal
https://orcid.org/0000-0003-4674-581X
Taylor's University, Malaysia

Siti Norlizaiha Harun
https://orcid.org/0009-0004-5658-1155
Universiti Teknologi MARA, Malaysia

Shahrul Yani Said
https://orcid.org/0000-0002-7727-5438
Universiti Teknologi MARA, Malaysia

ABSTRACT

Historic buildings (HB) are invaluable embodiments of culture and identity. Retrofitting HB has become a strategic imperative in Malaysia, given their substantial presence of building stock, moreover these buildings grappled with contemporary challenges. Energy retrofit allows adaptation and reuse while prioritizing environmental sustainability. However, such notions are still in their infancy stage, hindered by the gap between understanding heritage preservation techniques and energy efficiency consideration. The lack of inclusive guidance is an obstacle, restricting widespread application of retrofit intervention in HB conservation. This chapter aims to identify current limitations of Malaysia's Conservation Guidelines (HBCG) through a comparative analysis with similar international guidelines. Findings reveal lack of interpretations for contemporary green issues, unlike the latter which allows broader interpretation. The paper suggests additional guidance for practitioners in navigating environmental challenges, emphasizing intangible considerations such as heritage values and identity.

DOI: 10.4018/979-8-3693-1650-4.ch003

INTRODUCTION

Historic Buildings (HB) stand as iconic representations of Malaysia's rich historical and cultural heritage, reflecting the nation's background and character and generate significant economic benefits through urban vibrancy (Besen, 2018). Hence, it is vital to adapt and preserve their values for future generations. Heritage building conservation is not new to the Malaysian built environment industry. According to Harun (2011), heritage buildings represent the values of tangible cultural heritage, which have their own uniqueness and characters that need to be preserved. There are several importance and contributions of historical buildings in Malaysia, including, among others, preserving the history and beliefs of a society, maintaining the importance of architectural and cultural aspects, psychological aspects, and values of historic buildings, providing an image and identity to the city, providing continuity and stability in the built environment, as interpretational elements and tourism image, and providing heritage education to the young generation (Harun, 2021). In current record by the Malaysian Heritage Department, more than 125 historic buildings that are gazetted as national heritage (Jabatan Warisan Negara, 2023), with many potential future submissions. Since the addition of Georgetown and Malacca to the UNESCO World Heritage List on July 7, 2008, there has been a surge in demand for heritage conservation projects in Malaysia. In a preliminary study, Syed Zainol et al. (1995) reported that there are more than 39,000 historic buildings identified in Malaysia that were built between 1800 and 1948 in 235 towns and cities within Malaysia. Ali et al. (2018) highlights the distribution of historical buildings in Malaysia and reveal that over 20,000 buildings can be considered heritage-significant and deserving of conservation and preservation. Within established cities such as Kuala Lumpur, Georgetown, and Ipoh combined, a total of over 7500 historic buildings of various categories are recorded within the core areas alone (Ipoh, 2020; Georgetown, 2021; KLCP, 2020).

Hence, this signifies that historic buildings represents a considerable portion of the Malaysian building stock, and immediate actions need to be provisioned to protect such heritage assets, prolonging their survival from various threats and challenges (Hasan et al., 2019). Unfortunately, such buildings faced many challenges of the 21st century, which has alarmed the global heritage-built practice to rethink its current processes (Khalaf, 2021). With many of these buildings facing threats of defects, negligence, deterioration, and loss of performance over time, crucial adaptation measures are needed to be in place to ensure the survival of historic buildings, especially in contemporary settings (Ali et al., 2018; Harun, 2011). The effects caused by climate change have drastically degraded the building energy performance (BEP) and thermal comfort in Malaysia (Campagna & Fiorito, 2022). Studies have indicated traditional historic buildings in contemporary settings generally recorded poor environmental performance with high projections of energy consumption, contributed a high amount of carbon dioxide (CO_2) emissions, and need appropriate measures to improve their environmental performance (Khalil et al., 2021; Tang, 2019; Yau & Hasbi, 2017). These effects are widely held due to various conditions faced, being under threats of deterioration, were predominantly built in a time where energy efficiency regulation was limited, not being built sustainably, the types of materials used, and having less consideration for maintenance (Yazdani et al., 2018; Fillipi, 2015; Jagarajan et al., 2015; MacDonald et al., 2009). However, historic buildings could potentially achieve much more energy and carbon reduction compared to new construction, as it would take new buildings many years to achieve optimum sustainability impact as compared to historic properties.

Realizing the impact, Malaysia has been apportioned the importance of fostering green sustainability to address the environmental challenges in the recent years. The threats of climate change, the exhaustion of finite natural resources, and occupational health demands have driven countries to make immediate efforts towards environmental sustainability. Amongst the salient objectives was to meet the United Nations Sustainable Development Goals (UNSDG), among which the key measures are to reduce carbon emissions in the building sector by 45% by 2030 and to reach net-zero emissions by 2050 (UN, 2022; UNSDG, 2015; INDC, 2015). According to a report by the International Energy Agency (IEA), the building sector has a significant impact on energy intensity, which accounted for around 36% of global energy consumption and 39% of energy-related CO_2 emissions in 2017 (Abergel et al. 2017). Subsequently, in alignment with the determination as set out by the agenda and to mitigate climate change and non-renewable fuel dependency, the global energy intensity in existing building stock needs to be improved to at least 30% lower than the current level by the year 2030, with policies and actions initiated by the Malaysian government ranging from adopting sustainable material optimization, adaptation, resilience strategies, building operations, energy efficiency, renewable energy, green technology adoption, green renovation, and retrofitting (Sharma et al., 2022; International Energy Agency, 2017).

Hence, to promote building use and enhance the sustainability of historic cities, it would be advantageous to incorporate this recognized architectural heritage, into energy efficiency measures, even if HB in Malaysia is not covered by these plans. However, in order to create an energy retrofit plan that preserves the buildings' historical values and significance, a number of studies are needed. As there are many countries that are taking early steps and measures in this direction, for instance, in the United Kingdom (UK), there is a significant shift towards environmental sustainability concerning HB to achieve energy efficiency and carbon reduction, which has become an essential part of major policies, political, social, and economic agendas in the region (Historic Scotland, 2021; Historic England, 2018). These countries pledge to reduce carbon emissions, climatic impact and increase energy efficiency in historic buildings were manifested in a number of guidelines to enhance historic building's environmental performance, led by accountable heritage-built organizations.

Malaysian Heritage Building Conservation Guidelines

In responding to such phenomena, heritage bodies need to realize the impact, and this could be done through immediate initiatives. Policies and guidelines serve as an important tool to outline the scope and process for HB conservation practitioners, with the aim of ensuring that care processes take place in an evidence-based and structured manner (Harun, 2011). In Malaysia, there are several guidelines that detail the steps required for proper conservation of historic building elements, from early to post-conservation policies and statutory documents. The main piece of legislation for the protection and management of historic buildings is illustrated by the National Heritage Act 645 of 2005 (Act, 2005), which aims preserve and preventing heritage and culture from inappropriate use, subdivision, and development (Hariry, 2021). This has influenced the creation of the Historic Building Conservation Guideline 2016 (HBCG). Within the parameters of established government policy, the guidelines offer comprehensive information on conservation techniques and historic environment management. Several guidelines on heritage conservation practice and development control were developed in the structure of the HBCG dedicated to government conservation projects (Public Works Department, 2018) and local authorities of urban areas such as Kuala Lumpur, Ipoh, Georgetown, and Malacca. There are consultations as to whether historic buildings in city areas will be protected or not by the proposed 'Draft Special Area

Plan', which raised concerns about valued heritage and character areas being threatened by demolition (Jabatan Warisan Negara, 2016; Ipoh, 2020; Georgetown, 2021).

However, with the alarming issues of environmental sustainability, there is insufficient guidance concerning the energy retrofit works of HB within Malaysia. To cultivate effective practices for energy-efficient retrofitting of historical buildings, regulatory policies and frameworks must be introduced at the country level or even region-specific level (Jahed et al., 2020). The intersection of energy and conservation philosophies is a broader concern. Hence, developing new methodological frameworks in this realm would be possible with an appraisal of limitations and opportunities in various regulatory frameworks in different contexts. In recent years, several analyses have already been carried out to assess Malaysia's current heritage conservation policy and guidance (Azis et al., 2022; Zaid et al., 2021); however, none addresses the lack of technical and practical intervention frameworks for improving the energy efficiency of historic buildings. Through the issues discussed, the research aim is to identify the current limitations of the HBCG pertaining to historic energy retrofit practice and to provide context through discussions surrounding the issue. In relation to that, this paper identified two main research gaps to be studied, which consist of the following: (1) issues concerning the low energy performance of Malaysian historic buildings; and (2) the infancy of preservation guidance on energy efficiency aspects of historic buildings. Furthermore, these inquiries can aid conservation professionals in gaining a further understanding of the current state of energy retrofit practice in Malaysia.

LITERATURE REVIEW

Environmental Sustainability Challenges on Historic Buildings

At this point, it is important to understand that historic buildings are a finite resource that resembles the significant values, spirit, and identity of a country and place. However, as time eventually progresses, such historical assets are under constant pressure and challenges to improve their environmental sustainability performance (Soares et al., 2019). The present issues and debate on climate change, natural resource depletion, and global demands for energy security underscore the need for a more sustainable, heritage-built environment. The act of degradation, deterioration, and loss of performance in heritage buildings is caused by a variety of factors, which have resulted in a list of concerns and threats facing heritage buildings in Malaysia. In UNESCO world heritage sites, such as Penang and Malacca, heritage buildings have been cyclically adapted for new uses over time; one example is the Penaga Hotel building in Penang, whose state of building could be improved to meet the Green Building Index (GBI) criteria (Dewiyana et al., 2016; Harun, 2021). This cycle is part of the history of these buildings, which must be preserved. However, to guarantee their use in the future, these buildings must also be adapted, repaired, rehabilitated, and retrofitted to face future environmental challenges and user expectations regarding thermal comfort and energy efficiency.

Developing countries like Malaysia rely mainly on fossil fuels to supply the energy demand, which has many negative impacts on the environment (Chan et al., 2022; Moey et al., 2020).

Malaysia has a rising energy demand due to the growing development in the industrial, transportation, and agriculture sectors. The growth demand for electricity in Peninsular Malaysia from 2016 to 2021 was 21,773 MW to 23,344 MW (Malaysia Energy Statistics Handbook, 2021) and is expected to reach peak demand by the end of 2035. According to the Energy Commission Malaysia (2021), the total amount of installed energy capacity in 2021 will be 36,182.8 MW, with the biggest contributor to electricity generation coming from fossil fuel sources (82.1%), which come from sources of coal, gas, diesel, and oil (Energy Commission, 2021). Balaras et al. (2004) highlight that a modern building built after the year 1990 recorded an average energy use intensity (EUI) of around 40 kWh/m2, while buildings built before the year 1930, which are mostly considered old historical buildings, consumed an average of 170 kWh/m2. This is diminished further by the anticipated quick shift in climatic conditions, which might pose further threats to future heritage buildings. Furthermore, according to the United Nations (UN) Office for Disaster Risk Report (UNDRR, 2020), Malaysia is expected to experience up to 10% more annual rainfall owing to climate change, with mean annual surface temperatures rising by approximately 2° Celsius by 2050. In a study by Tang (2019), projected simulations predicted a temperature rise of less than 4°C till the end of the 21st century and a lack of a definitive pattern for rainfall, with increasing rainfall towards the end of the 21st century.

As for historic buildings, Yau & Hasbi (2017) have projected that from the baseline of the year 2000, the cooling demand is expected to gradually increase by 2.96 percent, 8.08 percent, and 11.7 percent in 2020, 2050, and 2080, respectively. With the maximum cooling load of 297,000 KJ/h generated in the year 2000, the increase is translated up to 330,000 KJ/h by the year 2080. The increase in maximum cooling demand is linked to a projected increase in external temperature, which causes more heat to penetrate spaces through the building fabric. Several case studies concluded substantial benefits can be gained through the practice of HB energy retrofit. For example, in the United Kingdom (UK), energy retrofit is capable of optimizing the building thermal value by as much as 70% with a significant improvement to Band B of the Energy Performance Certificate (Historic Scotland, 2020), and in Turkey, 37% of energy reduction was recorded through retrofit plans (Cho et al., 2022). Therefore, this signifies actions need to be undertaken to improve the performance and resilience of historic buildings to become adaptable to the challenges of the 21st century (Dias et al., 2021), which impose unpredictable weather behavior, higher temperatures, and increased flood risk as climate change takes effect.

Malaysian Policies and Initiatives Towards Energy Efficiency

The Malaysian government has undertaken many initiatives and measures to ensure the pursuit of environmental sustainability within the heritage-built environment to fulfill global climate aspirations, leveraging technological and innovative efforts. Since the time when the government's implementation of green policies such as the National Policy of Climate Change 2009 (NPCC, 2010) and the National Green Technology Policy 2009 (NGTP, 2009), has driven and established explicit directions and a set of expectations towards environmental sustainability and energy retrofit of existing buildings in the twenty-first century, The key expectations involved retrofitting existing buildings with efficient ventilation and cooling systems and lighting systems, the adoption of green technology, including energy efficiency (EE) features and renewable energy systems (RES), and enhancing energy conservation practices in buildings. Subsequently, the Malaysian Low Carbon Cities Framework Revision 2 highlighted specific performance criteria to reduce carbon emissions through retrofitting by extending the life cycle of existing building stock and enhancing the building's energy performance (LCCF, 2015; NPCC, 2010).

In the 12th Malaysia Plan, it is highlighted that the development and utilization of renewable energy (RE) resources will be intensified in many aspects to meet the target of 31% renewable energy of the total installed capacity. However, with such planning in place, the government has also recognized the impact and obstacles of implementing environmental sustainability practices in existing HB. Although there seems to be an awareness of conserving existing buildings for energy retrofit, there are still major barriers that hinder its application, especially at the implementation stage, which revolves around people, policies, and process aspects. According to the Green Technology Masterplan Malaysia 2017-2030 (GTMM, 2017), although retrofitting is recognized as an important approach to low energy and carbon reduction, it appears that none of the national or state-level regulatory instruments and codes of assessment are prepared to meet the needs of retrofitting; additionally, current green building certification has progressed primarily for new building categories, with limited uptake by existing and historic buildings. As a result, there is a lack of enforcement for retrofitting existing buildings and understanding among industry stakeholders about implementing responsible retrofit projects (Dewiyana et al., 2016; Al-Obaidi et al., 2017; Zaid et al., 2021; Che Husin, 2017; Mansir et al., 2016; Harun, 2011).

An Overview of Historic Building Retrofit

The current national-level heritage conservation guideline does not include a specific definition of retrofitting. However, the process of updating and reusing building components, which is similar to retrofitting characteristics, is expressed through a number of terms, including refurbishment, restoration, rehabilitation, and adaptation. These terms can be confusing, as such an approach can usually be defined similarly to retrofit (Sopery & Ayob, 2023). However, this has added a plethora of terms used to cover retrofit, which, as described by Wilkinson (2011), often exist in a 'state of pleasant confusion'. Hence, a distinction between these terms should be identified and clearly defined to establish a clear context for retrofitting. The terms retrofit and refurbishment are often used as if they are interchangeable. However, they have different meanings. Fundamentally, the term retrofit, as derived from Douglas (2006), is defined as "any work to a building above and beyond maintenance to change its capacity, function, or performance" or "any intervention to adjust, reuse, or upgrade a building to suit new conditions or requirements." According to Eames et al. (2014), retrofit means to provide something with a component or accessory not fitted during manufacture or to add something that did not exist when first constructed. It is frequently related to the installation of new building systems, but it may also refer to the fabric of a building, such as retrofitting insulation, walls, or windows. To further extend,

Historic England (2018) defined retrofit as the process of improving the energy and environmental performance of buildings through technical interventions, although to achieve benefits, this process often needs to encompass occupants' lifestyle changes and involves a systematic program of upgrades, repairs, and maintenance. The key focus of historic building retrofit is to achieve energy savings while maintaining satisfactory service levels and acceptable indoor thermal comfort suitable for contemporary lifestyles. (Almeida., 2014; Moran et al., 2013; Ma et al., 2012). For instance, in the building fabric, energy efficiency measures can be assessed and applied to modifications of building components such as walls, doors, windows, floors, and roofs, thereby cutting heating and cooling costs, energy use, and carbon emissions (Tobias, 2004). However, there are other important aspects of interventions to reduce these impacts, which include modifications to building systems, including natural daylighting and artificial lighting controls, indoor air quality, energy performance, ventilation, and using renewable energy.

METHODOLOGY

The study adopted a documentary analysis aligning to comparative study analyse the data and interpretations from the existing literature which consists of grey literatures and relevant scientific publications. The paper suggested three primary grey documents to be studied, namely the Guidelines on Heritage Building Conservation 2016 (Jabatan Warisan Negara, 2016), Conservation Principles, Policies and Guidance for the Sustainable Management of Historic Environment (Historic England, 2008) and The Secretary of the Interior's Standards for the Treatment of Historic Properties with Guidelines for Preserving, Rehabilitating, Restoring & Reconstructing Historic Buildings (National Park Service, 2017), was selected as the main documents for this study with the following reasons.

1. Although there are many guidelines related to HB conservation, the paper intends to place the scope to HBCG (Jabatan Warisan Negara, 2016), as it acts as the principal guideline in historic building conservation and is adopted in much guidance in Malaysia. Recognizing the existing content, along with its constraints, will enable the research to pinpoint its deficiencies and propose new criteria for enhancing the document with energy retrofit.
2. Historic England (2008) is a prominent guideline that focused on sustainable management of the historic environment. This document outlines principles, policies, and guidance aimed at conserving and managing historic sites, buildings, and landscapes in a way that preserves their cultural significance while also promoting their sustainable use. Such guidance places a focus on beneficial change and adaptability in novel treatments, which is lacking from other guidelines.
3. National Park Service (2017) is an established technical preservation brief providing a framework for the preservation, rehabilitation, restoration, and reconstruction of historic properties in the United States. These standards serve as guidelines for professionals, such as architects, historians, and preservationists, who are involved in the care and maintenance of historic buildings (Azis et al., 2022).

Furthermore, a comparative analysis technique is applied to analyse the content from the works of literatures. The study also adapted within the scope of refurbishment and adaptation, which also shares similar characteristics with energy retrofit.

MAIN FINDINGS

Although the comparison of guideline documents from other countries selected differs in their conservation practices from Malaysia, this paper aims to capture different interpretations, understandings, and approaches to dealing with historic energy retrofit, as shown in Table 1 below.

Table 1. Comparative analysis of energy retrofit through guidelines in Malaysia, United Kingdom (UK) and United States of America (USA)

Guidelines / Elements	Historic England (2008)	Jabatan Warisan Negara (2016)	National Park Service (2017)
Key Principles	The conservation principles are aligned with comprehensive framework for the sustainable management of the historic environment, under six headlines: The historic environment is a shared resource. Everyone should be able to participate in sustaining the historic environment. It is important to understand the significance of places. Significant places should be managed to sustain their values. Decisions about change must be reasonable, transparent and consistent. Documenting and learning from decisions are essential.	Be able to extend the building's lifespan. Respect the history and intrinsic quality of the property. Prioritise the original form, design, materials, and authentic craftsmanship. Be cautious and comprehensive in carrying out investigations and documentation before any work that may be disruptive is done. Tolerance in the interpretation and usability of the building. Minimum risk for the loss of parts of the important and unexpected damages during tests on building materials. Ensure that any disturbance or disruption can be reversed or made good. Changes or amendment to the building fabric should be minimized. Able to distinguish between old and new material. Provides maximum reinforcement to the original structure. Provides documentation throughout the conservation work.	Conservation approach that is based on the level of significance. Consideration based on the physical condition of the historic building. Prioritize clear documentation in determining conservation treatments. Regulations must be followed regardless of the course of treatment.
Conservation Approach	High level terms in conservation approaches	Preservation Prevention Consolidation Restoration Rehabilitation Reproduction Reconstruction Adaptation/Adaptive Reuse Maintenance	Recognizes four main approaches namely: Preservation Rehabilitation Reconstruction Rehabilitation
Conservation Concept	Recognizes beneficial change and adaptation to contemporary settings to reveal and reinforce heritage values, but understanding must depend on values and significance.	'Authenticity in Heritage Conservation' as the key concept and approach in the aspects of materiality, design, building works/ craftsmanship and layout.	Allowing adaptation with contemporary settings, depending upon the 'Level of Significance' and taking account its historic values and characters.

continued on following page

Table 1. Continued

Guidelines / Elements	Historic England (2008)	Jabatan Warisan Negara (2016)	National Park Service (2017)
Specific Requirement Related to Energy Retrofit	Following notes related to energy retrofit works or design: Conservation proposals are designed not to prejudice alternative solutions in the future. Actions taken to counter harmful effects of natural change, or to minimize the risk of disaster, should be timely, proportionate to the severity and likelihood of identified consequences, and sustainable. Changes to improve energy efficiency in historic buildings are capable of being reversed, in order not unduly to prejudice options for the future. Innovation is essential to sustaining cultural values in the historic environment for present and future generations but should not be achieved at the expense of places of established values. Energy efficiency, sustainable sourcing of materials, and environmental good practice should guide all new work, but not to the extent of causing harm to the heritage values of the place.	No specific guidelines but the following note related to energy retrofit works or design: Prioritizing original features and materials to preserve authenticity. Sympathetic in use and interpretations. Minimal risks of critical components being lost and unknown harm occurring when building materials are tested. Modifications to the building fabrics should be minimal. Able to distinguish between new addition and old materials.	The following notes related to energy retrofit works or design: Allowing for new technical improvements that can be employed to improve a historic building's efficiency. Consider sustainability aspects in the preservation and context of energy improvements. Allows energy improvements to a historic building, with proper evaluation of the existing energy-efficient characteristics of the building. In the context of a preservation project, sensitive upgrading of the building fabric and systems, as well as other code-required work, is appropriate. New sustainability treatments should generally be limited to updating existing features and systems and to have the least impact on the historical character of the building.
Further Preservation Guidance Specific on Energy Retrofit Works and Green Intervention	Further guidance on energy retrofit was extended in following: • Historic England: Energy Efficiency and Historic Buildings. • How to Improve Energy Efficiency and Application of Part L of the Building Regulations to Historic and Traditionally Constructed Buildings.	Not stated	Further preservation guidance on the context of environmental sustainability is addressed in following: • The Secretary of the Interior's Standards for Rehabilitation & Illustrated Guidelines on Sustainability for Rehabilitating Historic Buildings.

From the comparative analysis in Table 1, all of the guidelines put main emphasis on preserving authenticity, maintaining values and significance, however there are several areas within the Historic England (2008) and National Park Service (2017) guidance that highlights on the considerations related to energy retrofit are discussed below;

A thorough framework for the sustainable management of the historic environment is provided by Historic England (2008). In terms of energy considerations, the guideline allows historic building's adaptation to contemporary settings by taking account of environmental sustainability aspects such as energy, materials, and resources; however, this needs to be strictly put in place with the effects on authenticity and integrity. Although there is not much explicit guidance on energy retrofit, several areas of energy preservation and interventions were generically mentioned in sub-chapters on managing change in historic significant places. The guidance also emphasizes high-level terms in conservation approaches, which allows flexibility in interpreting actions on preservation. The guideline also extends further supporting guidance to cover more detailed interventions to enhance energy efficiency in historic buildings through retrofitting approach (McCaig et al., 2018; Pickles et al., 2018).

National Park Service (2017) provides an advisory guidance for preserving, rehabilitating, restoring and reconstructing historic buildings. In terms of energy retrofit, the guideline allows flexibility for traditional buildings adaptation with the contemporary settings, however it depends upon the 'Level of Significance' and taking account its historic values and characters. Several measures were discussed in sustainability topic of the guideline, with areas concerning energy improvements, new technologies adaptation and the need to conduct energy assessment for any decisions for conservation however, the detailed interventions are further covered in The Secretary of the Interior's Standards for Rehabilitation & Illustrated Guidelines on Sustainability for Rehabilitating Historic Buildings (Grimmer, 2011).

Unlike previous guidelines, the Guidelines on Heritage Building Conservation 2016 (JWN, 2016) was framed towards the fundamental ideology of heritage conservation itself, in which it is mainly positioned towards preserving the cultural heritage significance of histor. Upon investigation, the guidance emphasizes the detailed process of building conservation from the initiation to implementation. Contrasting with the new construction approach, where decisions are largely based on design intent and the need statement of the client, the conservation approach in HBCG enforces detailed analysis and assessment of the building's tangible and intangible conditions before any decisions towards conservation methods and techniques are made. This includes performing early building investigations, such as defects and dilapidation surveys of the building (Harun, 2011; Jabatan Warisan Negara, 2016). However, the implied assessments in HBCG place a strong emphasis on taking physical measures to preserve the fabric and materials with the main concept of maintaining the originality and "authenticity" of historic buildings (Khalaf, 2021; Harun, 2011). Lacking the coverage on prioritizing the energy performance, which significantly dissuades potential retrofitting considerations. The National Heritage Act and the Burra Charter, which underpin HBCG, both emphasize the preservation of cultural heritage values as the core objective of any historic buildings' conservation approaches (Azis et al., 2022). Consequently, this initiates additional conversations regarding the specific evaluation of cultural significance throughout the energy retrofit procedure to guarantee that these novel treatments are consistent with the conservation objectives. The exclusion of such criteria imposes a wider opportunity for historic buildings to address energy efficiency and innovative aspects, such as optimizing daylighting efficiency, thermal and energy performance, indoor air quality, and renewable energy strategies (McCaig et al., 2018).

Discussion on the Limitations of the Present Malaysia HCBG Guideline

The process approach to the energy retrofit of HB has been increasingly discussed in recent years (Filippi, 2015; Almeida, 2014; Jagarajan et al., 2015). In a study by Khalaf (2021), it was highlighted that to make heritage building conservation more adaptable to the demands of the 21st century, it is imperative that contemporary settings reconsider abandoning authenticity assessments and instead embrace a unified approach to preserving heritage integrity. This approach should prioritize continuity and compatibility, treating heritage assets as an evolving process that can effectively address global challenges in our modern era. This is important as tackling historic buildings to be mitigated and adaptable to the challenges of climate change, the ability to rebuild after conflict, disasters, and pandemics, and achieving sustainable development goals have become key priorities (UNSDG, 2015; UNDRR, 2020; UN, 2022). Thus, there is a need for the historic building conservation sector to put a specific focus on improving environmental sustainability through adaptation, as it is much more efficient than constructing new buildings and would also have an impact on the environment as historic buildings constitute a considerable portion of the building stock in Malaysia. Nonetheless, the majority of Malaysia's construction industry players are more concerned with creating new buildings, ignoring historic buildings' potential (Husin et al. 2019).

Moreover, this agrees with several previous studies where the content of the present heritage-built conservation guideline was deemed too prescriptive and lacked guidance on new interventions related to greening adaptation to prolong the lifespan of historic buildings. Hence, this signifies a gap because there seems to be a lack of understanding of the specific process and scope of how such an energy retrofit could be strategically positioned in the present national heritage guideline (Azis et al., 2022; Mohd-Isa, 2011; Bait et al., 2013; Harun, 2011; Ali et al., 2018; Hamzah et al., 2019). As energy retrofit practice has been widely sought in many countries, such as the UK and US, with extensive guidelines, case studies, and technical papers (BSI, 2017; ASHRAE, 2019; Historic Scotland, 2021), in the case of Malaysia, such greening intervention aspects for HB are yet to be widely discussed, especially in the current conservation policies, standards, and guidelines (Dewiyana et al., 2016; Hamzah et al., 2019). Furthermore, the present local authority conservation framework did not mention specifically on how to conduct historic retrofits aiming on environmental sustainability aspects such as energy efficiency and occupants' comfort (Che Husin, 2017).

Hence, given the lack of awareness, guidance, and legal protections on retrofitting, many historic buildings would consequently face threats of uncontrolled renovations, alteration, and conversion of spaces (Al-Obaidi et al., 2017). Although the National Heritage Act 2005 regulation gives emphasis on the social and economic aspects of sustainability, a lack of focus has been apportioned to the environmental sustainability aspects; hence, this would indirectly influence the absence of exploration of greening interventions in the present historic building conservation guidelines (Zaid et al., 2021). Hence, it is important for the practice to define the process of historic retrofit alongside the recent technical advancements for a much more inclusive heritage-built conservation practice (Ali et al., 2018; Halim et al., 2012). The challenges listed and the comparative analysis presented in Table 1 depict the essence of discourse on heritage conservation practice. This underscores the need for augmenting the existing HBCG with supplementary guidance tailored to address new interventions, particularly in the realm of energy retrofit, to ensure the preservation of its significance.

CONCLUSION

The Malaysian HBCG is a comprehensive document that guides the methods and procedures to preserve the fabrics of heritage buildings. Although there are some similarities with the listed countries historic building guidelines, there are obvious distinctions that limit the procedures to guide heritage building conservation practitioners in dealing with the context of energy retrofits in relation to environmental sustainability objectives. Findings from the comparative study identified the main research gap in the lack of consideration of energy preservation and intervention aspects at the current national level, in which the main concept, principles, and methods are more centric towards preserving cultural significance, specifically in the aspects of maintaining authenticity and originality in HB, putting aside considerations of energy preservation. In a world grappling to confront of environmental sustainability, a comprehensive and collaborative strategy to enhancing the energy efficiency of existing buildings, particularly historic buildings, is urgently required. Consequently, this comparative study and review reveal new ideas and areas of research to improve the existing guidelines and regulatory procedures with the goal of intensifying the value of historic buildings that involve greening and energy efficiency interventions in future studies. As the scope of energy retrofit is still in its infancy state, more studies can be aligned with such practices in Malaysia, including in aspects of industry's practical implementation, regulatory factors, and recent innovative and technological applications that would advance the productivity of retrofit projects in future heritage building conservation practice.

REFERENCES

Abergel, T., Dean, B., & Dulac, J. (2017). *Towards a Zero-Emission, Efficient, and Resilient Buildings and Construction Sector: Global Status Report 2017*. UN Environment and International Energy Agency.

Al-Obaidi, K. M., Wei, S. L., Ismail, M. A., & Kam, K. J. (2017). Sustainable Building Assessment of Colonial Shophouses After Adaptive Reuse in Kuala Lumpur. *Buildings*, 7(4), 87. 10.3390/buildings7040087

Ali, M., Ismail, K. M., Hashim, K. S. H. Y., Suhaimi, S., & Mustafa, M. H. (2018). Heritage Building Preservation through Building Information Modelling: Reviving Cultural Values through Level of Development Exploration. *Planning Malaysia Journal, 16*(6).

Almeida, S. L. G. D. (2014). *Retrofitting and Refurbishment Processes of Heritage Buildings: Application to Three Case Studies*. [Doctoral Dissertation].

ASHRAE. (2019). *ASHRAE terminology. ASHRAE searchable terminology tool. American Association of Heating, Refrigeration and Air-conditioning Engineers*. ASHRAE.

Azis, I. A. A., Khaidzir, K. A. M., & Rashid, A. M. (2022). Identifying Guidelines Limitation of Historic Interior in Adaptive Re-Use of Heritage Building in Malaysia. *Journal of Tourism Hospitality and Environment Management*, 7(27), 69–76. 10.35631/JTHEM.727006

Balaras, C. A., Dascalaki, E., & Kontoyiannidis, S. (2004). Decision support software for sustainable building refurbishment. *Transactions-American Society of Heating Refrigerating and air Conditioning Engineers, 110*(1), 592-601.

BSI. (2017). *Conservation of Cultural Heritage*. Guidelines for Improving the Energy Performance of Historic Buildings. BSI Standards Publication.

Campagna, L. M., & Fiorito, F. (2022). On the Impact of Climate Change on Building Energy Consumptions: A Meta-Analysis. *Energies*, 15(1), 354. 10.3390/en15010354

Chan, N., Go, T. F., Moey, L. K., & Chia, C. M. (2022). Short Review on Renewable Energy Policy and Energy Consumption of Buildings in Malaysia. *Journal of Engineering & Technological Advances*, 7(1), 64–82.

Che Husin, S. M. (2017). *Retrofitting Existing Building to Improve Energy Efficiency*.

Cho, H. M., Yun, B. Y., Kim, Y. U., Yuk, H., & Kim, S. (2022). Integrated retrofit solutions for improving the energy performance of historic buildings through energy technology suitability analyses: Retrofit plan of wooden truss and masonry composite structure in Korea in the 1920s. *Renewable & Sustainable Energy Reviews*, 168, 11280. 10.1016/j.rser.2022.112800

Dewiyana, E., Ibrahim, N., & Hajar, N. H. (2016). The Green Aspects of Adaptive Reuse of Hotel Penaga. *Procedia: Social and Behavioral Sciences*, 222, 631–643. 10.1016/j.sbspro.2016.05.220

Dias Pereira, L., Tavares, V., & Soares, N. (2021). Up-to-date challenges for the conservation, rehabilitation and energy retrofitting of higher education cultural heritage buildings. *Sustainability (Basel)*, 13(4), 2061. 10.3390/su13042061

Douglas, J. (2006). *Building adaptation*. Routledge. 10.4324/9780080458519

Eames, M., Dixon, T., Lannon, S., Hunt, M., Laurentis, C., Marvin, S., Hodson, M., Guthrie, P., Georgiadou, M. (2014). *Retrofit 2050: Critical Challenges for Urban Transitions*. Springer. 978-1-899895-12-0.

Filippi, M. (2015). Remarks on the Green Retrofitting of Historic Buildings in Italy. *Energy and Building*, 95, 15–22. 10.1016/j.enbuild.2014.11.001

Georgetown. (2021). Rancangan Kawasan Khas Tapak Warisan Dunia UNESCO: Jilid 2 Kawalan Pembangunan.

Grimmer, A. E. (2011). *The Secretary of the Interior's Standards for Rehabilitation & Illustrated Guidelines on Sustainability for Rehabilitating Historic Buildings*. US Government Printing Office.

Halim, A. H. A., Harun, S. N., & Hamid, M. Y. (2012). Peranan Sains dan Teknologi Dalam Kerja-Kerja Konservasi Bangunan Bersejarah.

Hamzah, N., Lou, E., Kamaruzzaman, S. N., Wong, P. F., Che-Ain, A. I., & Khoiry, M. A. (2019). An Environmental Study on Development of Refurbishment Assessment Themes for Heritage Non-domestic Buildings in Malaysia. *Ekoloji*, 28(107), 3–15.

Hariry, N. N. N. (2021). National Heritage Act 2005 (Act 2005): Problems and Solutions from the Perspective betweenanother Countries. [TURCOMAT]. *Turkish Journal of Computer and Mathematics Education*, 12(3), 2920–2925.

Harun, S. N. (2011). Heritage Building Conservation in Malaysia: Experience and Challenges. *Procedia Engineering*, 20, 41–53. 10.1016/j.proeng.2011.11.137

Harun, S. N. (2021). Pemuliharaan Bangunan Warisan. Dewan Bahasa dan Pustaka.

Hasan, H. S. M., Ab Wahab, L., & Ismail, D. (2019). Authenticity in Adaptive Reuse of Heritage Building in Malaysia. *International Journal of Cultural Heritage*, 4, 33–39.

Historic England. (2008). *Conservation principles: Policies and guidance for the sustainable management of the historic environment*. English Heritage.

Historic England. (2018). *Energy Efficiency and Historic Buildings: How to Improve Energy Efficiency*. Historic England.

Historic Scotland. (2020). *Historic Environment Scotland Refurbishment Case Study 37. Holyrood Park Lodge, Edinburgh: Thermal upgrade works to a 19th century lodge house*. Historic Scotland.

Historic Scotland. (2021). *Guide to Energy Retrofit of Traditional Buildings*. Historic Scotland. https://www.historicenvironment.scot/archives-and research/publications/publication/?publicationId=47c9f2eb-1ade-4a76-a775add0008972f3

Husin, S., Zaki, N., & Husain, M. (2019). Implementing Sustainability in Existing Building through Retrofitting Measures. [IJCIET]. *International Journal of Civil Engineering and Technology*, 10(01), 1450–1471.

INDC. (2015). (Submitted to). Intended Nationally Determined Contribution (INDC) of the Government of Malaysia. *the United Nations Framework*.

International Energy Agency. (2017). Towards a Zero-Emission, Efficient, and Resilient Buildings and Construction Sector. *UN Environment and International Energy Agency*, 2017, 2017.

Jabatan Warisan Negara. (2016). *Garis Panduan Pemuliharaan Bangunan Warisan*. Jabatan Warisan Negara.

Jagarajan, R., Abdullah, M. N., Asmoni, M., Lee, J. Y., & Jaafar, M. N. (2015). An Overview of Green Retrofitting Implementation in Non-Residential Existing Buildings. *Jurnal Teknologi*, 73(5). 10.11113/jt.v73.4324

Jahed, N., Aktaş, Y. D., Rickaby, P., & Bilgin Altınöz, A. G. (2020). Policy Framework for Energy Retrofitting of Built Heritage: A Critical Comparison of UK and Turkey. *Atmosphere (Basel)*, 11(6), 674. 10.3390/atmos11060674

Khalaf, R. W. (2021). World Heritage on the Move: Abandoning the Assessment of Authenticity to Meet the Challenges of the Twenty-First Century. *Heritage*, 4(1), 371–386. 10.3390/heritage4010023

Khalil, A., Stravoravdis, S., & Backes, D. (2021). Categorisation of Building Data in the Digital Documentation of Heritage Buildings. *Applied Geomatics*, 13(1), 29–54. 10.1007/s12518-020-00322-7

KLCP. (2020). *Kuala Lumpur City Plan 2020*. KLCP. http://klcityplan2020.dbkl.gov.my/sme/

LCCF. (2017). Low Carbon Cities Framework Version 2. Ministry of Energy, Green Technology and Water Malaysia (KeTTHA). Springer. ISBN 978-967-5893-10-0.

Ma, Z., Cooper, P., Daly, D., & Ledo, L. (2012). Existing Building Retrofits: Methodology and State-of-the-Art. *Energy and Building*, 55, 889–902. 10.1016/j.enbuild.2012.08.018

MacDonald, E., Cairns, P., Cormack, D., Máté, I., & McLaren, D. (2009). *Scottish House Condition Survey: Key Findings for 2008*.

Malaysia Energy Statistics Handbook. (2021). Department of Energy Management and Industrial Development Suruhanjaya Tenaga (Energy Commission), 1, 86. www.st.gov.my

Mansir, D., Muhammad, J. A., & Kasim, N. (2016). Reviewing the Need for Historic Building Information Modelling (HBIM) in the Conservation of Heritage Buildings in Melaka World Heritage City. *Social Sciences*, 11(11), 2777–2782.

McCaig, I., Pender, R., & Pickles, D. (2018). *Energy Efficiency and Historic Buildings: How to Improve Energy Efficiency*. Historic England.

Moey, L. K., Goh, K. S., Tong, D. L., Chong, P. L., Adam, N. M., & Ahmad, K. A. (2020). A Review on Current Energy Usage and Potential of Sustainable Energy in Southeast Asia Countries. *Journal of Sustainability Science and Management*, 15(2), 89–107.

Moran, F., Blight, T., Natarajan, S., & Shea, A. (2013). The Use of Passive House Planning Package to Reduce Energy Use and CO2 Emissions in Historic Dwellings. *Energy and Building*, 75, 216–227. 10.1016/j.enbuild.2013.12.043

National Park Service. (2017). *The Secretary of the Interior's Strandards for the Treatment of Historic Properties with Guideline for Preserving*. Rehabilitating, Restoring and Reconstructing Historic Buildings.

NGTP. (2009). *National Green Technology Policy 2009*. Ministry of Energy, Green Technology and Water.

NPCC. (2010). *National Policy on Climate Change 2009*. Ministry of Natural Resources and Environment Malaysia.

Pickles, D., & McCaig, I. (2017). *Energy Efficiency and Historic Buildings: Application of Part L of the Building Regulations to historic and traditionally constructed buildings*. Historic England.

Public Works Department. (2019). *Heritage Building Conservation Need Statement. Jabatan Kerja Raya*. JKR.

Sharma, S. K., Mohapatra, S., Sharma, R. C., Alturjman, S., Altrjman, C., Mostarda, L., & Stephan, T. (2022). Retrofitting Existing Buildings to Improve Energy Performance. *Sustainability (Basel)*, 14(2), 666. 10.3390/su14020666

Soares, N., Martins, C., Goncalves, M., Santos, P., da Silva, L. S., & Costa, J. J. (2019). Laboratory and In-Situ Non-Destructive Methods to Evaluate the Thermal Transmittance and Behavior of Walls, Windows, and Construction Elements with Innovative Materials: A review. *Energy and Building*, 182, 88–110. 10.1016/j.enbuild.2018.10.021

Sopery, L. H. M., & Ayob, M. F. (2023). An Investigation on Sustainable Refurbishment Practice of Building Works in Malaysia. *Journal of Architecture. Planning and Construction Management*, 13(1), 29–39.

Syed Zainol Abidin, I. (1995). *Pemeliharaan Warisan Rupa Bandar*. Badan Warisan Malaysia.

Tang, K. H. D. (2019). Climate Change in Malaysia: Trends, Contributors, Impacts, Mitigation and Adaptations. *The Science of the Total Environment*, 650, 1858–1871. 10.1016/j.scitotenv.2018.09.31630290336

Tobias, M. (2004). *Elements of Sustainable Architecture*. NY Engineers. https://www.ny-engineers.com/blog/elements-of-sustainable-architecture

UN. (2022). *Transforming our World: The 2030 Agenda for Sustainable Development*. UN. https://sustainabledevelopment.un.org/post2015/transformingourworld

UNDRR. (2020). Disaster Risk Reduction in Malaysia: Status Report 2020. Bangkok, Thailand, United Nations Office for Disaster Risk Reduction (UNDRR), Regional Office for Asia and the Pacific.

UNSDG. (2015). *UN General Assembly, Transforming our world: the 2030 Agenda for Sustainable Development, 21 October 2015, A/RES/70/1*. UN. https://www.refworld.org/docid/57b6e3e44.html

Wilkinson, S. (2011). *Sustainable Retrofit Potential in Lower Quality Office Stock in the Central Business District*. CIB Management and Innovation in the Sustainable Built Environment.

Yau, Y. H., & Hasbi, S. J. E. P. (2017). A Comprehensive Case Study of Climate Change Impacts on the Cooling Load in an Air-Conditioned Office Building in Malaysia. *Energy Procedia*, 143, 295–300. 10.1016/j.egypro.2017.12.687

Yazdani, S., Azandehi, S. K., Ghorbani, A., & Shakerian, S. (2018). Explaining the Process of Choosing Clinical Specialties in General Medical Graduates: A Grounded Theory. *Electronic Journal of General Medicine, 15*(6).

Zaid, N. S. M., Ghazali, M. I., Azmi, N. F., Kamarulzaman, P. N., & Akashah, F. W. (2021). Environmental Sustainability in Malaysian Heritage Building Conservation: Review of Heritage Legislation and Policy. *Malaysian Construction Research Journal (MCRJ)*, 203.

Chapter 4
A Review of Critical Factors of Successfully Implementing IBS in China Based on Scopus Database

Xiaonuo Kang
Taylor's University, Malaysia

Kenn Jhun Kam
https://orcid.org/0000-0003-4927-4611
Taylor's University, Malaysia

ABSTRACT

This abstract presents a comprehensive literature review on the critical factors driving China's successful adoption of industrialized building systems (IBS). Aiming at improving project efficiency and quality, this paper drawing insights from global experiences by utilizing the sysmatic literature review (SLR) method based on 78 IBS-related papers from Scopus database. The study emphasizes the significance of cognitive incentives, regulations, education, technology, construction management, the performance of IBS, culture, and tradition in the IBS implementation process. Ultimately, this chapter provides guidance for effective IBS implementation, distilled from regulatory, workforce, technological, collaborative, and cultural dimensions.

INTRODUCTION

The definition of IBS (industrial building system) is slightly different in different countries, but all countries believe that IBS mainly include the following several aspects: the use of industrialized standardized design, industrialized mass production, assembly and on-site construction (Wang Run, 2023). It is the overall assembly of the main body of the project, computer-aided assembly, the main components of the building are produced in the factory, and reliable connections are used. This method can achieve high efficiency, environmental friendliness and energy saving throughout the building's life cycle(Wang Run, 2023).In recent years, IBS has been widely promoted and applied in China and other

DOI: 10.4018/979-8-3693-1650-4.ch004

countries (Qiang Du et al., 2022), but it also faces some challenges and obstacles. Through the analysis of 78 papers on IBS by SLR, this article summarizes factors from cognitive incentives, policies, education, technology, construction management, performance, and cultural perspectives that influence the successful implementation of IBS. In addition, this article also discusses implementation strategies for each factor. It highlights the current research deficiencies and future directions at the end of the paper.

Current Status of China Construction Industry

With construction activity and growth in China, the housing market has also become active. The Chinese government focuses on developing infrastructure projects while promoting sustainable buildings and industrialized building systems (IBS) (Ying Rui Wang et al., 2020). In China, the construction industry provides many employment opportunities for society, leading to fierce competition within the industry.

According to data released by the National Bureau of Statistics of China, the GDP in 2022 reached 120,472.40 billion yuan, and the total output value of the domestic construction industry reached 30,793.538 billion yuan in 2022, of which the total output value of the construction industry accounted for about 25% of the GDP (National Bureau Statistics of China, 2022). However, with the rapid development of my country's construction industry, many disadvantages should be considered. For example, the waste and shortage of energy resources are prominent, environmental pollution and ecological damage are serious, and artificial rising costs drive construction costs to continue to rise(Jiahao Li, 2023). Construction Industry Institute research (CII) shows that 57% of non-value-added of work (inefficient work and waste) is in the construction industry, while at the same stage of development this proportion of manufacturing is 26%(Jiahao Li, 2023).

Moreover, China has been devoted to promoting infrastructures, like bridges, highways, light rail transit systems, etc. The implementation of the projects propels the construction industry to grow and also creates employment opportunities (Wang et al., 2020).

As a critical component of national development, the Chinese construction industry plays an essential role in the rapid and stable development of the economy.

Development History of IBS in China

The development stage of China's prefabricated buildings can be roughly divided into four stages: the initial stage, the development stage, the trough stage, and the further development stage(Jin Wu, 2022). In the early 1950s, the concept of prefabricated construction was first introduced to China. The technology at that time was mainly learned from the Soviet Union and Eastern Europe, and the housing construction model of standardization, and prefabricated construction was mainly promoted. From the 1960s to the early 1980s, prefabricated buildings experienced a period of stable development. At this time, the market demand for prefabricated buildings is single and the earthquake resistance requirements are not high. A large number of houses are suitable for construction in standardized methods(Jin Wu, 2022). Since the late 1980s, the construction market has had increasingly higher requirements for house design, seismic performance, and building scale. Unified construction methods can no longer meet these needs. At the same time, a series of quality problems have also appeared in a large number of prefabricated buildings. Since then, the development of prefabricated buildings has stagnated. From around year of 2008 to the present, the second industrial revolution has taken off, and foreign prefabricated construction technology has become very mature. With the loss of a large amount of cheap labor, labor costs are rising. At the same time, as environmental pollution becomes increasingly serious, prefabricated buildings have the advantages of saving resources and protecting the environment. Therefore, with the vigorous promotion of national competent authorities, prefabricated buildings have gradually developed.

China's Relevant Policies on IBS

The Chinese government has been promoting the development of IBS and has formulated relevant policies and programs to encourage the construction industry to adopt more industrialized construction methods. These policies help inspire builders and developers to adopt IBS technology.

On September 22nd, 2021, the Central Committee of the Communist Party of China and the State Council released a guidance document titled "Opinions on Complete, Accurate and Comprehensive Implementation of the New Development Concept and Carbon Implementation to Peak Carbon Neutrality." The document emphasizes the necessity for China to incorporate green and low-carbon considerations into all aspects of urban and rural planning, construction, and management. It advocates for the adoption of green construction practices throughout the entirety of project construction, the enhancement of building demolition management systems, and the cessation of large-scale demolition and construction activities. Furthermore, the document sets forth the objective for China to achieve comprehensive green transformation and significant progress by the year 2030 (Central Committee of the Communist Party of China and the State Council, 2021).

On October 21st, 2021, the "Opinions on Promoting Green Development of Urban and Rural Construction" was released by the General Office of the CPC Central Committee and the General Office of the State Council. This directive emphasizes the importance of advancing prefabricated building technologies, particularly in the realm of steel structure prefabricated residences. It also underscores the need to enhance the standardization of building components, foster a comprehensive industrial chain, and facilitate the harmonized progress of intelligent construction and construction industrialization (General Office of the CPC Central Committee and the General Office of the State Council, 2021).

Subsequently, on October 24th, 2021, China's State Council issued a notice on the "Carbon Peak Action Plan before 2030", which once again clarified the Chinese government's goal of strengthening the promotion of green construction methods and vigorously developing prefabricated buildings (China's State Council, 2021). In addition to these guiding opinions issued by the Chinese government, local governments have also given corresponding policies, including Bejing City, Tianjin City, Shanghai City, Hebei Province, Sichuan Province, etc. For instance, in Beijing, ten departments, including the Beijing Municipal Housing and Urban-Rural Development Commission (2021), issued "The Beijing Green Building Action Implementation Plan (2020-2022)," which clarifies that Beijing should steadily promote the development of prefabricated concrete structures, vigorously promote steel structure construction methods, and use steel structures in principle for new public buildings. Besides, the documents set the clear goal that in 2022, the proportion of prefabricated buildings in new construction areas will reach more than 35%.

The construction industry development plan outlined in the "14th Five-Year Plan" by Sichuan Province articulates a target whereby prefabricated buildings are projected to constitute 40% of new construction in the province by 2025 (Sichuan Provincial Department of Housing and Urban-Rural Development, 2021). This initiative will be actively advanced in municipal infrastructure endeavors, including municipal bridges, comprehensive pipe corridors, and urban rail transit projects (Sichuan Provincial Department of Housing and Urban-Rural Development, 2021).

Current Status of IBS in China

According to data published by the Ministry of Housing and Urban-Rural Development in 2022, the area of newly initiated Industrialized Building System (IBS) projects across China in 2020 amounted to 630 million square meters, with the utilization of prefabricated buildings showing a consistent growth trend in 2021. Projections indicate that by 2025, the majority of provinces and cities will mandate that prefabricated buildings constitute over 30% of their total construction area, with some regions requiring up to 50%. Notably, Tianjin and Shanghai are expected to enforce a 100% prefabricated building requirement. The statistics further reveal that the prefabricated modular building market in China has expanded from 29.79 billion yuan in 2015 to 114.26 billion yuan in 2021, with anticipated continued growth in the forthcoming years. The country is presently undergoing a rapid advancement in prefabricated building development, with several prominent prefabricated projects already in existence.

The China Antarctic Great Wall Station, established in 1985, represents an early example of a prefabricated steel structure that incorporates innovative materials and technologies, such as polyurethane composite panels and rapid-setting concrete (Sohu, 2018). The architect oversaw various aspects of the project, including construction, structural design, and organizational planning, while incorporating prefabricated steel elements. Subsequent to the initial construction of the Great Wall Station, it underwent multiple expansions, resulting in a complex comprising 25 buildings, including main structures, offices, dormitories, and research facilities, covering a total area exceeding 4,000 square meters. Regarded as a significant achievement in the realm of prefabricated construction in China, the Great Wall Station stands as a prominent landmark in the country (Sohu, 2018).

The Hong Kong-Zhuhai-Macao Bridge, spanning a total length of 55 kilometers, holds the distinction of being the lengthiest steel structure bridge globally. Various components such as bridge piers, bridge decks, steel box girders, and other prefabricated elements are initially manufactured in facilities located in Zhongshan, Dongguan, and other regions (Sohu, 2018). These components are then assembled using prefabricated construction techniques when favorable sea conditions permit, thereby enhancing operational efficiency and promoting environmental sustainability. Furthermore, the primary structure situated on the western artificial island of the Hong Kong-Zhuhai-Macao Bridge, encompassing a total construction area of 20,000 square meters across three levels, also leverages prefabricated construction methodologies (Sohu, 2018).

Although China has many prefabricated construction projects, there are still major difficulties in developing prefabricated buildings (Ping Cao & Xiatong Lei, 2023). The education system in China is not complete. For instance, as IBS is a new construction technology and concept, few workers on the construction site are willing to learn IBS construction technology, which directly leads to obstacles in the construction process of IBS.

The IBS specification system and contract conditions for IBS stakeholders in China also lack competency. Cooperation and synergy at all levels of the Chinese construction industry is an important factor driving the development of IBS. The joint efforts of all parties can promote the innovation and adoption of technology and solve the challenges that IBS may face in practical applications. Cooperation among different parties remains essential since efforts from various perspectives must still come together to reach a state of optimality.

The Importance of Studying Factors Affecting the Successful Implementation of IBS in China

Understanding why China successfully implements IBS (Industrialized Building Systems) is indispensable for its building sector and financial development. By exploring the factors for its successful application, it is possible to promote economic growth and support sustainable construction practices in Chinese (Houchao Sun et al., 2023).

In essence, the efficiency and criteria in the construction domain can be improved by gaining insight into the reasons for the successful implementation of IBS. Enhancing the construction efforts can significantly accelerate the construction schedules and ensure better construction caliber. Concurrently, the added employment opportunities are offered by successful implementation of the IBS (Su, 2020). Moreover, implementing the IBS prominently contributes to reducing the construction costs for the projects, due to the improved construction processes and enhanced resource utilization. Moreover, implementing innovative techniques can be heightened by investigating the influencing factors. Generally speaking, implementing IBS consists of pioneering building techniques and digital instruments. Hence, examining influencing factors can stimulate the construction sector to adopt new techniques and improve its degree of technology proficiency. Analysis of factors contributing to the successful implementation of IBS can give governments invaluable information for crafting policies and regulations that support its advancement. Governments can facilitate IBS adoption with incentives and policy support (Houchao Sun et al., 2023). Chinese modernization will aid its competitiveness by adopting advanced construction methods to attract international investment and become involved in global supply chains.

Overall, studying the influencing factors of the successful implementation of IBS in Chinese will help promote the development of the construction industry in a more sustainable and efficient direction, positively impacting the country's society, economy, and environment.

SYSTEMATIC LITERATURE REVIEW

According to the definition provided by Mark Petticrew and Helen Roberts (2008), a systematic review is characterized by literature reviews that rigorously follow scientific methodologies with the explicit goal of minimizing systematic errors, primarily through the comprehensive identification, evaluation, and synthesis of all pertinent studies, regardless of their design, to address specific inquiries or a series of questions.

PSALSAR

Maria J. Grant and Andrew Booth (2009) brought out a framework of SLR, by Search, Appraisal, Synthesis, and Analysis (SALSA). This guarantees methodological accuracy, systematization, completeness, and reproducibility. Most of the scientific works applied this methodological approach to reduce the risk of biased publication and increase the work's acceptance (Wondimagegn Mengist et al., 2020). And the method rectified and improved by Wondimagegn Mengist et al. (2020), SALSA method has developed to (PSALSAR) six-step framework which is associated with Protocol and Reporting results with Search, Appraisal, Synthesis, and Analysis framework.

The PSALSAR approach is a clear, adaptable, and repeatable process for carrying out systematic reviews. It aids in evaluating both quantitative and qualitative content analysis in literature reviews. This method includes two additional fundamental steps (establishing a protocol and reporting results) to the well-known SALSA framework(Wondimagegn Mengist et al., 2020).

Following the PSALSAR steps, the author modified its process as PSALSAR as to make it applicable to this study and aimed to do a systematic literature review and meta-analysis to gain a comprehensive and less biased result.

Table 1. The rectified frame work for IBS factors systematic and meta-analysis studies

	Steps	Outcomes	Methods
PPSALSAR FFRAMEWORK	Protocol	Define the Study Scope and Objective	**Research Objective:** to Identify the Factors Influencing the Implementation of IBS **Research Scope:** Define the Concept of Scope Including Research Area, IBS Type, Fields, etc
	Search	Define Search Strategy and Target Database	Define Search Strings and Use on Scopus Database
	Appraisal	Define Selecting Criteria of Resources	Define Selecting Inclusion and Exclusion Criteria
	Synthesis	Extract data	
	Analysis and Report	Analyse and Discuss the Outcome	

Systematic Literature Review Step 1: Protocol

For gaining the outcome with transparency and disclosure, systematic review is needed by exploring exhaustive literature searches. The first step is to define the study's detailed scope and objective. As the study scope is clear determined, further research questions and keywords, searching strings, research criteria can be produced.

The scope of the research is exploring the thesis, articles, conference papers, review, and short surveys about IBS all over the world, with a limitation of time from January 1st 2017, to 31th of December of 2022.

Table 2. IBS implementation factors SLR research scope

Concept of scope	SLR application
Area	this review target on IBS development with no limitations to the region; countries worldwide are accepted
Type	this review targets all kinds of IBS, including Steel structures, wood structures, concrete structures, etc
Fields	this review targets IBS application in different fields, including residential, commercial, medical, etc
Academic Outcome	previous research results, the future development trend, and the direction of IBS

Systematic Literature Review Step 2: Search

Scopus is used as the primary database, and the criteria is listed in Table 3.

Table 3. IBS implementation factors SLR eligibility criteria

Years Considered	January 1st 2017 - 31 st Dec 2022
Language	English and Chinese
Publication Status	Published
Document Type	Article/Conference Paper/Review/ Short Survey/Thesis

The number of target documents may vary based on the strings. If the strings are accurate and reasonable, the paper search results will be more accurate. The strings of the thesis should be based on the research scope of the SLR in this part. Since the ultimate goal of this SLR is to explore the influencing factors of IBS implementation as broadly as possible, the area scope includes a wide range and without limitation to region.

Before conducting the actual systematic review search, a pilot literature search should be undertaken to refine the search terms and cover the intended study objectives (Wondimagegn Mengist et al., 2020).

The author's goal is to analyse all the literature related to IBS implementation factors from January 1st 2017 - 31 st Dec 2022, so according to the advanced search syntax specified in the Scopus database, the determining right strings had four trials:

Table 4. Trials of string

Number of Attempts	Syntax Used	Outcome of Number of Journals	Date of Trial
1st trial	TITLE-ABS-KEY ((“PRECAST BUILDINGS” OR “PREFABRICATED BUILDINGS” OR “IBS” OR “INDUSTRIAL BUILDING SYSTEM”) AND (develop*)) AND (LIMIT-TO (PUBYEAR, 2022) OR LIMIT-TO (PUBYEAR, 2021) OR LIMIT-TO (PUBYEAR, 2020) OR LIMIT-TO (PUBYEAR, 2019) OR LIMIT-TO (PUBYEAR, 2018) OR LIMIT-TO (PUBYEAR, 2017))	3477	March 26th 2022
2nd trial	TITLE-ABS-KEY ((“PRECAST BUILDINGS” OR “PREFABRICATED BUILDINGS” OR “IBS” OR “INDUSTRIAL BUILDING SYSTEM”) AND (BUILDING CONSTRUCTION)) AND (LIMIT-TO (PUBYEAR, 2022) OR LIMIT-TO (PUBYEAR, 2021) OR LIMIT-TO (PUBYEAR, 2020) OR LIMIT-TO (PUBYEAR, 2019) OR LIMIT-TO (PUBYEAR, 2018) OR LIMIT-TO (PUBYEAR, 2017))	750	March 26th 2022
3rd trial	TITLE-ABS KEY ((“PRECAST BUILDINGS” OR “PREFABRICATED BUILDINGS” OR “IBS” OR “INDUSTRIAL BUILDING SYSTEM”) AND (“BUILDING CONSTRUCTION”)) AND (LIMIT-TO (PUBYEAR, 2022) OR LIMIT-TO (PUBYEAR, 2021) OR LIMIT-TO (PUBYEAR, 2020) OR LIMIT-TO (PUBYEAR, 2019) OR LIMIT-TO (PUBYEAR, 2018) OR LIMIT-TO (PUBYEAR, 2017))	96	March 26th 2022
4th trial	TITLE-ABS KEY ((“PRECAST BUILDINGS” OR “PREFABRICATED BUILDINGS” OR “IBS” OR “INDUSTRIAL BUILDING SYSTEM”) AND (“BUILDING CONSTRUCTION”)) AND (LIMIT-TO (PUBYEAR, 2022) OR LIMIT-TO (PUBYEAR, 2021) OR LIMIT-TO (PUBYEAR, 2020) OR LIMIT-TO (PUBYEAR, 2019) OR LIMIT-TO (PUBYEAR, 2018) OR LIMIT-TO (PUBYEAR, 2017)	119	March 3rd 2023

Select all documents of 119 papers in 4th trial for the next step.

Systematic Literature Review Step 3: Appraisal

Appraisal step is the phase in which the selected paper of Step 2 is assessed based on the SLR objectives. The selection process involves screening literature to identify relevant papers for review. The criteria of the inclusion and exclusion of screening papers is listed in Table 5.

Table 5. IBS implementation factors SLR selecting inclusions and exclusions criteria

Criteria	Decision
Keywords including "IBS" or "industrial building system" or "prefabricated buildings" or "precast buildings" should be included in title, keywords or abstract of the paper	inclusion
The paper should be written in the English or Chinese / Scopus database	inclusion
Papers are duplicated within the outcome of search results	exclusion
papers that are not accessible, including no access or fee papers	Exclusion
papers published before 2017	exclusion

Grey papers, pseudo-scientific literature, low-quality publications, retracted or withdrawn literature, unauthorized or copyright-infringing literature, presentations, and keynotes are rejected. Considering papers without keywords "IBS" "industrial building system" "prefabricated buildings" or "precast buildings" included in the title or the abstract may cover the content of IBS implementation factors as well, this SLR work, regardless of the works as the scope is too broad to be considered. The process of screening and the flow chart selection of papers is provided in Figure 1.

Figure 1. The flow diagram for the database search of publications for IBS implementation factors SLR

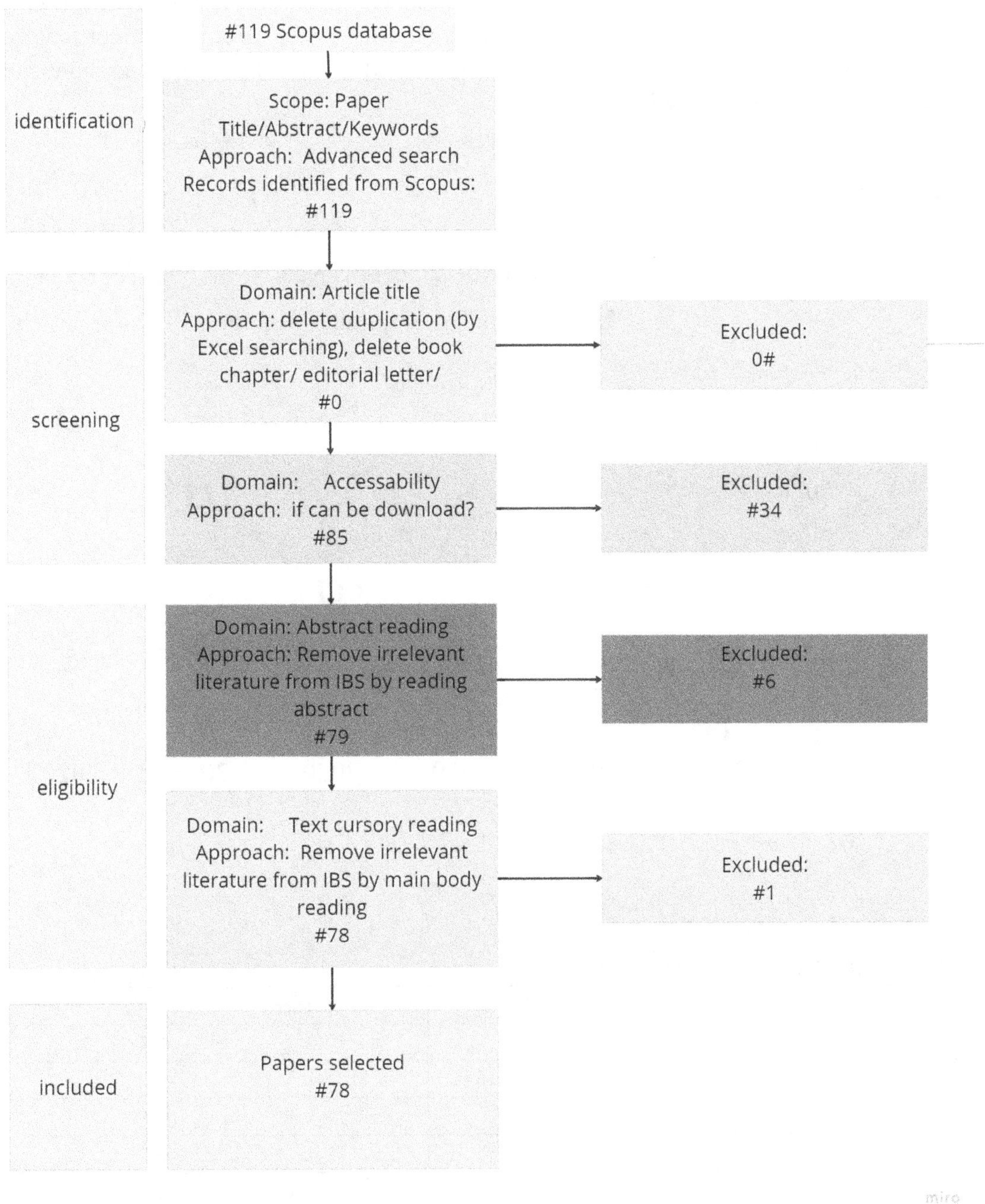

Systematic Literature Review Step 4: Synthesis

Summarize the Basic Information of the Selected Papers. As indicated from the line chart in Figure 2, Starting from 2017, the number of publications on IBS has increased almost proportionally, reaching a peak of 23 articles in 2021 and slightly decreasing to 19 articles in 2022. This trend shows the increasing attention paid to IBS in the scientific field.

Figure 2. Publication by year

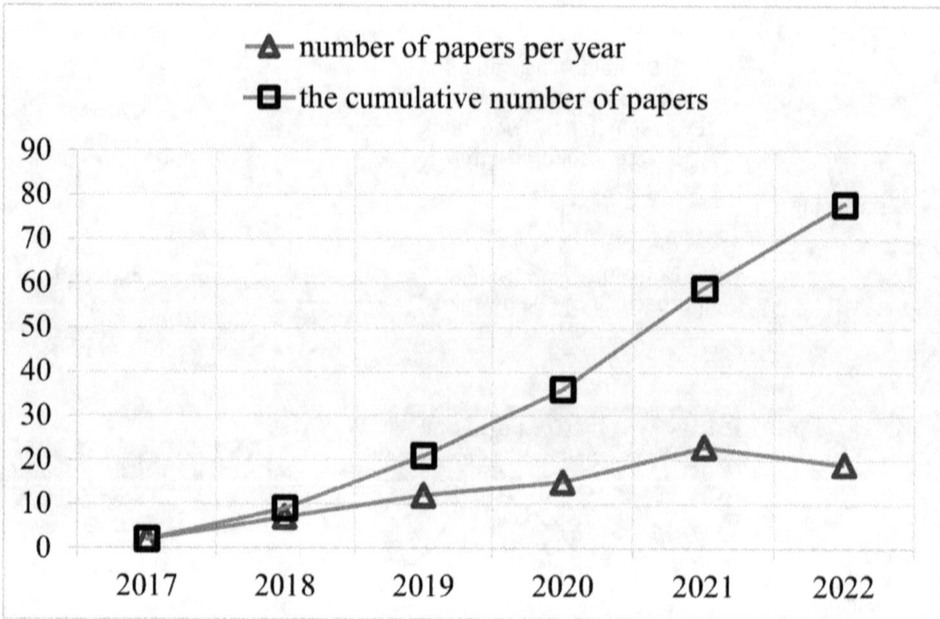

Table 6. Country of papers selected

Country	Number of paper
China	42
Malaysia	11
Australia	3
Australia; China	2
China (Hong Kong) Hong Kong; China United States; China Other countries	2 2 2 15

Figure 3. Country of papers selected

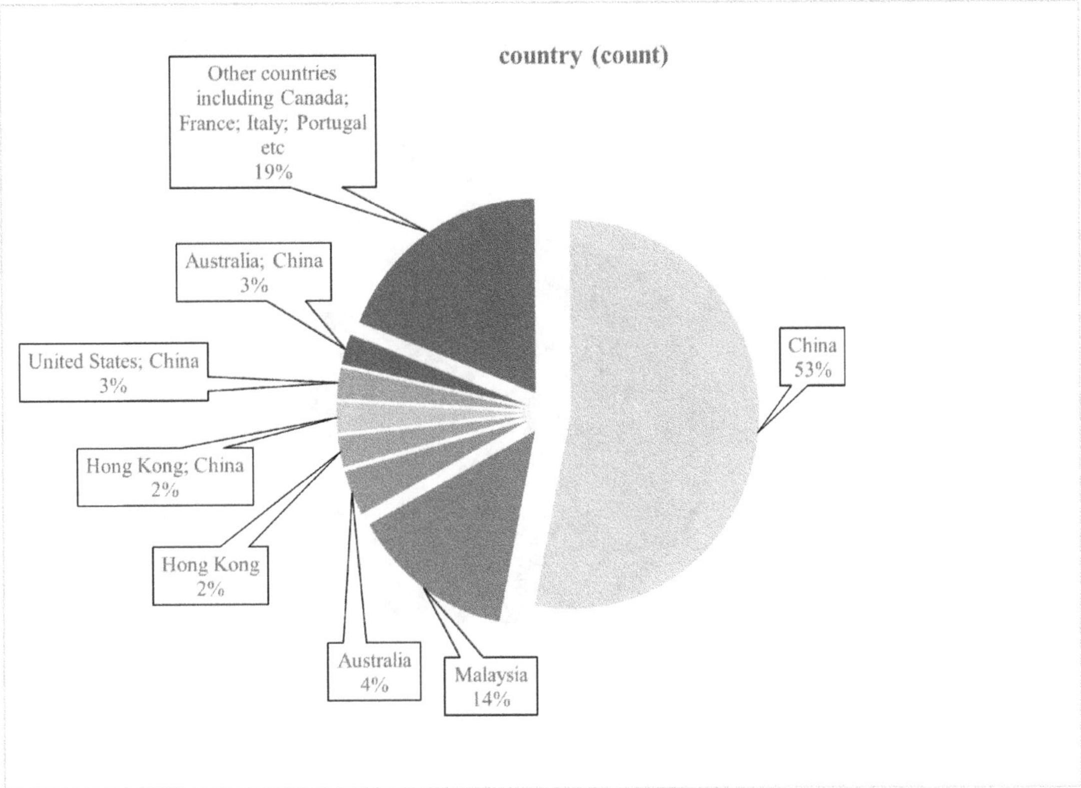

Table 7. Source of publication (sort the top five)

Source Publication	Count
IOP Conference Series: Earth and Environmental Science	6
Journal of Physics: Conference Series	6
Buildings	4
ACM International Conference Proceeding Series	3

Figure 4. Source of publication (sort the top five)

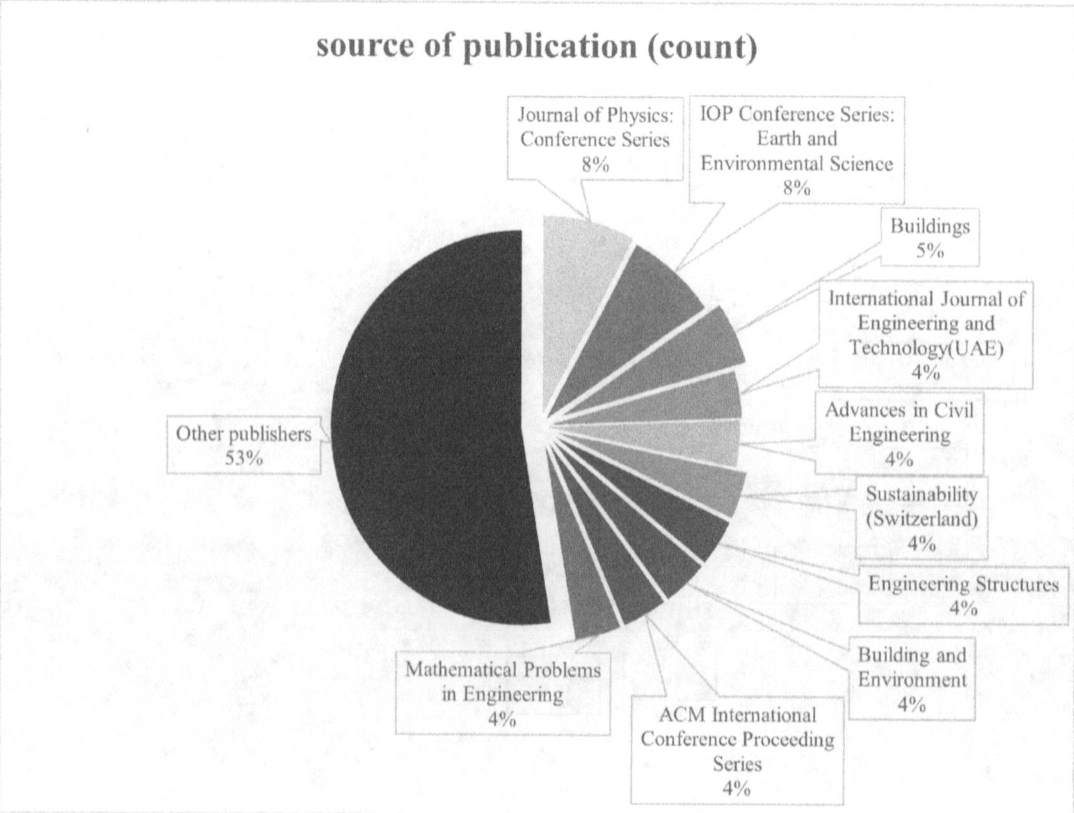

Systematic Literature Review Step 5: Analysis and Report

78 selected Paper contexts were classified and analyzed by factors.

1. Cognitive Incentives

The advantages of IBS can be regarded as stimulating factors that inspire people to implement IBS. After the advantages being accepted by social cognition, society will be more willing to adopt this new technology.

The advantages of IBS are mainly reflected in energy saving, emission reduction, cost reduction, quality improvement, safety, and productivity (Massofia Md Ali et al., 2018; Yuan Chang et al., 2018; Wahid Ferdous et al., 2019; Tan Let Hui & Ng Chee Khoon, 2020; Satheeskumar Navaratnam et al., 2019; V. Tavares et al., 2021; Yue Teng & Wei Pan, 2019). IBS can reduce the time, workforce, materials, and waste of on-site construction, thereby lowering carbon emissions and ecological footprint(Ting Han, 2023; Yingbo Ji et al., 2020; Shaozhe Wang & Rajib Sinha, 2021; Han Zhu et al., 2018). The IBS has excellent potential to improve building quality and precision (Dong Chen et al., 2021; Yadi Duan

& Guanlei Li, 2020; Yankun Guo, 2021). At the same time, implementing IBS helps promote building safety by decreasing incidents (Yuan Chang et al., 2018; Zul Atfi Ismail, 2019; Ye Shen et al., 2022; Bin Yang et al., 2021).

At present, IBS is not widely adopted in China. Thus, it is urgent to increase promotion to improve the public comprehension and trust (Mohd Nasrun Mohd Nawi et al., 2019). What's more, cognitive obstacles to the implementation of IBS have been mentioned in lots of studies, such as the deficient perception of changes from conventional methods, lack of trust in its durability, and aesthetic diversity problems (Massofia Md Ali et al., 2018; Hamizah Liyana Binti Tajul Ariffin et al., 2018; Ali Tighnavard Balasbaneh & Abdul Kadir Bin Marsono, 2017). Hence, the crucial stakeholders (e.g., designers, contractors, and suppliers) must share reliable information about IBS. It is also of great significance to display successful cases of the adoption to demonstrate its potential (Rafael E. López-Guerrero et al., 2022).

2. Regulations by Government (Legislations, Laws, Policy Support, Technical Standards or Conditions of Contract)

Meanwhile, there are some challenges for IBS in terms of market needs, governmental political support, supply chain coordination, quality control, etc (Massofia Md Ali et al., 2018; Hamizah Liyana Binti Tajul Ariffin et al., 2018; Ali Tighnavard Balasbaneh & Abdul Kadir Bin Marsono, 2017; Satheeskumar Navaratnam et al., 2019).

IBS also needs governmental policies and incentive measures to propel its design and use (Rafael E. López-Guerrero et al., 2022). In order for effective interaction and collaboration, all stakeholders, like suppliers, contractors, and designers, need to be combined (Bataglin et al., 2020; Wang et al., 2018). When the IBS is implemented, design standards and regulations should be developed in an attempt to monitor its utilization (Liu & Cui, 2020). In some researches, it is showed that the IBS is confronted with several policy obstacles, such as tax incentives, insufficient economic subsidies, and preferential loans (Ali et al., 2018; Ariffin et al., 2018; Balasbaneh & Bin Marsono, 2017).

Beyond that, there are several technological barriers for implementing IBS, such as insufficient consistent design norms and guidelines (Fernanda S. Bataglin et al., 2020; Yan Liu & Jialin Cui, 2020; Shaoli Wang et al., 2018).

3. Education

Given that IBS is not widely recognized and applied in China, it is important to increase promotion so as to amplify public understanding (Mohd Nasrun Mohd Nawi et al., 2019). Apart from that, some proficient professionals should be employed to elevate the building quality and efficiency (Riduan Yunus et al., 2020).

As indicated by some research, the educational barriers in the IBS domain include deficient formal training schemes, courses, etc. At the same time, the qualified instructors are deficient, while students and practitioners have a low level of awareness (Massofia Md Ali et al., 2018; Hamizah Liyana Binti Tajul Ariffin et al., 2018; Ali Tighnavard Balasbaneh & Abdul Kadir Bin Marsono, 2017). In this context, efforts should be made on the educational initiatives and resources to strengthen the competency of the IBS workforce. For example, IBS centers can be set up, and IBS courses can also be provided (Nawi et al., 2019; Yunus et al., 2020).

4. Technology

Technology of Design and Manufacture. As for the technology related to the IBS design mainly includes the procedure of producing, transporting, and installing certain elements. BIM, 3D printing, and so on are crucial techniques (Yangyang Guo, 2021; Stelladriana Volpe et al., 2021; Shuhai Wu & Qing Liu, 2020; Bin Yang et al., 2022; Yuhong Zhao et al., 2022). What's more, BIM promotes cooperation among lots of disciplines and stakeholders of the IBS projects (Fernanda S. Bataglin et al., 2020; Shaoli Wang et al., 2018). What`s more, BIM supports to integrate other techniques (e.g., Digital Twins) in order to fulfill the data exchange (Volpe et al., 2021; Zhao et al., 2022).

Through enhancing the three dimensions of prefabricated components and compatibility with the software, the effectiveness of the IBS design can be enhanced by BIM (Feng, 2021; Peng et al., 2021; Song, 2022; Yang, 2021).

By means of 3D printing, speedy IBS generation and individualization is enabled, by which digital design is transformed into tangible components and costs and material wastage are reduced (Volpe et al., 2021).

Laser scanning techniques are adopted in order to evaluate and locate the prefabricated components accurately in the process of building. Accordingly, it is possible to detect parameters such as the dimension and geometry (Yang et al., 2022).

Advanced Technology of IBS Structure. As for prefabricated components, the structure, functionalities, and materials chiefly show the IBS performance. Besides, the new material technology, innovative connections, and hybrid structures are also imperative (Chen et al., 2021; Menegon et al., 2020; Wu et al., 2020).

In terms of the IBS steadiness and seismic resistance, hybrid structures are indispensable. Additionally, the hardness and ductility of the components are reinforced by means of the technologies, such as Reinforced Concrete (RC), Steel-Concrete Composite Structures (SRC), and Timber-Concrete Composite Structures (CLT) (Loss et al., 2018; Menegon et al., 2020; Wu et al., 2020).

Additionally, the IBS demountability and dependence are enhanced by means of efficient connections. Whether to install or not is more simple by these technologies, such as adhesive connections, pin connections, and bolted connections. Thus, the structural integrity is defended (Duan & Li, 2020; Huo et al., 2021; Wu et al., 2020). These connection methods provide sufficient shear transfer and bending resistance can be offered adequately by means of these connection approaches. Thus, the adjustment and replacement are promoted (Yadi Duan & Guanlei Li, 2020; Tingting Huo et al., 2021).Bolted connections are another prevalent approach of connecting steel-concrete compound structures, which achieves the speedy assembly and disassembly of those components (Shuhai Wu & Qing Liu, 2020).

Materials. IBS durability and crack resistance are improved through new materials, Polymer Cement (PC), Glass Fiber Reinforced Cement (GRC), and nanomaterials, to name a few. Of the prefabricated components, weather-resistant and mechanical attributes are thus improved (Dong Chen et al., 2021; D. Vasovic et al., 2021).

5. Construction Management

IBS management is primarily part of risk management, quality control (QC), cost control, schedule control, and production planning. Under this background, these measures play an irreplaceable part, such as intelligent mechanism, simulation models, together with optimization algorithms (Chunguang

Chang et al., 2020; Yuncan Jin et al., 2019; Maohua Yang, 2022; Yisong Yuan et al., 2021; Wei Zheng et al., 2022).

IBS production effectiveness and resource applications are improved by means of the optimization algorithms. In doing so, the multipurpose, multi-constraint, and multi-modal scheduling issues are coped with for the purpose of determining the optimal production plans (Yisong Yuan et al., 2021).

The safety risk management of IBS is enhanced by intelligent systems. Through the use of AI and data mining technique, these kinds of systems can identify possible hazards in IBS building process and put forward prevention tactics (Chunguang Chang et al., 2020; Yuncan Jin et al., 2019; Bin Yang et al., 2022).

The Digital Twins can promote the real-time monitoring and management of IBS. Specifically, the Digital Twins can integrate IoT, sensors, and cloud computing techniques to link physical construction with virtual patterns, and propel data gathering and analysis (Zhansheng Liu et al., 2021; Yuhong Zhao et al., 2022).

6. Performance of Implementing IBS

Lots of papers have mentioned the assessment method of IBS. Thus, the performance of IBS is listed as a shadow factors influencing the adoption of IBS.

As for the assessment of IBS, it mainly includes the financial, environmental, and social aspects of prefabricated elements, in which methods such as the lifecycle assessment are of great significance (Yuan Chang et al., 2018; Yan Liu & Jialin Cui, 2020; Qiong Shen & Yaodan Liu, 2019; Shaozhe Wang & Rajib Sinha, 2021; Wei Zheng et al., 2022).

Cost-benefit analysis, which is used for evaluating the financial performance of IBS, involves comparison of the preliminary expenditure, operation expenditure, and maintenance expenditure of IBS with conventional construction methods to identify the investment return of IBS (Yuan Chang et al., 2018; Shaozhe Wang & Rajib Sinha, 2021).

The lifecycle assessment and optimization of IBS can be enhanced by the simulation models. Through the establishment of system dynamics, these tools can not only examine the energy use, carbon emissions, and cost indicators of IBS at different phases but also put forward improvement measures (Yan Liu & Jialin Cui, 2020; Qiong Shen & Yaodan Liu, 2019; Wei Zheng et al., 2022).

The social performance of IBS is evaluated using the multi-criteria decision-making. Beyond that, it determines the social benefits of IBS by considering its impact on user satisfaction, comfort, and security (Wei Zheng et al., 2022).

7. Culture and Tradition

The social-cultural impact of IBS has been analyzed in some research. To be specific, the user demands and preferences for IBS are considered in the diversity, health, security, etc. At the same time, methods such as multi-criteria decision-making are used to assess the social benefits of IBS(Wei Zheng et al., 2022). Besides, the preferences of different stakeholders of IBS projects (e.g. the contractors and suppliers) are also incorporated (Wei Zheng et al., 2022). It has been proposed that more social-cultural analyses should be made to gain an insight into the local norms for IBS (Rafael E. López-Guerrero et al., 2022).

Conclusion of Systematic Literature Review

Based on the above analysis, the influencing factors appearing in these 78 articles can be mainly divided into 7 categories: "cognitive incentives," "regulation by governments," "education," "technology," "construction management," "performance of implementing IBS" and "culture and tradition." Among them, 18 articles mentioned "cognitive incentives," indicating that researchers generally believe that IBS has advanced advantages and can promote the development of the construction industry through the implementation of IBS. At the same time, 18 articles mentioned the "technology" factor, indicating that in the current research, there is considerable room for improvement in the materials, design, component production, and construction technology of IBS. Secondly, 8 articles mentioned the factor "regulation by governments", indicating that government measures or opinions have a decisive impact on the implementation of IBS. Seven articles mentioned the "construction management" factor, indicating that scientific engineering management greatly impacts the successful implementation of IBS, and engineering management technology is also in the process of continuous development. There are 5 articles mention "education" and "performance of implementing IBS". Education is a relatively important link that the government should make efforts, should be paid attention to. Many scholars have also discussed the performance of IBS throughout its life cycle. If IBS has excellent performance throughout its life cycle, promoting the implementation of IBS will be one step closer to success. Only 2 articles mentioned information about "culture and tradition", but this factor is also an important factor affecting IBS that cannot be ignored. Implementing IBS should be consistent with local culture and tradition so that the public can easily accept it.

SUMMARY

Current Shortcomings in IBS Research

To summarize the overall situation of the 78 papers, IBS is a prospective building method that can enhance the efficiency and sustainability of the building sector. Nonetheless, the implementation of IBS in China and other nations has some challenges. Thus, it is essential to carry out more research. At present, the shortcomings of research areas are mainly reflected as follows:

The deficient systematic definition of IBS brings about differences in the comprehension and use of IBS in various fields.

The long-term performance of IBS lacks evaluation, leading to insufficient confidence in its lifespan and durability.

There is a lack of innovative design research for IBS, which can not satisfy various demands of IBS.

The social-cultural impacts of IBS lack research, so that there is deficient understanding and respect on the users' demands for IBS.

Future Research Direction

Thus, it is essential to conduct future research areas as follows:

Establishing a unified and standardized definition and classification system for IBS to facilitate communication and comparison between different countries and regions.

Conducting more experiments and field surveys to validate and evaluate the long-term performance and reliability of IBS under various environmental conditions.

Utilizing more advanced technologies, such as BIM, Digital Twins, and 3D printing, to achieve innovative and optimized designs, enhancing IBS's aesthetic appeal and diversity.

Carrying out more social-cultural analyses to understand and meet the needs and preferences of different user groups for IBS, increasing its social benefits and acceptability.

Resolution for IBS successfully Implementing in China-Based on Factors Extract from 78 Papers

1. Resolution for Factor "Cognitive Incentives"

For the Chinese society to fully understand the advantages of implementing industrialized building systems (IBS), a comprehensive series of publicity and education measures are needed. Through the use of publicity and media activities, demonstration projects and open days, information brochures and guides, online resource platforms, industry cooperation, community engagement and incentive programs, establish a comprehensive IBS publicity and education framework, thereby promoting its wider application in Chinese. Specific measures include the following aspects:

Publicity and media activities. Utilize various media channels, including television, radio, Internet and social media, to produce promotional videos, news reports, articles and blogs to convey the concepts, advantages and successful implementation cases of IBS to the public to Increase public awareness of IBS. Governments, construction trade associations and businesses can work together to engage in these campaigns (Mao Ran Che, 2022).

Demonstration projects and open days. Organize IBS demonstration projects and open days in different regions to allow the public to experience the implementation effects of IBS firsthand. Such events can provide practical demonstrations and increase public understanding and confidence in IBS.

Produce information manuals and guides. Produce easy-to-understand information manuals and guides to explain IBS's concepts, working principles and advantages. These materials can be distributed to construction industry practitioners, government agencies, schools and the general public.

Establish an online resource platform. Create a dedicated online platform to provide detailed information, research reports, case studies and related resources on IBS. This can serve as a centralized source of information for those interested to obtain detailed information.

2. Resolution for Factor "Regulation by Government"

This factor means legislations, laws, policy supports, technical standards or contract conditions which may be offered and rectified by the government authorities or related departments with the aim of guiding and regulating the industry.

In order for the improvement of industrialized building systems (IBS) in China, a range of supportive policies and steps can be taken in an effort to develop the extensive IBS application:

Develop incentive policies. The incentive policies are developed to adopt the IBS, like loan interest exemptions, tax benefits, or other economic incentives, with the objective to stimulating construction corporations to utilize the IBS technology (Che, 2022; Wang, 2019). Accordingly, corporations' initial investment costs can be decreased in the IBS implementation.

Establish standards and certification systems. The well-defined technical norms and certification systems are developed to safeguard the IBS dependability and quality. The regulations can be enacted by the government with the aim of requiring to use the IBS technology that meets standards and offers certification support. The first step has been taken by the Chinese government. As an illustration, the government has gradually revised and published IBS-associated norms and technical standards. For instance, the Department of Housing and Urban-Rural Development has launched a range of documents, like Technical Regulations for Prefabricated Concrete Structures (2010), "Assembly "Technical Regulations for Integral Concrete Structures" (2011), "Technical Regulations for Prefabricated Concrete Structures" (2014), "Technical Standards for Prefabricated Concrete Buildings" (2017), "Technical Standards for Prefabricated Steel Structure Residential Buildings" (2019). However, the unification and comprehensiveness of norms and technologies nationwide still demand to be deeply enhance.

Collaborate with industry associations. Construction industry associations work together with the professional groups for the sake of mutually formulating and boosting the plans of IBS advancement (Wang, 2019), which can stimulate more corporations to engage in and This can encourage more enterprises to participate in and stand up for the IBS technology.

Examine policies regularly. The related policies are regularly examined and updated, which guarantees that they coincide with market demands and technological progressions. By state`s macro-economic control, the IBS can be promoted better.

Collaborative research and development. The government can support research institutions and universities to conduct research projects on IBS to promote technological innovation and solve potential problems.

International experience exchange. Share the successful experience of IBS with other countries, participate in international seminars and cooperation projects, and refer to some normative policies for IBS in IBS developed countries to obtain global best practices and technological innovation.

By considering the above recommendations, the government can create an environment that supports the development of IBS in China and promote its widespread application in the construction industry. Government leadership and support are crucial for promoting IBS, which will help improve the construction industry's efficiency, sustainability and international competitiveness. In addition, technology and policy can be combined as key development directions. The technology- and policy-led linkage path has a wide range of applicable conditions and is the most applicable among the high-level development paths of 4 prefabricated buildings (Xiao Duo Cheng & Dongfang Wang, 2023).

3. Resolution for Factor "Education"

To promote the implementation of Industrialized Building Systems (IBS) in education, China can take the following measures:

Further integration of IBS courses to the education system. Integration of IBS-related courses in higher education institutions and vocational training (Jiang Yu Shi, 2023), including majors in architecture, civil engineering, construction technology and management. Ensure that students can learn about IBS technology and applications in the classroom.

Training faculty. Provide professional training to teachers and lecturers in educational institutions to ensure they understand the latest IBS technology and industry trends (Jiang Yu Shi, 2023). Establishing collaborations with industry professionals enables educational institutions to maintain links with the actual construction industry.

Practical projects and internship opportunities. Cooperate with construction companies and engineering projects to provide students with practical projects and internship opportunities (He Hui Zhao & Wen Qin, 2023), so that they can experience and understand the practical application of IBS first-hand. This helps combine theoretical knowledge with practical practice.

Develop a scholarship program. Establish a scholarship and funding program to encourage students to choose to study majors related to IBS (He Hui Zhao & Wen Qin, 2023). This can increase the demand for IBS technical talents and push students to choose academic majors in related fields.

By promoting IBS in the education system, more professional talents and technical experts can be cultivated to support the widespread application of this technology in the construction industry. This will also help create a more technologically advanced and innovative construction industry.

4. Resolution for Factor "Technology"

This factor has three sub-factors which are technology of design and manufacture, advanced technology of IBS structure, and materials.

In general, the following measures can be taken to improve IBS technology from design, manufacture and material aspects:

Establish a materials and technology research and development center. Establish an IBS materials and technology research and development center to bring together industry experts, scientific researchers and business representatives. This center can become the core of IBS materials and technology innovation and research, promoting the entire industry's progress.

Industry-university-research cooperation. Promote industry-university-research cooperation and encourage enterprises to establish close cooperative relationships with universities and research institutions. Such cooperation can promote knowledge sharing, technology exchange, and accelerate the progress of IBS materials and technology research.

Technical exchanges and seminars. Organize technical exchange meetings and seminars to provide a platform for experts (He Hui Zhao & Wen Qin, 2023), scholars and practitioners to share the latest IBS materials and scientific and technological research results. This helps facilitate the flow of information and collaboration within the industry.

Apply of digital modeling and simulation in IBS projects. In order to optimize the design and performance of IBS materials and simulate the practical application of new technologies (Qiu Ju Deng, 2024), digital modeling and simulation technology should be promoted. This promotion can be achieved using tools such as Building Information Modeling (BIM) (Qiu Ju Deng, 2024).

Expand the application of life cycle analysis. Conduct a life cycle analysis of IBS materials to assess their environmental and economic impacts throughout their life cycle (Rafael E. López-Guerrero et al., 2022). This helps promote the application of sustainable development concepts in IBS materials research (Rafael E. López-Guerrero et al., 2022).

Create a database. Create an IBS performance and research database so that designers, engineers, and architects have access to it (Yu Gao, 2023). This facilitates the standardization and broad use of IBS materials (Yu Gao, 2023).

Through these initiatives, the research and development of IBS materials and technology can be further deepened in China, and more advanced, efficient and sustainable solutions can be provided for the construction industry.

5. Resolution for Factor "Construction Management"

In China, there are still some aspects of engineering management in industrialized building systems (IBS) that need to be improved, including:

Talent training and skills improvement. Although IBS technology has been promoted in China, engineering management talents' training and skills improvement still need to be strengthened (Jiang Yu Shi, 2023). Training programs and education courses can place greater emphasis on the specific needs of IBS project management and ensure practitioners have the necessary skills and knowledge.

Standardization and Certification. Further develop the standards and certification system for IBS engineering management to ensure the quality and consistency of projects. Standardization can provide clear guidance for projects, reduce risks and improve the entire industry's sustainability.

Application of digital technology. Strengthen the application of digital technology in IBS project management (Yu Gao, 2023), including building information modeling (BIM), project management software and other intelligent tools. This helps improve project collaboration, transparency and efficiency.

Supply chain integration. Further integrate the IBS supply chain, including design, manufacturing, transportation and construction (Yan Liang Shang, 2023). Collaborative cooperation can help reduce the time and errors in information transmission and improve overall project execution efficiency (Yan Liang Shang, 2023).

By improving these aspects, the development of IBS engineering management in China can be further promoted and its application level and effect in the construction industry can be improved.

6. Resolution for Factor "Performance of Implementing IBS"

In order to improve the performance of IBS implementation in China, the following measures can be taken:

Improve construction project efficiency. Use advanced project management tools and techniques such as BIM (Building Information Modeling) to optimize project plans and schedules (Yan Liang Shang, 2023). Reduce fabrication and installation time of building elements through advanced planning and design optimization.

Reduce costs and improve economic benefits. Reduce manufacturing costs of IBS elements through large-scale procurement and supply chain optimization (Dong Mei Zhao & Gao Xian Li, 2023). Encourage innovation and find more cost-effective IBS technologies and solutions.

Improve safety performance. Provide professional safety training to ensure that practitioners understand the best safety practices during IBS construction (Ting Han, 2023). Design and implement an effective site safety plan to reduce potential construction risks.

Improve sustainability performance. Optimize building design to minimize environmental negative impact (Ting Han, 2023). Promote the use of renewable and environmentally friendly materials to reduce waste generation.

Subsequent maintenance and management. Consider maintenance needs during the design phase to simplify subsequent maintenance (Ting Han, 2023). Provide training and support to ensure maintenance personnel understand and adapt to the special requirements of using IBS.

Learning curve and improvements. Implement a culture of continuous improvement and encourage the team to continuously learn and optimize the implementation of IBS. Collect feedback and learn from each project to drive continuous improvement.

By taking these measures together, the performance of IBS in China can be continuously improved and its wider application in the construction industry promoted.

7. Resolution for Factor "Culture and Tradition"

Culture and tradition not only influence how people view architecture and engineering, but also profoundly impact the implementation, acceptance and final results of IBS.

Cultural adaptability. Different countries and regions have unique cultural backgrounds and traditions. These cultural factors deeply influence people's understanding and practice of architecture and engineering. For instance, residents in some regions may attach more importance to harmonious coexistence with nature, where those in other areas are likely to focus more on the practicality of the architecture. Understanding these cultural differences can not only facilitate the implementation of IBS but also gain support from lots of stakeholders.

Conventional architectural values. As a carrier of culture, the architecture manifests the values of different areas. In the implementation of IBS, it is essential to respect these traditional architectural values, and ensure the integration of new construction methods with traditional architectural arts. While enhancing people's sense of identity with IBS, this can maintain the valuable architectural and cultural heritage.

"Culture and tradition" factors play an important role in the implementation of IBS. Respecting and integrating local culture and traditions, and promoting community participation and cultural inheritance are the keys to implementing IBS successfully. Through cross-cultural exchanges and cooperation, education and training, IBS can be better combined with culture and tradition to create architectural works that are both modern and culturally distinctive.

SHORTCOMINGS OF THE STUDY

This study screened 78 articles related to the implementation of prefabricated buildings through a systematic literature review. The relevant factors affecting the successful implementation of prefabricated buildings in China mentioned in these 78 articles are summarized and corresponding countermeasures are presented. Since the study theme is about systematic literature review,factors extracted based on only literature, there is no verification and consideration at the practical level. No empirical analysis and questionnaire survey has been conducted. Therefore, further empirical analysis or questionnaire surveys are needed in future research to more deeply verify the reliability of factors, the relationship and importance between factors.

OUTLOOK AND CONCLUSION

To summarize, this article studied the influential factors of implementing IBS involved in 78 IBS-related papers from the Scopus database and found that 7 factors, including "cognitive incentives," "regulation by governments," "education," " technology," "construction management," "performance," and "culture and tradition" are crucial factors. Accordingly, through extensive reading of domestic and foreign literature, this article provides some implementation strategies suitable for China's national conditions. In addition, this article also proposes shortcomings in the current research on IBS, such as the lack of research on long-term performance, design innovation in IBS, and IBS culture and traditional cognition. These deficiencies need to be further improved in future research.

REFERENCE

Ali, M. M., Abas, N. H., Affandi, H. M., & Abas, N. A. (2018). Factors impeding the industrialized building system (IBS) implementation of building construction in malaysia. *International Journal of Engineering and Technology(UAE)*, 7(4), 2209–2212. 10.14419/ijet.v7i4.17863

Ariffin, H. L. B. T., Lynn, B. Y. H., Shukery, N. B. M., Rahiman, N. B. A., Mahmud, S. H. B., & Raslim, F. B. M. (2018). Innovative procurement adoption for industrialised building system (IBS) projects. *International Journal of Engineering and Technology(UAE), 7*(29), 887–892. 10.14419/ijet.v7i2.29.14277

Balasbaneh, A. T., & Bin Marsono, A. K. (2017). Strategies for reducing greenhouse gas emissions from residential sector by proposing new building structures in hot and humid climatic conditions. *Building and Environment*, 124, 357–368. 10.1016/j.buildenv.2017.08.025

Bataglin, F. S., Viana, D. D., Formoso, C. T., & Bulhões, I. R. (2020). Model for planning and controlling the delivery and assembly of engineer-to-order prefabricated building systems: Exploring synergies between lean and BIM. *Canadian Journal of Civil Engineering*, 47(2), 165–177. 10.1139/cjce-2018-0462

Beijing Municipal Housing and Urban-Rural Development Commission. (2021). *Notice on the issuance of the "Beijing green building creation action implementation plan (2020-2022)"*. Beijing Municipal Housing and Urban-Rural Development Commission. https://zjw.beijing.gov.cn/bjjs/gcjs/jzjnyjcjg/tzgg/tzgg98/325777657/index.shtml

Cao, P., & Lei, X. (2023). Evaluating risk in prefabricated building construction under epc contracting using structural equation modeling: A case study of Shaanxi Province, china. *Buildings*, 13(6), 1465. 10.3390/buildings13061465

Central Committee of the Communist Party of China and the State Council. (2021). *Opinions on complete, accurate and comprehensive implementation of the new development concept and carbon implementation to peak carbon neutrality*. Central Committee of the Communist Party of China and the State Council. https://www.gov.cn/zhengce/202110/24/content_5644613.htm

Chang, C., Wu, X., & Yan, X. (2020). Multiobjective optimization of safety risk of prefabricated building construction considering risk correlation. *Mathematical Problems in Engineering*, 2020, 1–11. 10.1155/2020/3923486

Chang, Y., Li, X., Masanet, E., Zhang, L., Huang, Z., & Ries, R. (2018). Unlocking the green opportunity for prefabricated buildings and construction in china. *Resources, Conservation and Recycling*, 139, 259–261. 10.1016/j.resconrec.2018.08.025

Che, M. R. (2022). *Research on incentive effect and optimization simulation of government policy on prefabricated buildings* [Master's thesis, Chongqing Jiaotong University]. CNKI. https://link.cnki.net/doi/10.27671/d.cnki.gcjtc.2022.000779doi:10.27671/d.cnki.gcjtc.2022.000779

Chen, D., Deng, J., Cheng, B., Wang, Q., & Zhao, B. (2021). New anticracking glass-fiber-reinforced cement material and integrated composite technology with lightweight concrete panels. *Advances in Civil Engineering*, 2021, 1–17. 10.1155/2021/7447066

Cheng, X. D., & Wang, D. (2023). Research on the high-level development path of prefabricated buildings: csQCA analysis based on TOE framework. *Journal of Engineering Management*, (6), 1–6. 10.13991/j.cnki.jem.2023.06.005

China's State Council. (2021). *Notice of the state council on issuing the action plan for carbon peaking before 2030.* CSC. https://www.gov.cn/zhengce/content/2021-10/26/content_5644984.htm

Deng, Q. J. (2024). Research on the application of BIM technology in the quality chain system of prefabricated buildings. *Journal of Beijing Industrial Vocational and Technical College*, 23(1), 1–5.

Du, Q., Hao, T., Huang, Y., & Yan, Y. (2022). Prefabrication decisions of the construction supply chain under government subsidies. *Environmental Science and Pollution Research International*, 29(39), 59127–59144. 10.1007/s11356-022-19861-035380327

Duan, Y., & Li, G. (2020). Analysis on the quality problems and preventive measures of prefabricated building construction. *Journal of Physics: Conference Series*, 1648(3), 032141. 10.1088/1742-6596/1648/3/032141

Feng, Y. (2021). Research on visualization of prefabricated building process based on BIM technology. *ACM International Conference Proceeding Series*, (pp. 307–312). ACM. 10.1145/3510858.3510955

Ferdous, W., Bai, Y., Ngo, T. D., Manalo, A., & Mendis, P. (2019). New advancements, challenges and opportunities of multi-storey modular buildings – a state-of-the-art review. *Engineering Structures*, 183, 883–893. 10.1016/j.engstruct.2019.01.061

Gao, Y. (2023). Research progress on digitalization of prefabricated construction industry. *Chinese Building Metal Structure*, 22(11), 140–142. 10.20080/j.cnki.ISSN1671-3362.2023.11.046

General Office of the CPC Central Committee and the General Office of the State Council. (2021). *Opinions on promoting green development of urban and rural construction.* CPC. https://www.gov.cn/zhengce/2021-10/21/content_5644083.htm

Grant, M. J., & Booth, A. (2009). A typology of reviews: An analysis of 14 review types and associated methodologies. *Health Information and Libraries Journal*, 26(2), 91–108. 10.1111/j.1471-1842.2009.00848.x19490148

Guo, Y. (2021). The establishment and evaluation method of prefabricated building construction quality evaluation index system taking into account multimedia sensor network nodes. *Advances in Multimedia*, 2021, 1–6. 10.1155/2021/9613100

Guo, Y. (2021). Research on design method and construction key points of prefabricated building structure combined with machinery. *Journal of Physics: Conference Series*, 1744(2), 022091. 10.1088/1742-6596/1744/2/022091

Han, T. (2023). Comprehensive benefit analysis of prefabricated buildings under the full life cycle. *Stone*, (12), 96–98. 10.14030/j.cnki.scaa.2023.0535

Hui, T. L., & Khoon, N. C. (2020). Comparative study on precast building construction and conventional building construction for housing project in sarawak. *Jurnal Teknologi*, 82(1), 75–84. 10.11113/jt.v82.13776

Huo, T., Ye, C., & Zhang, X. (2021). Research on construction technology of external hanging scaffold for prefabricated concrete shear wall. *Journal of Physics: Conference Series*, 1952(3), 032073. 10.1088/1742-6596/1952/3/032073

Ismail, Z. A. (2019). Optimising the safety of road transport workers on IBS building construction projects: A review. *Social Responsibility Journal*, 15(6), 837–851. 10.1108/SRJ-09-2018-0240

Ji, Y., Qi, K., Qi, Y., Li, Y., Li, H. X., Lei, Z., & Liu, Y. (2020). BIM-based life-cycle environmental assessment of prefabricated buildings. *Engineering, Construction, and Architectural Management*, 27(8), 1703–1725. 10.1108/ECAM-01-2020-0017

Jin, Y., Zhang, J., & Sun, L. (2019). Safety risk assessment of prefabricated building construction based on bayesian network. *IOP Conference Series. Earth and Environmental Science*, 371(3), 032052. 10.1088/1755-1315/371/3/032052

Liu, Y., & Cui, J. (2020). Identification of hazard sources in prefabricated building construction by entropy weight method. *IOP Conference Series. Earth and Environmental Science*, 560(1), 012073. 10.1088/1755-1315/560/1/012073

Liu, Z., Meng, X., Xing, Z., & Jiang, A. (2021). Digital twin-based safety risk coupling of prefabricated building hoisting. *Sensors (Basel)*, 21(11), 3583. 10.3390/s2111358334064110

López-Guerrero, R. E., Vera, S., & Carpio, M. (2022). A quantitative and qualitative evaluation of the sustainability of industrialised building systems: A bibliographic review and analysis of case studies. *Renewable & Sustainable Energy Reviews*, 157, 112034. 10.1016/j.rser.2021.112034

Loss, C., Rossi, S., & Tannert, T. (2018). In-plane stiffness of hybrid steel–cross-laminated timber floor diaphragms. *Journal of Structural Engineering*, 144(8), 04018128. 10.1061/(ASCE)ST.1943-541X.0002105

Menegon, S. J., Wilson, J. L., Lam, N. T. K., & Gad, E. F. (2020). Experimental testing of innovative panel-to-panel connections for precast concrete building cores. *Engineering Structures*, 207, 110239. 10.1016/j.engstruct.2020.110239

Mengist, W., Soromessa, T., & Legese, G. (2020). Method for conducting systematic literature review and meta-analysis for environmental science research. *MethodsX*, 7, 100777. 10.1016/j.mex.2019.10077731993339

Navaratnam, S., Ngo, T., Gunawardena, T., & Henderson, D. (2019). Performance review of prefabricated building systems and future research in Australia. *Buildings*, 9(2), 38. 10.3390/buildings9020038

Nawi, M. N. M., Noordin, A., Tamrin, N., Nifa, F. A. A., & Lin, C. K. (2019). An ecological study on enhancing the Malaysian construction ecosystem: Readiness implementation factors in industrialised building system (IBS) projects. *Ekoloji*, 28(107), 545–552.

Peng, J., Zhou, J., Meng, F., & Yu, Y. (2021). *Analysis on the hidden cost of prefabricated buildings based on FISM-BN*. PLoS ONE. 10.1371/journal.pone.0252138

Petticrew, M., & Roberts, H. (2008). *Systematic reviews in the social sciences: A practical guide*. Blackwell Publishing., 10.1002/9780470754887

Run, W. (2023). *Research on safety risk assessment and control of prefabricated building construction based on SEM-FCEM*. CNKI. https://link.cnki.net/doi/10.27784/d.cnki.gahjz.2023.000336

Shang, Y. L. (2023). Research on collaborative optimization management and platform construction of prefabricated construction supply chain enterprises based on BIM technology. *Journal of Shijiazhuang Railway Vocational and Technical College, 22*(3), 1–4+58.

Shen, Q., Liu, Y., & Shen, Q. (2019). Predicted model of ecological footprint based on PLCA. *IOP Conference Series. Materials Science and Engineering*, 688(5), 055035. 10.1088/1757-899X/688/5/055035

Shen, Y., Xu, M., Lin, Y., Cui, C., Shi, X., & Liu, Y. (2022). Safety risk management of prefabricated building construction based on ontology technology in the BIM environment. *Buildings*, 12(6), 765. 10.3390/buildings12060765

Shi, J. Y. (2023). Prefabricated construction industrialization worker development path and transformation and upgrading measures and strategies. *Sichuan Architecture*, 43(6), 289–290.

Sichuan Provincial Department of Housing and Urban-Rural Development. (2021). *Sichuan province issued the "14th Five-Year Plan" construction industry development plan*. SPD. http://jst.sc.gov.cn/scjst/ghxx/article_list.shtml

Sohu. (2018). *All classics! 10 well-known prefabricated buildings in china*. Sohu. https://www.sohu.com/a/284605502_120049046

Song, W. (2022). Building construction design based on particle swarm optimization algorithm. *Journal of Control Science and Engineering*, 2022, 1–8. 10.1155/2022/7139230

Su, T. T. (2020). The current situation of prefabricated buildings in china and their application in residential buildings. *Building Materials and Decoration*, (11), 98–99.

Sun, H., Fang, Y., Yin, M., & Shi, F. (2023). Research on the restrictive factors of vigorous promotion of prefabricated buildings in yancheng under the background of "double carbon". *Sustainability (Basel)*, 15(2), 1737. 10.3390/su15021737

Tavares, V., Gregory, J., Kirchain, R., & Freire, F. (2021). What is the potential for prefabricated buildings to decrease costs and contribute to meeting EU environmental targets? *Building and Environment*, 206, 108382. 10.1016/j.buildenv.2021.108382

Teng, Y., & Pan, W. (2019). Systematic embodied carbon assessment and reduction of prefabricated high-rise public residential buildings in Hong Kong. *Journal of Cleaner Production*, 238, 117791. 10.1016/j.jclepro.2019.117791

Vasovic, D., Lelovic, S., & Okrajnov-Bajic, R. (2021). Development of new materials for prefabricated concrete elements for building construction in tropical climatic conditions. *IOP Conference Series. Earth and Environmental Science*, 798(1), e012015. 10.1088/1755-1315/798/1/012015

Volpe, S., Sangiorgio, V., Petrella, A., Coppola, A., Notarnicola, M., & Fiorito, F. (2021). Building envelope prefabricated with 3d printing technology. *Sustainability (Basel)*, 13(16), 8923. 10.3390/su13168923

Wang, P. W. (2019). *Research on the influence of technical and economic policy on technological innovation of prefabricated building* [Master's thesis, Shenyang Jianzhu University]. CNKI. https://cdmd.cnki.com.cn/Article/CDMD-10153-1019255628.htm

Wang, S., Mursalin, Y., Lin, G., & Lin, C. (2018). Supply chain cost prediction for prefabricated building construction under uncertainty. *Mathematical Problems in Engineering*, 2018, 4580651. 10.1155/2018/4580651

Wang, S., & Sinha, R. (2021). Life cycle assessment of different prefabricated rates for building construction. *Buildings*, 11(11), 552. 10.3390/buildings11110552

Wang, Y. R., Pan, H. P., & Cai, W. Q. (2020). Research on china's prefabricated construction industry policies, current situation and development trends. *Journal of Anhui Jianzhu University*, 28(4), 47–51.

Wu, C., Yu, S., Liu, J., & Chen, G. (2020). Development and testing of hybrid precast steel-reinforced concrete column-to- H shape steel beam connections under cyclic loading. *Engineering Structures*, 211, 110460. 10.1016/j.engstruct.2020.110460

Wu, S., & Liu, Q. (2020). Analysis on the application of BIM and RFID in life cycle management of prefabricated building. *IOP Conference Series. Materials Science and Engineering*, 780(3), e032012. 10.1088/1757-899X/780/3/032012

Yang, B., Dong, M., Wang, C., Liu, B., Wang, Z., & Zhang, B. (2021). IFC-based 4D construction management information model of prefabricated buildings and its application in graph database. *Applied Sciences (Basel, Switzerland)*, 11(16), 7270. 10.3390/app11167270

Yang, B., Fang, T., Luo, X., Liu, B., & Dong, M. (2022). A BIM-based approach to automated prefabricated building construction site layout planning. *KSCE Journal of Civil Engineering*, 26(4), 1535–1552. 10.1007/s12205-021-0746-x

Yang, M. (2022). Risk management of prefabricated building construction based on fuzzy neural network. *Scientific Programming*, 2022, 2420936. 10.1155/2022/2420936

Yang, Y. (2021). *Application analysis of BIM technology and prefabricated structure construction [Paper presentation].* The Sixth International Conference on Information Management and Technology, Jakarta. 10.1016/j.cie.2021.107316

Yunus, R., Handan, R., & Riazi, S. R. M. (2020). Case studies on sustainability factors for industrialised building system (IBS). *International Journal of Sustainabl e Construction Engineering and Technology*, 11(2), 65–71. 10.30880/ijscet.2020.11.02.007

Zhao, D. M., & Li, G. X. (2023). Prefabricated building supply chain risk management based on AHP-DEMATEL. *Journal of Luohe Vocational and Technical College*, 22(6), 32–36.

Zhao, H. H., & Qin, W. (2023). Research on the training of prefabricated construction talents based on the integration of industry and education. *Shanxi Architecture*, 49(17), 196–198. 10.13719/j.cnki.1009-6825.2023.17.052

Zhao, Y., Cao, C., & Liu, Z. (2022). A framework for prefabricated component hoisting management systems based on digital twin technology. *Buildings*, 12(3), 276. 10.3390/buildings12030276

Zheng, W., Ye, Y., & Zang, H. (2022). Application of BIM technology in prefabricated buildings based on virtual reality. *Computational Intelligence and Neuroscience*, 2022, 9756255. 10.1155/2022/975625536225548

Zhu, H., Hong, J., Shen, G. Q., Mao, C., Zhang, H., & Li, Z. (2018). The exploration of the life-cycle energy saving potential for using prefabrication in residential buildings in china. *Energy and Building*, 166, 561–565. 10.1016/j.enbuild.2017.12.045

Chapter 5
The Role of AI in Urban Design and Planning in Responding to Extreme Flooding Events in Trinidad and Tobago

Roshnie Anita Doon
https://orcid.org/0000-0002-3285-0355
Global Labor Organization, Essen, Germany

ABSTRACT

Caribbean islands are some of the most vulnerable countries in the world to experience extreme flooding resulting from tropical cyclones, storm surges, and excessive rainfall. Persistent changes in the global climate have made countries like Trinidad and Tobago hot spots of climate change. As a result, the occurrence of extreme flooding events is likely to persist and is made worse by the barriers to adaptation and mitigation, such as having limited knowledge of climate change and how to plan for its impact at the individual level, as well as limited financial resources and technical know-how. Using a secondary research methodology, this chapter seeks to investigate the role that artificial intelligence (AI) plays in urban planning and design in responding to extreme flooding events in Trinidad and Tobago. This chapter discusses not only the historical occurrence of flooding events, and its causes, but also how the possible use and integration of AI in the planning and design of urban spaces in Trinidad and Tobago can assist in reducing the ill effects of such events.

INTRODUCTION

Climate Change is one of the most problematic issues facing the world today. In the face of extreme climate-related changes such as the growing intensity of extreme temperatures and metrological events, rising sea levels, and oceanic acidification, it is usually the poorer segments of populations within Small Island Developing States (SIDS) and Developing countries that Greenhouse Gases to suffer the most from the actions of more wealthy countries who are major emitters of Greenhouse Gases (GHG). Most of these emissions (60%) are produced by 10 countries, i.e., China, the United States (US), the

DOI: 10.4018/979-8-3693-1650-4.ch005

European Union (EU), India, Indonesia, Russia, Brazil, Japan, Iran, Canada, Japan, and Saudi Arabia (Climate Watch, 2022).

As these large fossil fuel-burning countries continue to grow their economies at the expense of smaller nations, it is, unfortunately, the poorer countries and regions around the world such as Bangladesh, Sub-Saharan Africa, Central America as well as the Caribbean, who will experience the grimmest of consequences through the frequent passage of tropical cyclones, heavier rainfall, intense flooding, greater incidence of water-borne diseases, droughts and poor harvests (Conner & Krogstad, 2018). As the climate crisis intensifies around the world, the most vulnerable in our society are the ones who will feel its disastrous aftereffects. This is indeed the case as the Caribbean region is expected to see its vulnerable population becoming increasingly displaced by natural disasters such as extreme flooding.

Considering such dislocation, diversity, equity, and inclusion have also been drivers behind the unequal implementation of mitigation and adaptation strategies for climate change. While some of the very same GHG-emitting countries, may be wealthy enough to fund the implementation of mitigation and adaptation strategies to reduce their GHG emissions, other countries like Trinidad and Tobago may not be able to do the same due to a lack of technical expertise, and financial resources, to build and strengthen the adaptive capacity needed to address climate change problems (Lemos et al., 2013). Notwithstanding this, the intensification of climate change events, like extreme flooding has a significant impact on the livelihoods of the most vulnerable communities in Trinidad and Tobago.

An examination of the climate change literature for Latin America and the Caribbean, reveals that while there is a wide range of studies which (1) examines climate change in coastal areas, and the impacts of floods occurring from extreme rainfall and snowmelt for Latin American Countries like Costa Rica, Ecuador, Chile and Argentina, and (2) studies that examine flood control mechanisms, flood insurance and management/analysis systems, flood risk mapping, and the impact of freshwater flooding on mangroves, there is a dearth of knowledge that examines the role of Artificial Intelligence (AI) in Urban Design and Planning in responding to extreme flooding events in Trinidad and Tobago.

Owing to this lack of knowledge, this chapter will aim to examine the role that AI plays in Urban Planning and Design in responding to flooding events in Trinidad and Tobago. To do this, using a conceptual research methodology, this study has several objectives, first to examine the historical flooding events in Trinidad and Tobago, second the primary causes of such flooding events, third explore the level of vulnerability to such natural hazards, fourth discuss the importance of urban design and planning in preventing urban flooding, fifth to examine the role that AI plays in urban design and planning in avoiding such events, and sixth to discuss the challenges that small island developing states like Trinidad and Tobago may encounter in integrating AI in its urban planning and design model, after which the study is concluded.

RESEARCH METHODOLOGY

This chapter takes a conceptual approach to investigating the role that AI plays in the Planning and Design of Urban areas in Trinidad and Tobago, as a response to extreme flooding events. This methodology was suitable for this chapter because it explores the existing literature on AI about Urban Planning and Design to develop a conceptual framework to understand better how AI-based strategies can be used to address severe flooding events in the Twin Island State. Further to this, given financial and time constraints, this approach was appropriate because it required fewer resources while allowing for both a deeper understanding of the complexities surrounding AI in urban planning and design, as well as providing the foundation for future avenues of research.

The first step to constructing the framework was to decide upon the topic of investigation, i.e., the role of AI in Urban Design and Planning in Responding to Extreme Flooding Events in Trinidad and Tobago. Following this, the second step consisted of using a secondary research approach to sourcing relevant scholarly literature on this topic from reputable journals such as the Journal of Environmental Management, Studies in Computational Intelligence, Sustainable Cities and Societies, and others, as well as notable organizational reports from the Inter-American Development Bank (IDB), the International Monetary Fund (IMF), and the World Bank (WB).

In the third step, the information from these sources was used to first add context to the study by (1) providing a historical overview of flooding events in Trinidad and Tobago, (2) exploring its causes, and (3) examining its vulnerability to these types of meteorological events. Second, using the relevant information from the literature two frameworks were then developed to examine the various themes of not only the importance of Urban Design and Planning in Flood Prevention, but also the role that AI plays in such Prevention. In addition, the Challenges that Small Island Developing States (SIDS) like Trinidad and Tobago may encounter in using AI-based responses were also discussed.

To this end, using this approach, a literature map as shown below in Figure 1 was developed and highlights 5 major themes (adaptive strategies, flood adaptation models, flood resiliency systems, non-structural mitigation plans, and building regulations) that are important in Urban Planning and Design aspect of Flood Prevention, and 3 major themes (flood resilience, storm detention, and forecasting and prediction) in the role that AI plays in the in Urban Planning and Design aspect of Flood Prevention.

Figure 1. Literature map of the importance and role of AI urban design and planning in flood prevention

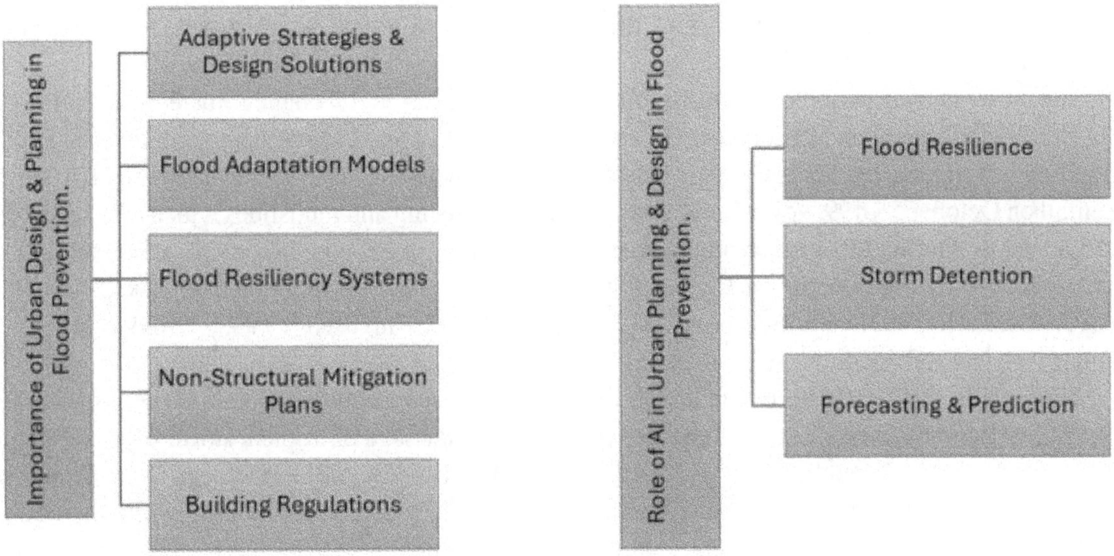

A HISTORY OF FLOODING EVENTS IN TRINIDAD AND TOBAGO

Flooding in Trinidad and Tobago, like the island nations across the Caribbean, has become an annual problem, which has intensified with the frequent occurrence of environmental and climate change events such as a rise in the sea level, stronger hurricanes, and tropical storms, as well as longer dry seasons and shorter wet seasons. Even though the Caribbean region makes a low contribution to global Green House Gas (GHG) emissions Climate Change events continue to pose a worrying hazard for the region's population. The increased occurrence of these hydro-meteorological events, together with a change in the flood plains, are likely to have contributed to changing patterns and intensity of precipitation in Trinidad and Tobago, so much so that even shorter periods of intense rainfall events may lead to frequent and widespread flooding (Fontes de Meira & Phillips, 2019). An inventory of the floods occurring between 1986-2006 where flooding has occurred 10 or more times within the last 20 years, revealed that based on the physical characteristics of the soil, the most flood-prone areas in Trinidad and Tobago remain in the counties of Caroni and Victoria (Aliasgar, 2012). However, while there has been a growth in the reports of flooding and landslides such as during the 2005 and 2006 dry season floods, flooding is now taking place in areas which were never historically prone to flooding.

An examination of the flood risk models for Trinidad and Tobago using remotely sensed data collected in 2007, revealed that while an increased frequency and intensity of rainfall events contribute to the flooding problem, so too does the growth of urbanization (Thongs, 2019). The expansion of urban and suburban developments, in Trinidad and Tobago's low-lying flood plains and areas where the land has become degraded, as well as unplanned development and nonsynchronous infrastructural road works, may have both caused natural drainage systems to change in these areas (Narinesingh, 2014). These changes would have contributed to the growth of flooding in Trinidad and Tobago because the increased sedimentation in streams and rivers, may have clogged waterways. The most at-risk flooding areas in Trinidad are the areas of St. Joseph, San Juan, and Beetham in North Trinidad, as well as Barrackpore in South Trinidad, while most areas along the East-West Corridor and Southwestern Areas of Trinidad remain highly susceptible to flooding.

The three known periods of extreme flooding in Trinidad and Tobago's meteorological history, i.e., 1993, 2017-2018, and 2022, were triggered by meteorological events. Although historical data on flooding in Trinidad and Tobago is quite sparse, the flooding events of 1993 saw between 60-100mm of rainfall on October 5th, 1993, over 2 hours causing heavy flooding and mudslides which affected a region of 1,1170km. During this time, ten persons were noted to have become homeless, and 5 persons died as a direct result of the flooding activity, while the economic damages amounted to US$70,000 (Udika, 2010). Earlier in 1993, with the passage of Hurricane Bret on August 7th, 1993, Trinidad experienced minor instances of flooding.

Throughout 2017-2018, Trinidad and Tobago experienced extraordinary rainfall events that led to five major flooding events, i.e., (1) June 19th-21st, 2017(the passage of tropical storm Bret), (2) October 18th-21st, 2017; (3) December 29th-31st, 2017; (4) October 18th-20th, 2018 (October Great Floods); (5) November 15th-19th, 2018 (November Floods). These instances of severe flooding while increasing in terms of severity were noted to be concentrated in the communities in Southern and Eastern Trinidad in 2017, and in North-Eastern and Southern Trinidad in 2018, affecting 150,000 persons (11% of the population of Trinidad and Tobago (Fontes de Meira & Phillips, 2019).

The aftermath of such floods in both years led to not only the loss of property, crops and livestock of persons residing in these communities, but also worsened the presence of flood-related natural disasters such as mass wasting which is a type of landslide which includes the slope movement of either bedrock, debris, or earth that is worsened by the presence of rainfall. Notwithstanding this, these regions are historically susceptible to flooding, during the period 1999-2011, communities such as Barrackpore, Caroni Village, Guayaguayare, and Penal, were high-risk areas, while communities such as Bamboo, Carenage, La Seiva, Maloney Gardens, Spring Village and Siparia, were at a medium risk of experiencing flooding events.

The final major flooding event to have occurred in Trinidad and Tobago, took place between November 25th- 29th, 2022, following weeks of excess rainfall between November 9th-11th, 2022 (UN-OCHA, 2023). During this time, the Ortoire, Caparo, South and North Oropouche, and the Caroni rivers all began reaching their maximum capacity. Heavy rainfall events concentrated in North and Central Trinidad, caused flash flooding in communities along the East-West Corridor, while riverine flooding in areas such as Bamboo #2, Madras Road, St. Helena, and Kelly Village were caused by the broken banks of the Caroni River (UN-OCHA, 2023). It was reported that there were 30 flooding incidents, and more than 51 landslides, which affected approximately 100,000 persons who resided in the areas of Mayaro, Rio Claro, Penal, Debe, San Juan, Laventille, Sangre Grande, Tunapuna and Piarco (IFRC, 2022). Following the rainfall event, the length of time that communities remained submerged in flood waters also posed a major problem, as several road networks in these areas were damaged having become either undermined and caved in or completely washed away.

THE CAUSES OF FLOODING IN TRINIDAD AND TOBAGO

Occurring on an annual basis, flooding has become one of the most common natural disasters to affect Trinidad and Tobago, brought upon by either natural or human-caused factors. In the case of Trinidad and Tobago, there are several factors which influence the occurrence and type of flooding that takes place.

Intensity and Duration of Rainfall

The intensity and duration of rainfall can significantly impact the flooding outcomes of communities. In the first case of duration, Prolonged Periods of Rainfall over several days in which water accumulated faster than the ability of the soil to absorb and carry it away, while in the second case of intensity, often results from Intense Periods of Rainfall over a shorter period. In both cases, rivers tend to overflow because the water control structure of the tributary is compromised. In Trinidad and Tobago, Flash flooding is common, as floods typically develop within at least 6 hours of rainfall. These types of floods are characterized by the rapid rise and swelling of water courses in low-lying areas, resulting from a copious amount of rainfall from severe thunderstorms, and tropical cyclones, which can also be associated with dam failure, and levee breaches (WHO, 2023). Instances of Flash flooding was noted to have occurred in the Penal/Debe area, Riverine and Urban flooding are becoming increasingly present in the area.

Saturated Soils and Topography

When debris, i.e., mineral, and organic sediment-rich floods take place in sharp waterways it destabilizes the banks of rivers causing them to become mobile. These types of floods can be categorized based on either their exceedance of shear stress, its initial cause, i.e., either being triggered by debris flow inside of the channel, or from the slanted impact of a debris flow-prone river and finally the outbreaking of floods from natural or artificial dams. The structure of the land and its geographical location also plays an important role in the incidence of flooding, as the topography and slope of a floodplain can regulate many aspects of a flooding event such as its velocity (flood flow velocity), depth (flood water depth), and duration (inundation duration of a flood) (Sendek et al., 2021). As a result, Roopnarine et al. (2018) in their risk mapping study of Trinidad show that 23% of the island's landmass is likely to be at a considerable risk of flooding, particularly in urban and suburban areas.

Meteorological Events

Tropical cyclones, Hurricanes, Storm surges, Tsunami and weather conditions associated with the passage of the Inter-Tropical Convergence Zone (ITCZ), can all contribute to periods of intense flooding and landslides that can occur either inland between valleys and depressions, or on the coast. Coastal flooding usually occurs with the landfalling of a tropical storm or hurricane, which produces not only large storm surges and waves that destroy buildings, roads, and homes in coastal communities, as well as the erosion of beaches, but also contributes to the development of caves and karsts environments (Arkle et al., 2017). Furthermore, the erosion and recession of the coast may cause the marine ecology of the beach to change so much so that leatherback turtles that visit the beaches on the eastern coast of Trinidad and Tobago to nest will no longer come to these traditional sites to lay their eggs. In the cases where there is freshwater flooding of Mangroves in Trinidad and Tobago, this can have a detrimental effect on the mangrove ecosystem as the excess sediments deposited by flood waters may cause the mangrove to die due to the lack of oxygen.

Deforestation and Land Use Practices

Throughout the period 2001/2002-2022, Trinidad and Tobago have lost 2.58kha of its humid primary forest, and 23.5kha of its tree coverage (GFW, 2023). During this time, 0.29% of this tree coverage loss was due to practices such as logging, slash/burning practices used in agricultural practices, forest fires and infrastructural developments such as urbanization and road building, which have resulted in deforestation (GFW, 2023). In areas where the land has reduced tree/plant coverage, floods can occur frequently as the water flows quickly on the surface of the soil rather than infiltrating the soil. Since the amount of foliage is sparse in these areas, periods of moderate-heavy rainfall guarantee that there will be a rise in soil erosion and deposition of sediments on lower slopes because of increased surface runoff as reflected by the heavy sediment load in the Caparo River Basin (Ramlal & Baban, 2008). For this reason, it is important to address the causes of land degradation present in Trinidad and Tobago, such as deforestation, loss of tree coverage, flooding, soil erosion/deterioration, coastal erosion, and pollution. Brookhuis and Hein, (2015) also explain that there is a rise in small-medium flooding events in Trinidad where there is decreasing forest cover, while larger floods are found primarily in catchment areas where there is less than 85% forest cover.

Improper Waste Disposal, Pollution and Illegal Quarrying

A poor waste disposal system can clog drains and prevent water from flowing freely, leading to flooding of the surrounding area during adverse weather events. This is especially important as a build of garbage that is not properly disposed of, especially if dumped into watercourses, can clog drainage systems, obstructing the free and natural flow of water in rivers (Canisius & Nancy, 2009). The practice of illegal quarrying involves the clearing away of land to produce aggregates, i.e., sand, and gravel, which are used in construction activities. The poor regulation of quarrying often leads to instances of illegal quarrying which leaves the land bare, as it removes all tree/plant coverage, which intrudes increased surface runoff when adverse weather events occur. The outcome of such illegal quarrying activities according to Scobie et al. (2023) are experienced far from these sites as sediments are deposited along rivers further inland.

Land Development and Urbanization

Illegal Hillside land development, Land development, squatting along flood plains, and urbanization in the upper segment of watersheds for housing and road infrastructure are all known to contribute to *urban flooding* because the permeable layers of soil are replaced by construction materials, i.e., concrete, bricks, stones, wood, asphalt and building coverage, that are impenetrable by water (Sangster et al., 2020). Apart from the changing land use cover, urbanization also causes the habitat of animal species to dwindle (biodiversity loss) and become extinct as the fragmentation of the land weakens the amount of ecologically habitual land (Mycoo, 2018). Further to this, although some catchment areas in Port-of-Spain may have better flood control mechanisms in place such as levees, terraces, and drainage ditches, if these mechanisms are either not supported by frequent maintenance, or these mechanisms may not be provided in other urban areas in Trinidad and Tobago, the vulnerability of flooding risk will tend to remain elevated in the City of Port-of-Spain and other urban communities (Brookhuis & Hein, 2015).

Lack of Information and Poor Management

The poor management and coordination of planned and unplanned development by ministerial bodies in Trinidad and Tobago, also contribute to instances of flooding, particularly in urban areas like the Chaguanas Borough (Narinesingh, 2014). These unplanned development activities, together with road works being done by the government ministry, without sufficient investigation and consultations, contributed to the flood waters of approximately 1.0 meters in the area, as new infrastructural works in old, planned developments were not required to meet design standards as there was no re-certification needed. The increased growth in unplanned development may also reflect both institutional and legislative weaknesses of ministerial bodies which supervise such projects. This is further compounded by the moderate level of preparedness and response by disaster response agencies, and community members who may have either little knowledge or resources to prepare for flooding events.

TRINIDAD AND TOBAGO'S VULNERABILITY TO EXTREME FLOODING

The concept of vulnerability is important in the discussion of extreme flooding in Trinidad and Tobago, not only because it indicates how likely persons and communities are to be harmed by extreme flooding events, but also because of how different groups of persons are likely to experience the different levels of vulnerability, based on the level of risk. While there are several indicators can be used to assist in determining the level of vulnerability associated with natural hazards and climate change, for purposes of this chapter only the Local Disaster Index (LDI), and the Risk Management Index (RMI) will be examined.

The LDI index detects the social and environmental risks associated with low recurrent events at the local and subnational level that have a significant impact on the more socially disadvantaged population. The use of this index is especially important in disaster risk and reduction because it not only evaluates the natural tendency of Trinidad and Tobago to experience small-scale natural disasters created by flash, riverine, and urban flooding events which are common occurrences, but also its cumulative impact on progress and development of local private and public sector enterprises. Such an index is of particular importance as it captures the impact of small-scale natural disasters like floods that may not be included in international disaster databases. In a 2008 Vulnerability Study undertaken by the IMF, the LDI of Trinidad and Tobago, was found to be 23, which implies that the country's risk management performance is not only unsatisfactory but implies that the country may not have the capacity to identify, reduce risk, provide financial protection, and respond and recover to natural disasters like extreme flooding (IMF, 2013).

The aim of the RMI index is typically used to measure the performance and effectiveness of risk management. According to the IDB (2010b, p. 19), the *"index is a qualitative measurement of risk based on pre-established levels (targets) or desirable references (benchmarking) towards which risk management should be directed, according to its level of advance,"* and is composed of four different areas of public policy, i.e., Risk Identification (RI), Risk Reduction (RR), Disaster Management (DM), and Finance Protection (FP). An average of the four policy areas is used to determine the RMI, where on a scale of 0-100, 0 implies minimum performance, and 100 is maximum performance. In the case of Trinidad and Tobago according to the IDB (2010a), from 1995 to 2008 the RMI of the twin island state although improving over time, has remained low between an RMI of 12 in 1995, to an RMI of 22 in 2008. This outcome implies that while Trinidad and Tobago have made little progress concerning risk management associated with natural disasters and climate change events like extreme flooding, risk management is poor and remains low.

THE IMPORTANCE OF URBAN DESIGN AND PLANNING IN FLOOD PREVENTION

Given that Trinidad and Tobago, has a high level of vulnerability to climate change, and natural hazards such as the extreme flooding events experienced as recent as 2017-2022, in highly populated urbanized areas in Northern and Southern Trinidad, the discussion on urban design and planning in preventing the occurrence of these floods is of utmost importance. Urban design and planning play a critical role in preventing urbanized flooding associated with meteorological events such as tropical cyclones, storms, thunderstorms, extreme rainfall, and storm surges which occur frequently in the Caribbean on an annual basis. For purposes of this chapter, this section will briefly discuss some of the primary ways in which urban design and planning can help prevent urban flooding in Trinidad and Tobago.

Adaptive Strategies and Design Solutions

To tolerate these metrological events, urban communities and cities in Trinidad and Tobago must create resilient urban infrastructural designs, which can reduce the vulnerability that persons in these areas experience. In this context, adaptive urban designs can be used to generate creative and innovative solutions to dealing with extreme flooding in urban areas in Trinidad and Tobago. Since few cities around the world are resilient when it comes to dealing with the ill effects of climate change, there are several design adjustments that urbanized areas can make, such as the adaptation of sustainable urban drainage systems, and the transformation of urban waterfronts by integrating riverbanks with urban infrastructure to deal with flooding problems. Decision aid tools such as Cost Benefit Analysis (CBA), Cost Effectiveness Analysis (CEA), and Life-Cycle Costing (LCC) can also be used with Sustainable drainage systems (SuDS) when assessing the sustainability of drainage systems (Zhou, 2014).

Further to this, Bakhtiari et al., (2023) have also explained that digital visualization tools such as virtual reality, and simulation models can be used to improve both the management of adaptation and mitigation strategies and to also design effective flood evacuation routes. This is especially important as the focus on flood protection in urban areas has begun shifting away from the construction of flood defence mechanisms like floodwalls, seawalls, dikes, and levees, to the development of urban areas in which adaptative strategies such as spatial and repurpose design strategies, and green infrastructure play a significant role in urban flood management (Hooimeijer et al., 2022). Focusing on the latter concept of flood-adaptive green infrastructure, such an approach can improve urban flood resilience because apart from controlling urban flooding from stormwater by reducing runoff and regulating the rate of flow and volume of water, it has many positive effects on the environment such as improving water and air quality, increasing biodiversity, reducing heat, and conserving water which can be used during periods of droughts (Lee et al., 2021).

Flood Adaptation Models

Building on the discussion of adaptive strategies and design solutions, alternative flood adaptation models can also be effective in promoting alternative flood adaptation standards that focus on flood prevention and building flood resilience in urban communities. Such alternative flood adaptation models that can be implemented in Caribbean countries like Trinidad and Tobago include water-sensitive urban design models, such as those implemented in Melbourne and Adelaide in Australia, which use rain gardens, and green roofs to manage urban water, while reducing flood risk and urban heat, as well as encouraging biodiversity. Another proposal by Miguez et al. (2018) on hydrodynamic models, and Lourenco et al. (2020) to support flood prevention in urban areas through the planning process can include initiatives to promote sustainable drainage and the restoration of rivers in city areas by using sustainable drainage concepts in multifunctional urban open spaces like natural and semi-natural areas, lakes, rivers, canals, wasteland, and derelict open land, to serve as an environmental agent that protects fluvial parks. The latter can mitigate against flooding, as these are urban open spaces that can absorb the excess water which prevents the sedimentation of rivers.

Flood Resiliency Systems

Further to this, flood resiliency approaches for urban planning can also assist with flood adaptative measures in urbanized areas in Trinidad and Tobago. One such example of a flood resiliency approach includes the Protect, Accommodate, Retreat, and Avoid (PARA) framework implemented by Canadian communities to build and sustain adaptability against climate change events. According to Doberstein et al. (2019), such an approach uses dikes and pumping stations as "protective" measures, flood relief programmes as "accommodation" measures, urban zoning, and land expropriation as "retreat" measures, and floodplain development as "avoid" measures, has been instrumental in building flood resilience and reducing flood risk. Flood resilience assessment frameworks that make use of the Robustness-Resistance-Recovery (3Rs) plan can assess the resilience of fluvial systems to flooding by looking at their ability to remain functional during extreme flooding events, by withstanding and absorb excess water, and their ability to adapt and recover after such flooding has taken place, unlike other flood resilience assessment methods that focus on social-ecological areas that take into account the relationship that human being have with natural systems (Yu et al., 2023).

Non-Structural Mitigation Plans

Another interesting way in which urban design and planning can prevent urban flooding in Trinidad and Tobago is by merging nonstructural mitigation plans with urban planning to help communities in urban areas become resilient to floods. In countries such as China, where a substantial proportion of the population resides in cities, the concept of a Sponge City was used to deal with flooding and water shortage problems. By integrating green infrastructure discussed earlier, with the current urban drainage system it can absorb excessive water during heavy rainfall events to assist in stormwater management in urban areas (Han et al., 2023).

Zeng et al., (2023) explain that for the sponge city approach to be effective in flood prevention, its construction must conform to the geographic environment in which it was situated while being guided by the appropriate technology and legislative guidelines. In addition to the sponge city concept, Middle Eastern countries in Southwest Asia and North Africa, that have integrated Nature-Based Solutions (NBS) like low-impact development mechanisms, i.e., bio-retention cells, and infiltration trenches in their urban design and planning process have found that they were effective in reducing flood runoff, which can aid policymakers in building and sustaining flood resistant cities (Mabrouk et al., 2023).

Building Regulations

The final area in which urban design and planning have played a significant role in flood prevention is in building regulation. Building regulations must deal with regulating the density of buildings constructed in flood-prone areas, preventing the construction of buildings close to or on flood plains, and creating urban sinks in Trinidad and Tobago. Looking at each area, enforcing building regulations in urban areas that are prone to flooding in Trinidad and Tobago, is critical for not only flood prevention but also for the preservation of buildings. The growth of unplanned urban communities, together with the construction of buildings, which may not have adhered to the building codes and standards may become a great risk to the lives of its occupants (Abdrabo et al., 2022). In that, these buildings may not be structurally sound, which increases the risk of building collapse as their structural integrity may be compromised by storm and flood waters, and poor construction, and as a result, persons occupying these buildings will be less likely to meet the requirements to apply for flood insurance and to make claims.

Furthermore, urban design and planning are also essential to help developers in Trinidad and Tobago design urban spaces around and at a safe distance away from flood plains, which are hotspots for urban flooding. The growth in the informal urban sprawl in floodplain areas is likely to increase the risk of flooding because building in these areas reduces the size of the floodplain itself, so when excessive rainfall does occur there is nowhere for the excess water to go (Abdrabo et al., 2022). This is because as the floodplains become lost to housing developments and agricultural activity, their ability to absorb excess water from storms is reduced, which causes land degradation, and severe flood damage. Thus, it is important to encourage urban planners to prioritize the urbanization of safe areas, to avoid building hazardous flood zones.

THE ROLE OF AI IN URBAN PLANNING AND DESIGN IN FLOOD PREVENTION

As discussed in the previous section, urban planning and design plays a great role in preventing floods in highly populated urban cities and communities in Trinidad and Tobago and around the world. With the advent of Industry 4.0, the integration of advanced technologies such as Artificial Intelligence (AI) has become quite important in the planning and design of urban spaces, as its use has the potential to revolutionize how cities operate, as well as how persons live and work, so that it may create a more sustainable form of living (Ghobakhloo, 2020).

To do this, the technologies associated with Industry 4.0, have enabled countries around the world like Brazil, to reshape the identities of their urban environments, which improves the quality of life of persons living in these spaces, by being sustainable, inclusive, and environmentally friendly, through waste reduction, improved resource allocation, and urban planning (Fichinelli et al., 2023). Considering that Industry 4.0, in particular, the use of AI has benefitted the design and planning of sustainable urban environments through the design and management of smart cities, it is also likely that its use will play a significant role in the flood prevention in urban flood-prone areas in Trinidad and Tobago, as based on the data available, it allows urban planners to come with innovative and novel solutions that are also cost-effective (Yigitcanlar et al., 2021).

Flood Resilience in Planning and Design

There are several ways in which AI can be used in urban design and planning to prevent flooding in countries like Trinidad and Tobago, one such approach is using AI-based platforms and urban big data analytics to aid in the design and planning of flood resilience programmes. These platforms apart from the planning and design aspect, can be used for multiple purposes such as promoting flood awareness amongst the public which closes the knowledge gap, as well as encouraging collaboration between stakeholders, i.e., private, and public sector entities located in urban areas in Trinidad and Tobago which allows the best flood resilience outcomes to be attained (Ye et al., 2021). One example of this is the merging of digital technologies such as Machine Learning (ML) with the mining of hydrological data has been used to create a platform to manage floods in countries like Iran (Akbarian et al., 2022). As well as AI-based water management and governance systems such as the Cyber-Physical Systems

(CPS), as discussed by Alexandra et al., (2023) can be used to manage and monitor the governance of hydrological systems in urban settings.

The use of such AI-based platforms together with the use of urban big analytics can then provide several benefits for the response and management of flooding in urban settings, in that it not only allows for the processing and updating of forecasts and systems in real-time, which enables environmental and disaster planning agencies in Trinidad and Tobago like the Office of Disaster Preparedness and Management (ODPM), and the Tobago Emergency Management Agency (TEMA) to respond to potential flooding events and risks in a more proactive in manner. Using such information to fuel a forward-looking approach to flood management in Trinidad and Tobago may then act as a stimulus for community members in urban areas to engage with these agencies as the use of the platform increases the likelihood, they will obtain assistance and support promptly.

Another useful AI-based tool that can assist planners in flood management in Trinidad and Tobago, is the implementation of the Flood Detection Secure System (FDSS) used in countries like Saudia Arabia. Based on Deep Learning (DL) techniques and blockchain technology, planners can use FDSS systems to train models that can be used to track the changes in water systems and estimate the risk of flooding in urban areas (Alsumayt et al., 2023). A computerized management advisory system such as the Decision Support System (DSS) when implemented can support planners when approaching flooding disasters in urban areas in Trinidad and Tobago. In that, the data collected by the DSS on areas such as flood simulations, historical flooding events, and rainfall forecasts through AI can be used to conduct disaster drills and provide greater public education programmes while addressing engineering and non-engineering issues linked to urban flood management.

Notwithstanding the multitude of benefits that DSS systems are likely to provide for environmental and disaster response agencies in Trinidad and Tobago, one of its main advantages is that the use of such AI-based systems may likely be quite cost-effective in the long run (Ye et al., 2021). Even though the initial development and implementation costs of a DSS may be high and may be a barrier to its use by countries like Trinidad and Tobago may not have the financial resources to support such a system, if implemented there are a great deal of financial and non-financial benefits to be accrued in the long run. In that, not only would the probability of the loss of lives to urban flood decline, but the costs associated with infrastructural and recovery efforts in Trinidad and Tobago, are all likely to decline. This is because the DSS provides agencies with the necessary flood management information and data to not only make more informed decisions but also facilitate a better flow of communication and understanding between stakeholders that can encourage greater collaboration, productivity, and innovation in addressing urban flooding in Trinidad and Tobago.

Stormwater Detention Facilities

Moreover, AI-based tools can assist in the design of urban stormwater detention facilities, that function as a storage mechanism, and provide support for the strategies developed for urban flood management, and the creation of water-sensitive cities in Trinidad and Tobago. According to Yang et al. (2019), the use of such stormwater facilities in cities like Taipei City, Taiwan are essential to flood mitigation systems, to preserve the infrastructure of buildings during and after typhoon and storm events, because the intelligent flood drainage system, can monitor the fluctuations in the amount of water held in storage ponds. Based on this information collected, the system can then determine when peak water levels will occur, and when and for how long pump switches can stay on, which encourages optimal use of stormwater detention, and flood drainage systems. Further to this, these stormwater facilities can also be supported by AI-based systems such as Centaur and DL algorithms that can be used for flood risk reduction in flood-prone urban communities, as they monitor both (1) the flow of water in the sewer system to reduce the risk of excrement overflow, (2) the detection of leakages and contamination, which helps sewage treatment facilities to operate more efficiently (Fu et al., 2022).

In the context of countries like Trinidad and Tobago which also experience tropical storm events of a similar magnitude to a Typhoon in Taipei, Taiwan, the construction of stormwater detention vaults integrated with AI-based systems, to manage the excess stormwater runoff in urban cities can play a vital role in not only the prevention of floods but also the erosion of major rivers, such as the Maraval and St. Ann's river in the City of Port-of-Spain. The use of an intelligent form of the stormwater management system, as well as the smart management of sewer overflows, in urbanized areas in Trinidad and Tobago like that implemented by Shanghai, China; Ulsan, South Korea and Taipei, Taiwan, to reduce the downstream peak flow, to use a more data-driven form of analysis using sensitivity analyses and remote sensing can all help support the development of more climate resilient cities in not only Trinidad and Tobago but also the Caribbean (Saddigi et al., 2022). Furthermore, the use of these systems is also likely to contribute to reduced maintenance costs in the long run as the AI-based system can detect exactly where repairs are needed in real-time, thus improving the life and efficiency of detention facilities.

Forecasting and Prediction of Flooding Events

One of the final ways in which AI can be used in urban planning and design is to forecast and predict the occurrence of floods. Using AI-based technologies together with geospatial analysis, planners in Trinidad and Tobago can obtain statistics and information to analyze in real-time that can be used to provide residents in urban cities and communities, together with the national disaster planning committee with the information needed to mitigate against, as well as to forecast and predict the frequency and location of urban floods. In addition to this, planners can use AI models, ML, DL, as well as Long Short-Term Memory (LSTM) and Resner-Unet Neural Networks as well as Numerical Models and Simulations to rapidly predict urban flooding risks which are integral to flood prevention and mitigation (Chen et al., 2023). The main benefit of using LSTM for flood prediction is that planners together with natural disaster planning agencies in Trinidad and Tobago can use data analytics to examine the relationships between past, current, and future urban flooding events that may be developing over time, to better understand and make more informed decisions regarding the optimization of flood mitigation strategies, as well as anticipate infrastructural needs.

Data collected on the different features of urban and flash floods such as their frequency, depth of water flow, the water logging depth of cities, and the volume of the water flow, can all assist in efficiently predicting its occurrence in urban communities where there are complicated land use patterns (Al Mehedi et al., 2022). In this regard, the data collected, and the leveraging of temporal dependencies can assist urban planners in Trinidad and Tobago in developing more adaptative strategies that encourage urban planners to consider the information provided through flood prediction models to ensure that the growth of urban spaces in Trinidad and Tobago is not only sustainable, but whose design mitigates against flood risks.

The amalgamation of AI with the Internet of Things (IoT), also known as the Artificial Intelligence of Things (AIoT) can also be used in the detection and prediction of floods. Using AIoT-based sensor nodes, when placed in water systems in urban areas in Trinidad and Tobago, the data collected can be used to drive the analytical estimates on how likely these systems are to flood, to design strategies such as drainage systems modelling, flood mapping, and rainfall forecasting (Chitwatkulsiri & Miyamoto, 2023). Such an outcome can be achieved because AIoT and cloud computing create a network that automatically forecasts the function and features of urban floods in cities in a structured and logical manner, without human input.

CHALLENGES OF USING AI IN URBAN PLANNING & DESIGN TO RESPOND TO FLOODING

The integration of AI in the planning and design of urban cities and communities in Trinidad and Tobago has the potential to transform the mitigation and adaptation strategies used by planners to deal with the ill effects of extreme flooding events that occur annually. While the use of AI-based technologies such as machine learning, deep learning, blockchain technologies, and neural networks can all help to develop and sustain the resilience of urban communities against flooding events, the use of AI in the planning and design process does present several issues that may hinder the ability of urban planners in Trinidad and Tobago to use the technology to deal with flooding problems. These issues can originate from a multitude of areas, however for purposes of this chapter, emphasis will be placed on four main challenges concerning, access to data and algorithm biases, technical ability and data privacy, and social inequality.

Limited Access to High-Quality Data and Algorithm Biases

Countries around the world like Brazil, Saudia Arabia, and China have effectively used AI-based technologies to create smart cities and use various tools to mitigate against extreme flooding events but were only able to do so because these technologies are driven by large datasets. Often the collection of this data is done in an intelligent manner, that sees information being gathered during an iterative process on various topics related to population demographics, traffic flow, the experience of persons residing in urban communities, their level of risk and vulnerability to flooding events, the types and extent of flooding events, and well as the characteristics of the floods themselves (Al Mehedi et al., 2022). Using this information, planners then then again to AI-based technologies to model the frequency and intensity of flood events in urban areas, and design strategies to deal with this problem. However, this could only have been made possible if the data used to power these AI-based technologies is of a high quality, and is relevant to issues of urban planning, design, and flooding events.

While the use of AI can amplify the ability of high-income countries to deal with flooding events from an urban design and planning perspective using AI, since the collection, and collation of large datasets and the training of AI-based models is extremely expensive and time-consuming, the same outcome may not be realized by countries like Trinidad and Tobago. This is primarily the case, because urban planners in Trinidad and Tobago, may not have sufficient financial and computational resources to fund the design and collection of the data needed to drive the AI-based technologies used in the planning and design process (Sharma et al., 2022). As a result, this may not only reduce the rate at which AI is implemented in urban planning and design in Trinidad and Tobago, but the use of poor-quality data, which is often restricted in terms of the type of information collected, may lead to algorithm biases, as well as the generation of inaccurate models, predictions, and policies, that may worsen the overall flooding outcomes of persons living in urban spaces.

Limited Technical Know-How of AI Tech and Privacy Issues

Another important challenge that Trinidad and Tobago may experience in its use and integration of AI, is that of limited technical expertise in AI technologies and how to use them for urban planning and design. Given that a vast majority of AI-based technologies are built on a variety of specialized areas used in the writing and training of the machine and deep learning algorithms, AI developers must have the technical expertise and competence to use, implement, and maintain these resources (Sanchez, et al., 2023). As a result, urban planners and AI developers who possess these skills can make more informed and accurate decisions regarding the possible flooding of urban spaces, and how to adapt their designs such as the designs of roads, drainage systems, and green spaces, to deal with these issues. Using the data collected by AI and their technical expertise, urban planners can also forecast the expansion of urban spaces, and using historical climate change data can predict how urban spaces in Trinidad and Tobago are likely to change over time.

Even though it is important to have a prominent level of expertise and training in not only the use of AI technologies but also how to apply them in urban planning and design, in many countries like Trinidad and Tobago such benefits may not materialize due to poor AI readiness. Although high-income and resource-rich countries are benefitting from the integration of AI in their industries, Poor AI readiness has been an issue experienced by small island states like Trinidad and Tobago, and resource-poor countries like Africa, because the growth and employment of AI start-ups have very little seed funding needed to invest in key AI infrastructure such as high-computing, networking, and storage capacity, as well as security (Arakpogun et al., 2021).

Furthermore, such poor AI readiness may also be because the government of Trinidad and Tobago, may not be ready to implement AI, in that according to Rogerson et al., (2022) in its Government AI Readiness Index 2022, Trinidad and Tobago is currently ranked 106th, out of 181 countries. Such an outcome implies that Trinidad and Tobago may not have the strong internal digital capabilities, innovative capacities, and human capital, as well as high-quality data, needed to develop and govern the use of AI in an ethical manner. Moreover, bearing in mind that the government of Trinidad and Tobago has a low rank in its AI readiness, it is the ability to secure the data collected is also a matter of concern.

Data privacy associated with the implementation of AI in urban planning and design particularly in countries like Trinidad and Tobago is a major challenge as the deployment of IoT in urban environments may cause divergences in privacy and security of urban residents in regards to city planning and housing, as the use of ICT such as security cameras and sensors, can be considered a violation of personal privacy since it may collect personal data related to the location of persons, as well as audio and visual information (Hassan & Awad, 2018). As a result, if the integration of AI with urban planning and design is used to reduce the ill effects of extreme flooding in Trinidad and Tobago, there will be a great need to put in place the technological infrastructure needed to prevent cyber-attacks and the illegal access to urban residents' data by malicious persons in the dark web.

Societal Inequalities

Another challenge that is likely to be encountered during and after the integration of AI in urban planning and development in responding to extreme flooding events in Trinidad and Tobago is that its use may also encourage the growth of inequality in society. In the context of urban planning and development, AI can worsen societal inequalities between urban and rural communities, as urban communities and cities are more likely to benefit from the technological advancements associated with urban planning and design such as reduced traffic, preferred parking and transportation routes, aesthetically pleasing and holistic infrastructural designs, than rural communities which reduces the inequity of infrastructure and other services such as healthcare experienced by persons living in these areas (Guo & Li, 2018). The limited use of AI in the planning and design of rural communities, may not only reduce the quality of public infrastructure in rural communities in Trinidad and Tobago, but also worsen their flooding outcomes during and after natural disasters, as they may be geographically isolated from technological advancements, and may not have the technological infrastructure needed to build smart rural communities (Li et al., 2020).

CONCLUSION

Exploring the relevance of AI in urban design and flooding as a mechanism to respond the extreme flooding events in Trinidad and Tobago is important because it can assist planners, institutional stakeholders and persons living in urban communities to gain insights into how flooding events emerge and how its associated risks evolve. AI in this context can help urban cities and communities to build resilience against natural disasters and improve their flood risk in the future, by prioritizing investments that can strengthen existing infrastructure in flood-prone urban areas.

A study of this nature is exceedingly important because Trinidad and Tobago like many small island developing states around the world, is highly vulnerable to the negative impacts of climate change, such as extreme flooding events which occur on an annual basis. Such a high vulnerability to extreme flooding events is linked to the high population densities, small land mass, and limited financial as well as technological resources to deal with climate change risks. The use of AI in urban planning and design to respond to extreme flooding in urban areas in Trinidad and Tobago can provide the insight needed to reduce flood risk by efficiently allocating resources to build and design urban communities that are resilient to future flooding risks.

A study of this nature helps build the literature on using AI in urban planning and design with climate change events like flooding in Trinidad and Tobago, where there has previously been limited information. Such a study is important because it highlights how AI-based technologies and applications used in planning and design can be applied to deal with the issue of extreme flooding events in urban areas in Trinidad and Tobago. For such technologies to become a cornerstone of urban planning and design, this study also acknowledges the challenges that Trinidad and Tobago must first overcome so that the integration of AI can become a reality.

REFERENCES

Abdrabo, K. I., Kantosh, S. A., Saber, M., Sumi, T., Elleithy, D., Habiba, O. M., & Alboshy, B. (2022). The Role of Urban Planning and Landscape Tools Concerning Flash Flood Risk Reduction Within Arid and Semiarid Regions. In Sumi, T., Kantoush, S. A., & Saber, M. (Eds.), *Wadi Flash Floods. Natural Disaster Science and Mitigation Engineering: DPRI reports*. Springer. 10.1007/978-981-16-2904-4_11

Akbarian, H., Gheibi, M., Hajiaghaei-Keshteli, M., & Rahamani, M. (2022). A hybrid novel framework for flood disaster risk control in developing countries based on smart prediction systems and prioritize scenarios. *Journal of Environmental Management*, 312, 114939. 10.1016/j.jenvman.2022.11493935338986

Al Mehedi, M.-A., Smith, V., Hosseiny, H., & Jiao, X. (2022). Unravelling the complexities of urban fluvial flood hydraulics through AI. *Scientific Reports*, 12(18738), 1–18. 10.1038/s41598-022-23214-9

Alexandra, C., Daniell, K. A., Guillaume, J., Saraswat, C., & Feldman, H. R. (2023). Cyber-physical systems in water management and governance. *Current Opinion in Environmental Sustainability*, 62(101290), 1–11. 10.1016/j.cosust.2023.101290

Aliasgar, K. (2012). *Developing a geoinformatics-based early warning system for floods in the Caribbean, Trinidad, and Tobago*. Southern Cross University.

Alsumayt, A., El-haggar, N., Amouri, L., Alfawaer, Z. M., & Aljameel, S. S. (2023). Smart Flood Detection with AI and Blockchain Integration in Saudi Arabia Using Drones. *Sensors (Basel)*, 23(5148), 1–30. 10.3390/s2311514837299876

Arakpogun, E., Elsahn, Z., Olan, F., & Elsahn, F. (2021). Artificial intelligence in Africa: challenges and opportunities. In *The Fourth Industrial Revolution: Implementation of Artificial Intelligence for Growing Business Success. Studies in Computational Intelligence* (pp. 375–388). Springer., 10.1007/978-3-030-62796-6_22

Arkle, J., Owen, L., & Weber, J. (2017). Trinidad and Tobago. In Allen, C. (Ed.), *Landscapes and Landforms of the Lesser Antilles. World Geomorphological Landscapes* (pp. 267–291). Springer., 10.1007/978-3-319-55787-8_17

Arnaud, L. (2021). Trinidad & Tobago's Mangrove Ecosystems suffocating in plastic waste: Recommendations moving forward. *Academic Letters*, 3745, 1–9. 10.20935/AL3745

Bakhtiari, V., Piadeh, F., Behzadian, K., & Kapelan, Z. (2023). A critical review of the application of cutting-edge digital visualisation technologies for effective urban flood risk management. *Sustainable Cities and Society*, 99(104958), 1–14. 10.1016/j.scs.2023.104958

Brookhuis, B., & Hein, L. (2015). The value of the flood control service of tropical forests: A case study for Trinidad. *Forest Policy and Economics*, 62, 118–124. 10.1016/j.forpol.2015.10.002

Canisius, F., & Nancy, C. (2009). GeoFIS flood insurance system for Trinidad: A case study for San Juan downstream. *Urban and Regional Information Systems Association (URISA). Journal*, 21(1), 1–11.19331084

Chen, J., Li, Y., Zhang, C., Tian, Y., & Guo, Z. (2023). Urban Flooding Prediction Method Based on the Combination of LSTM Neural Network and Numerical Model. *International Journal of Environmental Research and Public Health*, 20(1043), 1–12. 10.3390/ijerph2002104336673799

Chitwatkulsiri, D., & Miyamoto, H. (2023). Real-Time Urban Flood Forecasting Systems for Southeast Asia- A Review of Present Modelling and Its Future Prospects. *Water (Basel)*, 15(178), 1–18. 10.3390/w15010178

Climate Watch. (2022). *Historical GHG Emissions*. Climate Watch. https://www.climatewatchdata.org/ghg

Doberstein, B., Fitzgibbons, J., & Mitchell, C. (2019). Protect, accommodate, retreat, or avoid (PARA): Canadian community options for flood disaster risk reduction and flood resilience. *Natural Hazards*, 98(1), 31–50. 10.1007/s11069-018-3529-z

Fichinelli, A. C., Yigitcanlar, T., Sabtini-Marques, J., Cortese, T. T. P., Sotto, D., & Libardi, B. (2023). Urban Smartness and City Performance: Identifying Brazilian Smart Cities through a Novel Approach. *Sustainability (Basel)*, 15(10323), 1–24. 10.3390/su151310323

Fontes de Meira, L., & Phillips, W. (2019). An economic analysis of flooding in the Caribbean. The case of Jamaica and Trinidad and Tobago, *Studies, and Perspectives series-ECLAC subregional headquarters for the Caribbean, 78*. Santiago: Economic Commission for Latin America and the Caribbean (ECLAC).

Fu, G., Jin, Y., Sun, S., Yuan, Z., & Butler, D. (2022). The role of deep learning in urban water management: A critical review. *Water Research*, 223(118973), 1–16. 10.1016/j.watres.2022.11897335988335

Ghobakhloo, M. (2020). Industry 4.0, digitization, and opportunities for sustainability. *Journal of Cleaner Production*, 252(119869), 1–21. 10.1016/j.jclepro.2019.119869

Global Forest Watch (GFW). (2023). *Global Forestry Watch*. GFW. https://www.globalforestwatch.org/

Guo, J., & Li, B. (2018). The application of medical artificial intelligence technology in rural areas of developing countries. *Health Equity*, 2(1), 174–181. 10.1089/heq.2018.003730283865

Han, J., Wang, C., Deng, S., & Lichtfouse, E. (2023). China's sponge cities alleviate urban flooding and water shortage: A review. *Environmental Chemistry Letters*, 21(3), 1297–1314. 10.1007/s10311-022-01559-x

Hassan, A. M., & Awad, A. I. (2018). Urban Transition in the Era of the Internet of Things: Social Implications and Privacy Challenges. *IEEE Access : Practical Innovations, Open Solutions*, 6, 36428–36440. 10.1109/ACCESS.2018.2838339

Hooimeijer, F. L., Yoshida, Y., Bortolotti, A., & Iuorio, L. (2022). Integrated urban flood design in the United States and the Netherlands. In Brody, S., Lee, Y., & Kothuis, B. (Eds.), *Coastal Flood Risk Reduction: The Netherlands and the U.S. Upper Texas Coast* (pp. 241–254). Elsevier. 10.1016/B978-0-323-85251-7.00018-4

Inter-American Development Bank (IDB). (2010a). *Indicators of Disaster Risk and Risk Management Program for Latin America and the Caribbean Summary Report*. IDB. https://publications.iadb.org/publications/english/viewer/Indicators-of-Disaster-Risk-and-Risk-Management-Program-for-Latin-America-and-the-Caribbean-Summary-Report.pdf

Inter-American Development Bank (IDB). (2010b). *Indicators of Disaster Risk and Risk Management Program for Latin America and the Caribbean, Barbados*. IDB. https://www.imf.org/external/np/seminars/eng/2013/caribbean/pdf/barbados.pdf

International Federation of Red Cross and Red Crescent Societies (IFRC). (2022). *DREF Application-Trinidad and Tobago Floods-November 2022*. IFRC. https://reliefweb.int/attachments/4fb5022e-a0e1-4ea3-ab80-0415ba23ee2e/MDRTT002do.pdf

International Monetary Fund (IMF). (2013). *Measuring Disaster Risk*. IMF. https://www.imf.org/external/np/seminars/eng/2013/caribbean/pdf/measuring.pdf

Lee, H., Song, K., Kim, G. W., & Chon, J. (2021). Flood-adaptive green infrastructure planning for urban resilience. *Landscape and Ecological Engineering*, 17(4), 1–11. 10.1007/s11355-021-00458-7

Lemos, M., Eakin, H., Nelson, D., & Engle, N. (2013). Building adaptive capacity to climate change in less developed countries. In *Climate science for serving society* (pp. 437–457). Springer. 10.1007/978-94-007-6692-1_16

Li, R., Chen, K., & Wu, D. (2020). Challenges and opportunities for coping with the smart divide in rural America. *Annals of the American Association of Geographers*, 110(2), 559–570. 10.1080/24694452.2019.1694402

Lourenco, I. B., Beleno de Oliveira, A. K., Marques, L., Quintanilha Barbosa, A. A., Veról, A. P., Magalhães, P. C., & Miguez, M. G. (2020). A framework to support flood prevention and mitigation in the landscape and urban planning process regarding water dynamics. *Journal of Cleaner Production*, 227(122983), 1–17. 10.1016/j.jclepro.2020.122983

Mabrouk, M., Han, H., Fan, C., Abdrabo, K. I., Shen, G., Saber, M., Kantoush, S. A., & Sumi, T. (2023). Assessing the effectiveness of nature-based solutions-strengthened urban planning mechanisms in forming flood-resilient cities. *Journal of Environmental Management*, 344, 118260. 10.1016/j.jenvman.2023.11826037354590

Miguez, M. G., Raupp, I. P., & Verol, A. P. (2018). An integrated quantitative framework to support the design of resilient alternatives to manage urban flood risks. *Journal of Flood Risk Management, 12*(S2-e12514), 1-13. 10.1111/jfr3.12514

Mycoo, M. (2018). Urban Sustainability in Caribbean Small Island Developing States: A Conceptual Framework for Urban Planning using a case study of Trinidad. *International Development Planning Review*, 40(2), 143–174. 10.3828/idpr.2018.8

Narinesingh, P. (2014). An Assessment of River Maintenance Practices in Trinidad and Tobago. *The Journal of the Association of Professional Engineers of Trinidad and Tobago*, 42(2), 19–28.

Ramlal, B., & Baban, S. (2008). Developing a GIS-based integrated approach to flood management in Trinidad, West Indies. *Journal of Environmental Management*, 88(4), 1131–1140. 10.1016/j.jenvman.2007.06.01017714856

Rogerson, A., Hankins, A., Nettel, P. F., & Rahim, S. (2022). Government AI Readiness Index 2022. *Oxford Insights*. https://www.unido.org/sites/default/files/files/2023-01/Government_AI_Readiness_2022_FV.pdf

Roopnarine, R., Opadeyi, J., Eudoxie, G., Thongs, G., & Edwards, E. (2018). GIS-based flood Susceptibility and Risk Mapping Trinidad Using Weight Factor Modeling. *Caribbean Journal of Earth Science*, 49, 1–9.

Saddiqi, M. M., Zhao, W., Cotterill, S., & Dereli, R. K. (2023). Smart management of combined sewer overflows: From an ancient technology to artificial intelligence. *WIREs. Water*, 10(3), e1635. 10.1002/wat2.1635

Sanchez, T. W., Shumway, H., Gordner, T., & Lim, T. (2023). The prospects of artificial intelligence in urban planning. *International Journal of Urban Science*, 27(2), 179–194. 10.1080/12265934.2022.2102538

Sangster, N., Bisnath, J., Maharajh, A., & Sammy, A. (2020). Investigation of the Watercourses and Measures to Alleviate Flooding in the Penal/Debe Region. *The International Conference on Emerging Trends in Engineering and Technology (IconETech-2020)*. Faculty of Engineering, The UWI. 10.47412/CHIF9112

Scobie, M., Gopinathan, U., Kelman, I., & Myhre, S. (2023). The agency of community groups in health and climate change adaptation governance and policy in SIDS: The case of in Toco, Trinidad and Tobago. *Environmental Science & Policy*, 147, 116–125. 10.1016/j.envsci.2023.05.006

Sendek, A., Kretz, L., van der Plas, F., Seele-Dilbat, C., Schulz-Zunkel, C., Vieweg, M., Bondar-Kunze, E., Weigelt, A., & Wirth, C. (2021). Topographical features related to flooding frequency promote ecosystem multifunctionality of riparian floodplains. *Ecological Indicators*, 132(108312), 1–11. 10.1016/j.ecolind.2021.108312

Sharma, M., Luthra, S., Joshi, S., & Kumar, A. (2022). Implementing challenges of artificial intelligence: Evidence from the public manufacturing sector of an emerging economy. *Government Information Quarterly*, 39(4), 101624. 10.1016/j.giq.2021.101624

Thongs, G. (2019). Integrating risk perceptions into flood risk management: Trinidad case study. *Natural Hazards*, 98(2), 593–619. 10.1007/s11069-019-03720-2

Udika, R. (2010). *Flood Management: An examination of mitigation measures for flooding in Urban areas in Trinidad*. 46th ISOCARP Congress, Nairobi, Kenya. https://www.isocarp.net/data/case_studies/1758.pdf

United Nations. (2023). (UN) Office for the Coordination of Humanitarian Affairs (UN-OCHA). *Trinidad and Tobago: Floods - Nov 2022*. UN-OCHA. https://reliefweb.int/disaster/fl-2022-000374-tto

World Health Organization (WHO). (2023). *Floods*. WHO. https://www.who.int/health-topics/floods#tab=tab_1

Yang, S.-N., Chang, L.-C., & Chang, F.-J. (2019). AI-based design of urban stormwater detention facilities accounting for carryover storage. *Journal of Hydrology (Amsterdam)*, 575, 1111–1122. 10.1016/j.jhydrol.2019.06.009

Ye, X., Wang, S., Lu, Z., Song, Y., & Yu, S. (2021). Towards an AI-driven framework for multi-scale urban flood resilience planning and design. *Computational Urban Science*, 1(11), 11. 10.1007/s43762-021-00011-0

Yigitcanlar, T., Corchado, J. M., Mehmood, R., Li, R. Y. M., Mossberger, K., & Desouza, K. (2021). Responsible urban innovation with local government artificial intelligence (AI): A conceptual framework and research agenda. *Journal of Open Innovation*, 7(1), 71. 10.3390/joitmc7010071

Yu, S., Kong, X., Wang, Q., Yang, Z., & Peng, J. (2023). A new approach of A new approach of Robustness-Resistance-Recovery (3Rs) to assessing flood resilience: A case study in Dongting Lake Basin. *Landscape and Urban Planning*, 230, 104605. 10.1016/j.landurbplan.2022.104605

Zeng, C., Aboagye, E. M., Li, H., & Che, S. (2023). Comments and Recommendations on Sponge City-China's solutions to prevent flooding risks. *Heliyon,9*(e12745):1-11. doi: . 2022.e1274510.1016/j.heliyon

Zhou, Q. (2014). A Review of Sustainable Urban Drainage Systems Considering the Climate Change and Urbanization Impacts. *Water (Basel)*, 6(4), 976–992. 10.3390/w6040976

Chapter 6
Achieving Cultural Sustainability Through Museum Cultural and Creative Products:
Museum Cultural Heritage Conservation and Innovation

Hui Liu
Taylor's University, China

Siti Norzaini N. Z. Zainal Abidin
https://orcid.org/0000-0002-2637-4794
The Design School, Taylor's University, China

Filzani Illia I. Ibrahim
The Design School, Taylor's University, China

Cheng Shu An
The Design School, Taylor's University, China

Wang Lu
Wenzhou University, China

ABSTRACT

Museum cultural and creative products (MCCP) are seen as carriers of museum cultural resources and are important for the sustainable development of cultural heritage, Nevertheless, the cultural aspect is often overlooked within the realm of sustainable product design. Hence, a primary objective of this research endeavour was to address this gap by developing a comprehensive framework that clarifies the interconnectedness between CS and MCCP. Furthermore, an additional objective of this study was to put forth a conceptual framework and specific indicators for the assessment of the cultural sustainability of MCCP. The indicator sets were designed through the literature and analytic hierarchy process. This process has yielded a tool that will undergo additional testing and practical implementation. The findings of this study will enhance comprehension of cultural sustainability within the context of MCCP

DOI: 10.4018/979-8-3693-1650-4.ch006

and provide guidance for future MCCP design aligned with the principles of sustainable development.

INTRODUCTION

Museums possess the capacity to enhance cultural vibrancy and aid in the conservation of heritage, making them crucial in upholding and raising public consciousness about culture. (Aboulnaga et al., 2022). In recent times, museums have begun to adopt a new role in which they provide the public with tangible cultural and creative products (Nie & Li, 2020).In addition to the traditional functions of museums in the sense of collecting, displaying, and educating. Museum cultural and creative product(MCCP)refers to a product for sale based on museum collections, exhibitions, cultural characteristics, combined with creative design, aesthetics, fashion, and other factors(Tu et al., 2019). According to UNESCO, cultural and creative products (CCPs) are consumer goods that convey creative concepts, symbols, and ways of life(UNESCO, 2003). MCCPs are the best link between museums and the public (Hu, 2018). The recent appeal of these products to audiences of all ages and genders can be attributed to their unique combination of cultural inventiveness, deep cultural connotations (Song & Li, 2018). The cultural relics in museum collections contain endless cultural and artistic values, and through the use and regeneration of culture by MCCPs, they can promote museum culture into the real life of the public, enrich public life and art, and then satisfy the public's demand for culture and art, serving the purpose of educating people through culture.

The interrelationship between design and culture is often manifested in the creation of novel cultural products that integrate cultural elements into their design characteristics. Design is not only a science to make an aesthetic product, but also it could make a sustainability for a good caring (Armayuda et al., 2020). For MCCPs, preserving and expressing culture has become the most important consideration in the design process, which is the first element of their evaluation(Chen, 2021). In some articles, it has been discussed that CS can be achieved by developing cultural capital, which focuses mostly on buildings, cultural services, events, and heritage(Soini & Birkeland, 2014a), excluding cultural and creative products. It can be observed that the characteristics of a product and its constituent elements influence the way in which consumers perceive that product (Camilleri et al., 2023). In the recent literature on the topic of assessing the performance of MCCPs, no comprehensive methodology has yet been proposed. Meanwhile, there is a lack of quantitative methods that provide more precise results.

RESEARCH OBJECTIVES

Currently, there is a paucity of evaluative research on the design of MCCPs, as most research on MCCP focus on industrial development(Cascini et al., 2020), research and development strategies(Lu & Zhu, 2020; Jin, 2019; Chen & Yang, 2022), design methods, etc. However, this study can benefit from research on the evaluation of cultural and creative products in general. For example, Shen(Shen, 2022) presents a method for the evaluation and selection of cultural and creative product design solutions based on grey correlation multicriteria decision analysis and the indicators used are practicality, durability, energy efficiency, workability, loanability and innovation. Additionally, Cheng and Guo(Cheng & Guo, 2022) provide an analytic hierarchy process model for evaluating Yan'an Red CCPs. Nevertheless, the evaluation indicators employed by this model are constituted by constituent elements, functional experience, and cultural characteristics. The cultural characteristics include reference, integration, inheritance, commemoration, region, and region. These researchers acknowledged the influence of cultural factors on cultural and creative products but did not give a detailed analysis of the relationship between CS and MCCPs, nor did they explain the evaluation criteria in terms of achieving cultural sustainability. The director of the Development and Operation Department of Guangdong Museum mentioned in an interview that due to insufficient funds and the absence of a professional design director, the R$D of MCCPs had to be outsourced to a third party. However, their lack of understanding of the museum's collections has led to a serious homogenization of MCCPs. Meanwhile, museums lack evaluation indicators to screen creative programs that are culturally relevant. Ultimately, MCCPs of the Guangdong Museums are "distorted" and "unrefined" in terms of cultural symbols, which are not attractive to the market and do not truly reflect the characteristics of the museums' culture(Hu, 2018).

To fill this research gap, the following objectives have been adopted:

1.To review the relationship of MCCP and CS.
2.To propose a system of evaluation indicators to assess the CS performance of MCCPs
3.To examine the CS performance of MCCPs in Guangdong Museum

This paper is structured as follows: the 2nd section provides a review of the existing literature with the aim of identifying compatibility between cultural sustainability and MCCPs. Section 3 builds the research framework based on design thinking and clarify the research method used, namely analytic hierarchy process (AHP). Section 4 develops the evaluation model of MCCP in cultural sustainability dimension guidance of AHP. Section 5 examines the CS performance of MCCPs in Guangdong Museum. Ultimately, section 6 summarizes the new findings presented in the paper.

LITERATURE REVIEW

Design Evaluation of MCCP

Design evaluation (DE) is an evaluation of the quality and value of the design object and is a specific activity to objectively evaluate the results and benefits of design activities. The majority of DE methods are predicated on fixed theories, such as the analytic hierarchy process (AHP) (Ujwary-Gil et al., 2023),the technique for order performance by similarity to ideal solution(TOPSIS)(Quan et al., 2019; Tian et al., 2018),Based on the design thinking, it is crucial to classify the design criteria and calculate the weight of each criteria to improve the quality of a decision of the MCCPs design (Gao & Xu, 2021). There is little related research on the DE of MCCPs, but research on the DE of CCPs can be drawn upon. AHP is the dominant method utilized in the DE of CCPs, as demonstrated in Tabel 1. Besides,It also was applied to evaluate the competitive properties of industrial product design(Özsoy, 2020).Meanwhile, the indicators are also isolated and lack analysis from a specific theoretical perspective, such as the triple bottom line (TBL) framework. This finding is in line with the research conclusions in Camilleri et al (Camilleri et al., 2023). The research team revealed that a multitude of factors can influence consumer perception with regard to three dimensions of sustainability: environmental, social, and economic. Notably absent from the analysis was a consideration of cultural sustainability.

Table 1. Research about DE of CCPs

Product	Method	Indicator	Limitation	Reference
CCP	AHP	Innovation, Simplicity, Quality, Environmental Protection, Beauty, Refinement, Understanding, Functionality, Ease of Use, Reference, Integration, Heritage, Memorial, Regional and Resonance	The selection and weighting of indicators are subjective.	(Gao & Xu, 2021)
CCP	Delphi, Fuzzy logic theory	Cultural Images, Innovation and Creativity, Customer Demand and City Charm.		(Hsueh et al., 2022)
MCCP	Order relation method, Entropy weight method	Shape, Colour, Texture and Material	Need for distinct, non-conflicting design elements in museum cultural creation	(Zhao, 2022)
Red CCP	AHP-TOPSIS	Constituent Elements, Functional Experience and Cultural Characteristics		(Cheng & Guo, 2022)
Cultural Heritage	Comparative analysis	Investments, Tourism, Innovation and Quality of life	The scope of the study is limited by the methods utilized.	(Reher, 2020)
Tourist Souvenir	AHP	Commemorative Value, Commercial Value, Aesthetic Value and Symbolic Value	It has not been demonstrated whether there is a more advanced method of designing memes than described in this paper.	(Liu et al., 2022)
CCP	AHP, TOPSIS	Practicality, Durability, Energy efficiency, Workability, Load ability and Innovation	The indicators selected are limited	(Shen, 2022)

Cultural Sustainability

The concept of CS was previously defined by the World Commission on Culture and Development as the accessibility of cultural resources to individuals across all age groups and generations (Soini & Birkeland, 2014). An increasing number of people have come to recognise the importance of cultural factors in ensuring the long-term stability of any given society, alongside social, economic and environmental concerns. In fact, the post-2015 sustainability objectives of the United Nations made the incorporation of a cultural perspective a central objective (IFACCA 2013). S is founded on the principle that the present generation can modify and make use of cultural heritage, as long as it does not hinder future generations from understanding its significance and interpretations(Petti et al., 2020). Thus, Sustainability is primarily focused on the preservation of cultural values that establish a connection between the past, present, and future(Vries & De Vries, 2022).

Cultural Sustainability and Product Design

The principles of CS and product design are inherently interconnected. The incorporation of traditional cultural features into product design can serve as an important impetus for the promotion of cultural sustainability. This approach has the potential to safeguard and inherit traditional cultures, whilst fostering novel opportunities for traditional products(Qin et al., 2019) .The success stories of companies like Braun in Germany, Philips in the Netherlands, and Alessi in Italy exemplify the attainment of cultural identity through design. It may be concluded that culture has a significant influence on design, and conversely, design also has an impact on culture (Langeveld & Das, 2021). A study conducted on young Chinese consumers found that culturally innovative product designs with chinese traditional cultural properties had a significant positive impact on their perception of cultural sustainability and resilience.

In order to assist designers to use culture as a catalyst for designing innovative products, Moalosi et al. (2010) proposed a "culture--orientated model" based on the Botswana's context.The cultural product design model purposed by Qin and Ng(Qin & Ng, 2020),offers designers a valuable reference point for the successful design of a cross-cultural product.

Cultural Sustainability and Museum

In accordance with the International Council of Museums, museums serve society as educational, cultural, and entertaining institutions. They preserve, exhibit, and maintain the past and present human cultures and environments. For the first time in history, museums have been identified as a crucial contributor to the United Nations Sustainable Development Goals. Their unique collections and resources are invaluable assets in the achievement of these goals, as they facilitate the understanding and implementation of sustainable development principles(Richardson, 2022).

How to attract new audiences through creativity and innovation in order to promote cultural diversity and intercultural dialogue is also an issue to be considered in the process of helping museums achieve CSD(Villacañas de Castro et al., 2022). Innovative, diverse, and interesting cultural activities, as well as the development of cultural and creative products, are effective ways to attract potential audiences, thus raising museums' competitiveness (Tsai & Lin, 2018).

A review of the existing research indicates that museums have identified the objective of CS as the preservation and maintenance of the quality of the cultural heritage (Adams, n.d.; Loach & Rowley, 2022; Pop & Borza, 2016), but there is no measure developed to suggest ways museums can contribute to cultural sustainability beyond heritage conservation.

Cultural Sustainability and MCCP

The function of museums is to meet the cultural needs of individuals, communities, countries, humanity and the environment. In order to achieve this function, the exhibits become tools for their communication(Villacañas de Castro et al., 2022). The artworks played a crucial role in conveying the knowledge of local cultural history through individual interpretations(Härkönen & Stöckell, 2019). MCCP is an innovative use of museum cultural resources. A MCCP, according to Nadja Race, director of international affairs at the British Museum, allows customers to communicate with the museum, learn the story behind the relics, and re-experience the story(Pan, 2019). Each element of a MCCP represents its cultural temperament, from its appearance to its material style, structure and function, and decoration aesthetics. Aesthetics and the experience of artistic creation are a manifestation of cultural values, and their sustainability is critical in the work of cultural and artistic organizations(Brown & Vacca, 2022).

MCCP has the ability to 1) effectively convey museum culture, 2) challenge preexisting perceptions associated with museums, and ultimately, 3) attract a wider audience while yet maintaining the engagement of core viewers(Hong & Qianli, 2019).In addition, it has some social function such as education(Bu & Cao, 2020; Xie & Lai, 2020; Zheng et al., 2020). MCCPs with educational function are one of the ways to realize the educational function of museums(Bu & Cao, 2020), which should be readable, operable and inequitable(Zheng et al., 2020).

Many researchers study how to evaluate the effectiveness of MCCPs from a consumer perspective (Qin et al., 2019; Tu et al., 2019; Yang et al., 2021). In addition, the "Design Performance" is the most important consideration. Conversely, Tu et al.(2019) discovered that consumers attach significant importance to factors such as 'cultural connotation' and 'unique creativity' in their selection of MCCPs at the Palace Museum..

Form the experts perspective, four criteria suitable for the DFuzzy model, namely cultural image, innovation and creativity, customer demand, and city charm, could be used to evaluate the CCPs(Hsueh et al., 2022) .

METHOD

Research Framework

The selection and development of the CS indicators were supported by a framework shown in Figure 1. The research process is carried out under the guidance of design thinking (DT) process. It can be applied to social, economic, environmental, or business problems(Shrivastav & Joshi, 2021). More and more public and non-profit organizations, including museums, are integrating DT into their business practices. The website *designthinkingformuseums.net* was one of the first professional forums to promote and document design thinking approaches in cultural heritage organizations in the US and UK(Mason, 2022). *Given its potential to generate novel sustainability solutions, the application of DT has transcended its original remit as a tool for product design, becoming an effective methodology for addressing complex problems across various fields and sectors*(equity design, 2019).According to Dewberry and Sherwin (Dewberry & Sherwin, 2002), DT can be used to envision sustainability in the future(Dewberry & Sherwin, 2002). For more sustainable outcomes, Shapira et al.(2017) have developed and tested sustainability-oriented add-ins which are designed to complement DT processes.

The "double diamond" design process, established by the Design Council in 2005,is one of the 10 DT frameworks (Dam & Siang, 2020). This model involves collecting a large number of ideas and then narrowing them down until innovative solutions can be developed(Jauhari & Prayudi, 2023).It is composed of four stages, namely: discover, define, develop and deliver, as illustrated in Figure 1.

a. Discover: Understand the problem or challenge that needs to be solved
b. Define: Identify problems or challenges that need to be addressed
c. Define: Develop ideas and prototypes for solutions
d. Deliver: Implement design and test solutions

Figure 1. The research framework of this study based on the double diamond model

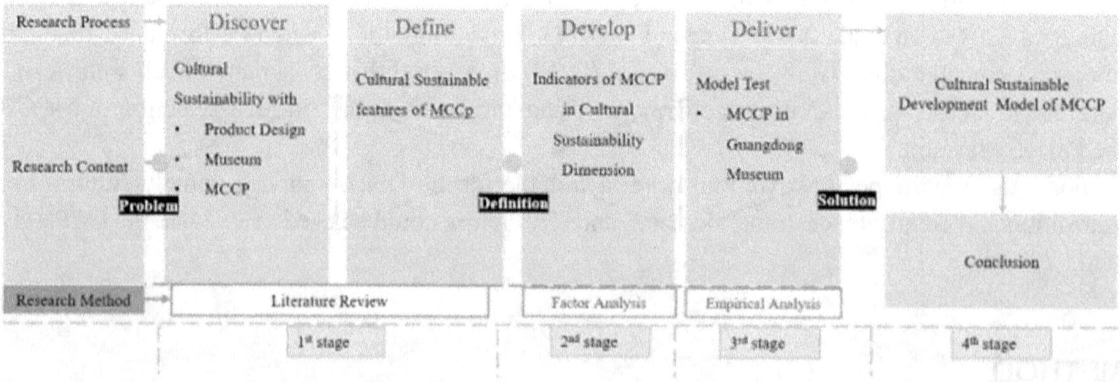

(Drawn by authors)

Analytic Hierarchy Process

The Analytic Hierarchy Process (AHP) was developed in the 1970s by T L Saaty. It is a scientific method that combines qualitative and quantitative techniques. This method yields a simplified evaluation problem. It allows the original complex evaluation problem to be decomposed into one quantifiable evaluation object, which is then summarized and converged(Ujwary-Gil et al., 2023). AHP has been widely used in various fields of social development due to their simplicity and effectiveness. In the field of design, scholars employed the AHP, the Kano Model and Quality Function Deployment (QFD), in order to enhance the design of a tractor seat(Hridoy et al., 2020). AHP involves the following steps: 1)define the problem,2)development of the hierarchical analysis structure model,3)employ a simple pair-wise comparison matrices for each of the lower levels,4)estimate relative weights of the components of each level and 5)undertake a consistency test(Wang & Hu, 2021). Then complete the comprehensive evaluation and make the decision (Wen & Zhang, 2022).

Evaluation Model of MCCP in Cultural Sustainability Dimension

Stage 1: Develop the Hierarchical Analysis Structure Model

Based on the literature related with CS integration of general products, museums, and MCCPs, the hierarchical structure of MCCP evaluation in CS dimension is established, which is shown in Figure 2. This structure is divided into three levels, i.e., goal, criterion, and factor/attribute(Tian et al., 2018; Jin et al., 2023). Goal Level(A) is the MCCP evaluation in CS dimension. Criterion Level(C) includes seven indicators, which are cultural trait (C1), cultural education(C2), cultural creativity and innovation(C3), cultural transmission(C4), cultural authenticity(C5), cultural identity(C6) and cultural vitality(C7). The detailed attributes of these seven metrics are listed in Table 2.

Figure 2. Hierarchical structure of MCCP evaluation in CS dimension

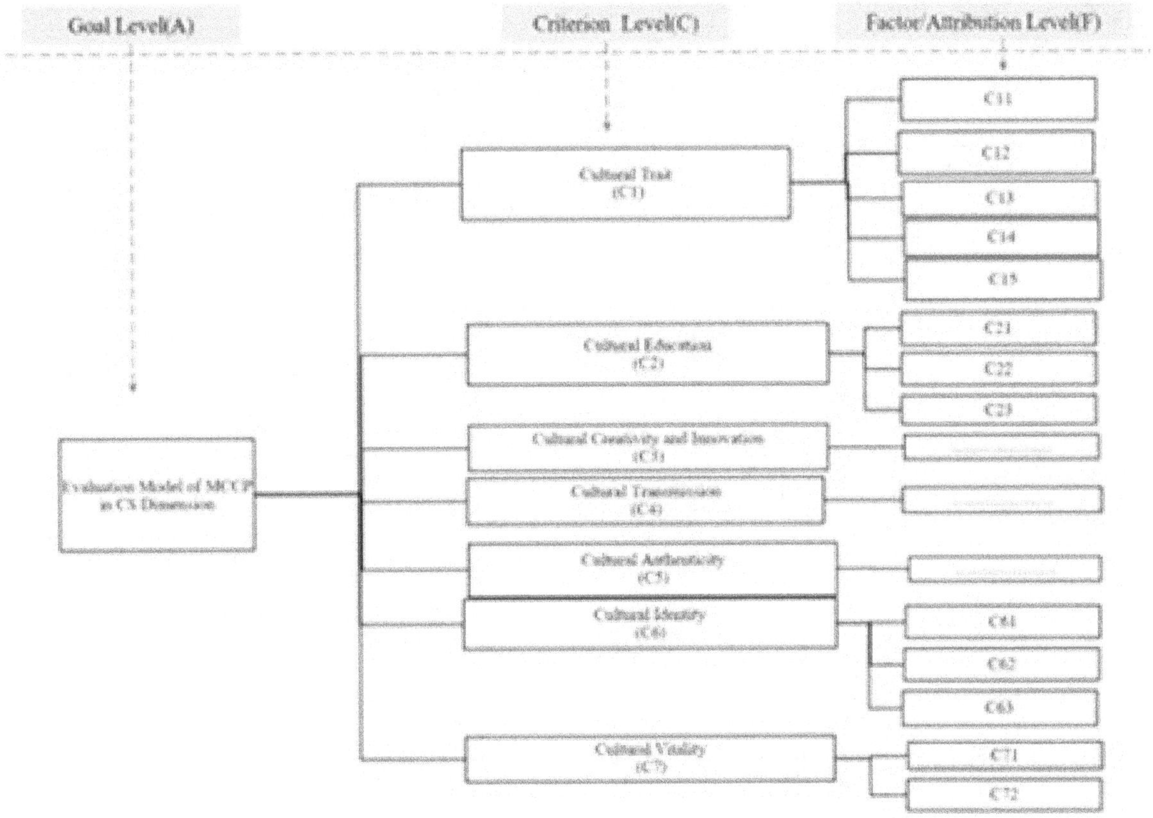

Table 2. Description of evaluation indicators of MCCP in CS dimension

Criterion Level(C)	Factor/Attribution Level(F)
Cultural Trait (C1)	C11. Colour, pattern, and appearance of MCCP can reflect museum's cultural haracteristics
	C12. Colour, pattern, and appearance of MCCP can reflect local cultural characteristics
	C13. Colour, pattern, and appearance of MCCP's package can reflect museum's cultural characteristics
	CT4. Colour, pattern, and appearance of MCCP's pakage can reflect local cultural characteristics
	C15. MCCP has the characteristics of storytelling
Cultural Education (C2)	C21. MCCP has a cultural educational function
	C22. MCCP stimulates consumers' reflection and imagination about culture
	C23.MCCP gives consumers a sense of long history and cultural connotations

continued on following page

Table 2. Continued

Criterion Level(C)	Factor/Attribution Level(F)
Cultural Creativity and Innovation (C3)	C31. Cultural uniqueness
	C32. uniqueness of innovation
	C33. Rebuild the local art and cultural history
	C34. Continuous product innovation and iteration
Cultural Transmission (C4)	C41. Unique brand image
	C42. Packaging and products complement each other and promote brand promotion
	C43. Create a multi-dimensional marketing system with new media
	C44. Have an online marketing platform
	C45. Encourage consumers to participate in the process of product design, R&D, and marketing
	C46. Rich product variety
	C47. Continuously attract new consumers
Cultural Authenticity (C5)	C51.Display authentic historical or regional culture
Cultural Identity (C6)	C61: Fulfill the need for consumer identity
	C62: Make consumers feel a sense of belonging
	C63. Have specific cultural value
Cultural Vitality (C7)	C71.Desig for different consumer groups
	C72. Encourage cultural interaction and exchange

Stage 2: Construct Judgment Matrix

At this stage, an evaluation scale ranging from 1 to 9 (Table 3) was adopted(Saaty, 1990) for each indicator, and the relative significance of the two indicators was assessed through a consultation with experts..

Table 3. 1~9 Evaluation scale (Saaty, 1990; Zhang & Li, 2022)

Intensity of weight	Definition(Comparison of x and y)
1	x and y are equally important
3	x is slightly more important than y
5	x is apparently more important than y
7	x is strongly more important than y
9	x is extremely more important than y
2, 4, 6, 8	Intermediate values
Inverse	If the importance ratio of x and y is a_{xy}, the ratio of y and x is $a_{yx} = 1/a_{xy}$

Stage 3: Calculate Weights

A hierarchical analysis questionnaire was sent to five experts in the cultural and creative design industries. The experts included two cultural and creative product designers, two product design teachers, and one museum manager. After several rounds of discussion, the judgement matrix of the indicator level was obtained, and then the matrix was normalised by column vector, and then by sum normalisation, and finally the weight vector ω and the weight value were obtained(Table 4).

Table 4. Judgement matrix and weight value

A	C1	C2	C3	C4	C5	C6	C7	ω	weight value (%)
C1	1	4	1	3	1	6	4	2.246	23.87
C2	1/4	1	1/5	1/2	1/5	2	5	0.652	6.93
C3	1	5	1	2	1/4	2	8	1.694	18.00
C4	1/3	2	1/2	1	0.2	2	3	0.877	9.32
C5	1	5	4	5	1	6	5	3.139	33.36
C6	1/6	1/2	1/2	1/2	1/6	1	3	0.521	5.54
C7	1/4	1/5	1/8	1/3	1/5	1/3	1	0.281	2.99

Stage 4: Calculating the Consistency

To make the data reasonable, a consistency test is needed. This was assessed using Consistency Ratio (CR), which was calculated using the formula:

CR=CI/RI

CI is the index for determining the consistency of the matrix.

$$CI = \frac{|\lambda \max - n|}{n - 1}$$

λ_{max} represents the largest eigenvalue in the matrix.
n represents the order of the judgement matrix.
RI(Table 5) is the average random consistency index.
If CR≤0.1, the consistency test is passed.
If CR>0.1, it means that the judgement matrix does not pass the consistency test and the previous evaluation data need to be corrected.

Table 5. Average Random Consistency Index Values

N	1	2	3	4	5	6	7	8	9
RI	0.00	0.00	0.58	0.90	1.12	1.24	1.32	1.41	1.45

According to the above method, the consistency of the data was tested and yielded CR=0.078<. 0.1, which passes the consistency test. It shows that the weights obtained in Table 3 are reasonable.

Stage 5: Result Analysis

Cultural authenticity (33.36%) is at the top of the list of indicators, indicating the need to truly reflect historical and cultural connotations in the design of MCCPs. Cultural trait (23.866%) followed closely. It is consistent with the connotation of MCCPs. A MCCP is a recreation based on a cultural resource (Yu, 2019). Cultural creativity and innovation have a weight of 18.00%, reflecting the creativity of MCCPs. Cultural transmission (9.32%), cultural education (6.93%) and cultural identity (5.54%) all have a weighting of less than 10%. Culture vitality (2.99%), which focuses on different groups of consumers and encourages cultural exchange, is considered the least important factor. This is in line with the findings of the *"2019 Museum cultural and creative products market data report"*. It shows that young urban women are the main consumer group of MCCPs. Furthermore, the development of products can be optimised to meet the specific needs of children, adolescents, and the elderly.

CASE STUDY

MCCP in Guangdong Museum

Guangdong Museum serves as one of the inaugural pilot units for the advancement of cultural and creative products intended for the benefit of nationally significant cultural and heritage sites in Guangdong Province. In 2010, it established a special Development and Operation Department to handle the development, promotion, and sales of MCCPs. Presently, it offers a wide range of MCCPs, including stationery, quality products for daily use, exquisite accessories, clothing and bags, treasured books, and limited gifts, which are categorized into several specific types of products, and more than 1,000 MCCPs have been developed and released. Meanwhile, it is expanding its distribution channels. Several experience shops have been established offline, including the Xinhua Bookstore and the Guangzhou Tower Store. Online, it opened its Yuebo Micro Shop in 2018, and in 2020, it opened its flagship store on Tmall. In view of the powerful communication power of the Internet, three MCCPs, shown in Table 6 on the Guangdong Museum's Tmall flagship shop were selected as study cases for this study.

Table 6. Study cases of MCCPs

No	Case 1	Case 2	Case 3
Picture			
Name	Underwater archaeology blind box "Nanhai I"	Chaozhou wood carving large shrine metal bookmark	Flower picture badge Magnolia brooch
Category	Gift	Stationery	Accessory
Design Source	In the 12th and 13th centuries AD, a large ship loaded with Chinese porcelain, gold jewellery and other commodities set sail at a port on the southeast coast, bound for Southeast Asia and other places. When passing through the South China Sea, it encountered a disaster and accidentally sank under the sea..In the 1980s, the Southern Song Dynasty ship was rediscovered and named "Nanhai I".	The shrine is a special artifact of the Chaoshan folk to worship the ancestral deities. The shape of the shrine is solemn, the production is exquisite, is one of the most exquisite workmanships of Chaozhou wood carving products, with gold lacquer wood carving decoration is the most common.	Magnolia, orchids, lilies in the "Hundred Flower Scroll" as the inspiration prototype, suitable for all kinds of bags, clothes and hats. Using paint and enamel technology.

Assessment of Cultural Sustainability

Following the definition of the program, ten experts in the fields of the cultural and creative industries (Table 7) evaluated the extent to which the project had achieved its cultural sustainability objectives. The result shows that case 1 performs best in terms of cultural identity. This indicates that its design can reflect the historical and cultural stories behind it well. However, case 1 performs poorly in cultural vitality, which is because the blind box is mostly used by teenagers and cannot adapt to the needs of a wider range of consumers. Case 2 is like case 1. Case 3 ranks the worst in cultural identity, which shows that it does not bring a sense of cultural belonging to consumers. This may be due to the fact that museums such as the Forbidden City and the Shanghai Museum (Liang, 2020; Hu, 2019) have similar jewellery designs. Concurrently, it fails to convey the intrinsic value of cultural education. As museums serve as a treasure trove for conducting inquiry-based learning, the content and appearance design of MCCP should consider how to stimulate consumers' interest in learning and have an explorable void(Zheng et al., 2020).

Table 7. Grading experts

No	Gender	Age	Working experience Related to MCCP
expert 1	Male	38	7
expert 2	Male	44	9
expert 3	Female	42	8

continued on following page

Table 7. Continued

No	Gender	Age	Working experience Related to MCCP
expert 4	Male	45	10
expert 5	Female	41	15
expert 6	Male	38	13
expert 7	Female	41	10
expert 8	Male	35	6
expert 9	Female	31	4
expert 10	Male	36	6

Figure 3. Expert evaluation results

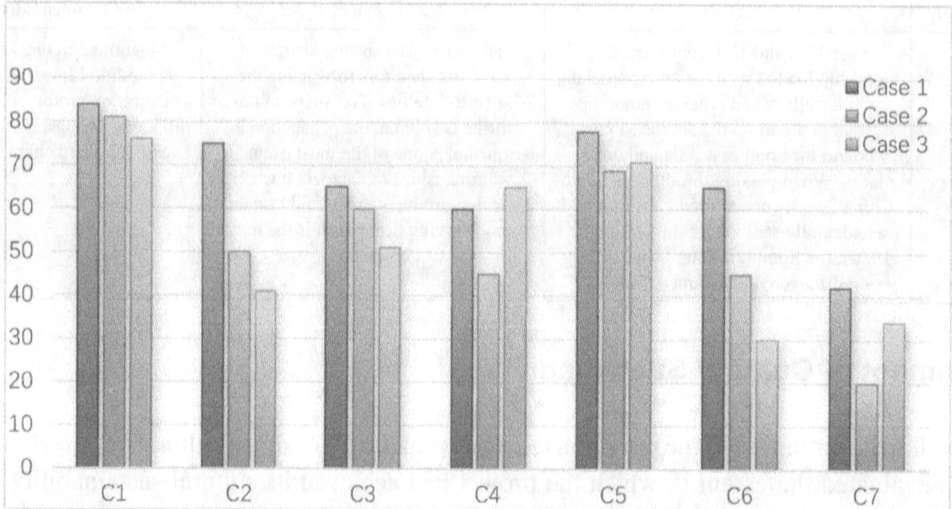

CONCLUSION

This study explores the relationship between museum cultural heritage and CS. MCCP, as a vehicle for innovative expression of museum cultural heritage, is one of the channels for museums to achieve cultural sustainability. To achieve the degree of CS, this study constructed the MCCP evaluation model in CS dimension in using the hierarchical analysis method. This model contains seven indicators, including *cultural trait, cultural education, cultural creativity and innovation, cultural transmission, cultural authenticity, cultural identity, and cultural vitality.* The findings of the study indicate that cultural authenticity is the most crucial factor in attaining cultural sustainability. Regardless of the design methodology and innovative thinking employed in the design of MCCPs, it is imperative to ensure that the cultural connotations are presented in their original form. This is in line with the views of Cao et al.(Bu & Cao, 2020) and Shao (Shao, 2022).

LIMITATION AND FURTHER RESEARCH

These evaluation indicators were identified from the literature, which means that they are secondary data. To improve the evaluation model, future studies should survey industry experts, such as museum experts, cultural research scholars, and designers of MCCPs. Furthermore, this study validated the model with only three products from Guangdong Museum, and the next study can expand the selection of cases for further validation.

REFERENCES

Aboulnaga, M., Puma, P., Eletrby, D., Bayomi, M., & Farid, M. (2022). Sustainability Assessment of the National Museum of Egyptian Civilization (NMEC): Environmental, Social, Economic, and Cultural Analysis. *Sustainability, 14*(20), 13080. Q3. 10.3390/su142013080

Adams, E. (n.d.). Towards Sustainability Indicators For Museums In Australia.

Armayuda, E.-, Lin, P. H., & Lin, R. (2020). Sustainability in Transforming Cultural Feature Into Cultural Creative Product Design (Case Study of Malangan Mask). *Journal of Arts and Humanities*, 9(3), 3. 10.18533/journal.v9i3.1842

Brown, S., & Vacca, F. (2022). Cultural sustainability in fashion: Reflections on craft and sustainable development models. *Sustainability: Science, Practice and Policy, 18,* 590–600. N/A. 10.1080/15487733.2022.2100102

Buyan & Cao, L. (2020). The road of R&D and thinking of educational cultural creativity in museums--Taking Shaanxi History Museum as an example. *Museum of China, 3,* 117-121.

Camilleri, M. A., Cricelli, L., Mauriello, R., & Strazzullo, S. (2023). Consumer Perceptions of Sustainable Products: A Systematic Literature Review. *Sustainability, 15*, 8923. Q3. 10.3390/su15118923

Cascini, G., O'Hare, J., Dekoninck, E., Becattini, N., Boujut, J.-F., Ben Guefrache, F., Carli, I., Caruso, G., Giunta, L., & Morosi, F. (2020). Exploring the use of AR technology for co-creative product and packaging design. *Computers in Industry*, 123, 103308. 10.1016/j.compind.2020.103308

Chen, B.-S. (2021). Research on cultural industry fund to support the development of cultural and creative industries in museums. *Cultural Industry*, 26, 159–161.

Chen, X., & Yang, S. (2022). Research on sustainable design strategy of creative products from the perspective of cultural consumption. *Packaging Engineering*, 43(14), 320–325. 10.19554/j.cnki.1001-3563.2022.14.039

Cheng, X., & Guo, Y. (2022). *A Quantitative Evaluation System of Yan'an Red Cultural and Creative Product Design Based on AHP-TOPSIS Method. Proceedings of the International Conference on Art Design and Digital Technology, ADDT 2022*, 16-18 September 2022, Nanjing, China. 10.4108/eai.16-9-2022.2324890

Dam, R. F., & Siang, T. Y. (2020, June 8). *10 Insightful Design Thinking Frameworks: A Quick Overview.* The Interaction Design Foundation. https://www.interaction-design.org/literature/article/design-thinking-a-quick-overview

Dewberry, E., & Sherwin, C. (2002). E., & Sherwin, C. (2002). Visioning Sustainability through Design. *Greener Management International*, 2002(37), 125–139. 10.9774/GLEAF.3062.2002.sp.00011

equityXdesign. (2019, January 30). Racism and inequity are products of design. They can be redesigned. *Medium*. https://medium.com/@equityXdesign/racism-and-inequity-are-products-of-design-they-can-be-redesigned-12188363cc6a

Härkönen, E., & Stöckell, A. (2019). Cultural Sustainability in Art-Based Interdisciplinary Dialogue. *International Journal of Art & Design Education, 38*(3), 639–648. Q2. 10.1111/jade.12246

Hong, P., & Qianli, S. (2019). Researches on Museum Cultural Creative Products Based on Domains Crossing. *IOP Conference Series. Materials Science and Engineering*, 573(1), 012032. 10.1088/1757-899X/573/1/012032

Hridoy, R. M., Parvez, M. S., & Mohsin, N. (2020). Joining Methods of Analytic Hierarchy Process (AHP), Kano Model and Quality Function Deployment (QFD) to Improve the Tractor's Seat Design for Tractor Drivers in Bangladesh. *American Journal of Industrial and Business Management*, 10(05), 05. Advance online publication. 10.4236/ajibm.2020.105071

Hsueh, S.-L., Zhou, B., Chen, Y.-L., & Yan, M.-R. (2022). Supporting technology-enabled design education and practices by DFuzzy decision model: Applications of cultural and creative product design. *International Journal of Technology and Design Education, 32*(4), 2239–2256. Q4. 10.1007/s10798-021-09681-7

Jauhari, M., & Prayudi, Y. (2023). IMPLEMENTASI METODE DOUBLE DIAMOND DALAM PERANCANGAN PROTOTIPE APLIKASI SISTEM ERP BERBASIS WEBSITE. *AKSELERASI: Jurnal Ilmiah Nasional*, 5(1), 85–98. 10.54783/jin.v5i1.699

Jin, Y., Zhou, X., & Zhang, Z. (2023). Design and evaluation of domestic baby bathtubs based on fuzzy hierarchical analysis. *Furniture and Interior Decoration*, 30(8), 72–75. 10.16771/j.cn43-1247/ts.2023.08.012

Langeveld, L., & Das, A. K. (2021). Impact of Culture on Design Versus Design on Culture. In Chakrabarti, A., Poovaiah, R., Bokil, P., & Kant, V. (Eds.), *Design for Tomorrow—Volume 2* (pp. 455–467). Springer. 10.1007/978-981-16-0119-4_37

Liang, S. M. (2020). Research on the development, design and promotion of cultural creations of the Forbidden City in Beijing under the background of "Internet+". *Packaging Engineering, 41*(8), 309-312+316. 10.19554/j.cnki.1001-3563.2020.08.050

Liu, F., Li, L., & Liu, C. (2022). Research on the design and evaluation model of Jing tourism souvenirs under the perspective of cultural genes. *Packaging Engineering*, 43(2), 333–340. 10.19554/j.cnki.1001-3563.2022.02.044

Loach, K., & Rowley, J. (2022). Cultural sustainability: A perspective from independent libraries in the United Kingdom and the United States. *Journal of Librarianship and Information Science, 54*(1), 80–94. Q3. 10.1177/0961000621992824

Lu, M., Wu, J., & Meng, Q. (2015). Comprehensive Development Evaluation System of Asian Infrastructure Investment Bank Based on Double Diamond Model. *American Journal of Industrial and Business Management*, 05(07), 07. 10.4236/ajibm.2015.57051

Lu, Q., & Zhu, J. H. (2020). A study on the design strategy of cultural and creative products in museums: The case of Yachou-axe. *Architecture and Culture*, 5, 46–48.

Mason, M. (2022). The Contribution of Design Thinking to Museum Digital Transformation in Post-Pandemic Times. Multimodal Technologies and Interaction, 6(9). *Multimodal Technologies and Interaction*, 9(9), 79. 10.3390/mti6090079

Moalosi, R., Popovic, V., & Hickling-Hudson, A. (2010). Culture-orientated product design. *International Journal of Technology and Design Education, 20*(2), 175–190. Q4. 10.1007/s10798-008-9069-1

Nie, H. T., & Li, N. (2020). Protection and Creation: An Analysis of Copyright Legal Issues of Museum Cultural and Creative Products. *Chinese Museum, 1*, 13-18.

Özsoy, H. Ö. (2020). Evaluating Competitiveness In Product Design By Using Analytic Hierarchy Process. *İstanbul Ticaret Üniversitesi Sosyal Bilimler Dergisi, 19*(37), 37.

Pan, Y. (2019). The British Museum's cultural and creative exploration: How do world-class museums develop cultural and creative industries? *PR World*, 22, 23–26.

Petti, L., Trillo, C., & Makore, B. N. (2020). Cultural Heritage and Sustainable Development Targets: A Possible Harmonisation? Insights from the European Perspective. Sustainability, 12(3). *Sustainability (Basel)*, 3(3), Q3. 10.3390/su12030926

Pop, I., & Borza, A. (2016). Factors Influencing Museum Sustainability and Indicators for Museum Sustainability Measurement. *Sustainability, 8*(1), 101. Q3. 10.3390/su8010101

Qin, Z., & Ng, S. (2020). Culture as Inspiration: A Metaphorical Framework for Designing Products with Traditional Cultural Properties (TCPs). Sustainability, 12(17). *Sustainability (Basel)*, 17(17), Q3. 10.3390/su12177171

Qin, Z., Song, Y., & Tian, Y. (2019). The Impact of Product Design with Traditional Cultural Properties (TCPs) on Consumer Behavior Through Cultural Perceptions: Evidence from the Young Chinese Generation. Sustainability, 11(2). *Sustainability (Basel)*, 2(2), Q3. 10.3390/su11020426

Quan, H., Li, S., Wei, H., & Hu, J. (2019). Personalized Product Evaluation Based on GRA-TOPSIS and Kansei Engineering. *Symmetry*, 11(7), 7. 10.3390/sym11070867

Reher, G. S. (2020). What is value? Impact assessment of cultural heritage. *Journal of Cultural Heritage Management and Sustainable Development, 10*(4), 429–436. Q4. 10.1108/JCHMSD-10-2019-0128

Richardson, J. (2022, March 27). *Museums and sustainable development goals*. MuseumNext. https://www.museumnext.com/article/museums-and-sustainable-development-goals/

Saaty, T. L. (1990). *The Analytic Hierarchy Process: Planning, Priority Setting, Resource Allocation*. RWS Publications.

Shao, L. (2022). Research on the visual design of non-heritage cultural and creative products based on sustainable design theory. *Art and Design (Theory)*, 2(4), 87–92. 10.16824/j.cnki.issn10082832.2022.04.020

Shapira, H., Ketchie, A., & Nehe, M. (2017). The integration of Design Thinking and Strategic Sustainable Development. *Journal of Cleaner Production, 140*, 277–287. Q1. 10.1016/j.jclepro.2015.10.092

Shen, Y. (2022). Evaluation and Selection of Cultural and Creative Design Solutions Based on Grey Correlation Multicriteria Decision Analysis. *Mathematical Problems in Engineering*, 2022, e1755728. 10.1155/2022/1755728

Shrivastav, S., & Joshi, R. (2021). Design Thinking for Long-Term Product Planning. In Chakrabarti, A., Poovaiah, R., Bokil, P., & Kant, V. (Eds.), *Design for Tomorrow—Volume 3* (pp. 387–398). Springer. 10.1007/978-981-16-0084-5_31

Soini, K., & Birkeland, I. (2014a). Exploring the scientific discourse on cultural sustainability. *Geoforum, 51*, 213–223. Q1. 10.1016/j.geoforum.2013.12.001

Song, Y., & Li, M. (2018). Research on Cultural and Creative Product Development Based on Museum Resources. *IOP Conference Series. Materials Science and Engineering*, 452(2), 022090. 10.1088/1757-899X/452/2/022090

Tian, G., Zhang, H., Zhou, M., & Li, Z. (2018). AHP, Gray Correlation, and TOPSIS Combined Approach to Green Performance Evaluation of Design Alternatives. *IEEE Transactions on Systems, Man, and Cybernetics. Systems*, 48(7), 1093–1105. 10.1109/TSMC.2016.2640179

Tsai, P.-H., & Lin, C.-T. (2018). How Should National Museums Create Competitive Advantage Following Changes in the Global Economic Environment? *Sustainability (Basel)*, 10(10), 3749. 10.3390/su10103749

Tu, J.-C., Liu, L.-X., & Cui, Y. (2019). A Study on Consumers' Preferences for the Palace Museum's Cultural and Creative Products from the Perspective of Cultural Sustainability. Sustainability, 11(13). *Sustainability (Basel)*, 13(13), Q3. 10.3390/su11133502

Ujwary-Gil, A., Florek-Paszkowska, A., & Kozioł, A. (2023). AHP/ANP Applications in Economic Policy, Business, and Management with the T. L. Saaty Decision Making Hackathon at ICEFM 2022. *International Journal of the Analytic Hierarchy Process*, 14(3). 10.13033/ijahp.v14i3.1071

UNESCO (Ed.). (2003). *Culture, trade and globalization: Questions and answers*. Unesco.

Villacañas de Castro, L. S., Moreno-Serrano, L. M., & Giner Real, C. (2022). Museum education, cultural sustainability, and English language teaching in Spain. *Pedagogy, Culture & Society, 30*(2), 201–223. Q1. 10.1080/14681366.2020.1794947

Vries, G. D., & De Vries, G. (2022). Culture and the sustainable development goals: Challenges and opportunities. *Economia Della Cultura*, 4/2022. 10.1446/107171

Wang, M., & Hu, Y. (2021). AHP-TOPSIS-based Attachment Product Design for Children with Autism Spectrum Disorder. *Packaging Engineering*, 42(18), 220–226. 10.19554/j.cnki.1001-3563.2021.18.025

Wen, D., & Zhang, Z. (2022). Furniture Design Strategies for Young People Living Alone Based on KANO and TRIZ Theories. *Furniture & Interiors*, 3, 6.

Xie, L., & Lai, X. (2020). The development of cultural and creative products in red culture memorial museums. *Culture Industry*, 8, 26–27.

Yang, J., Luo, J. M., & Lai, I. K. W. (2021). Construction of Leisure Consumer Loyalty from Cultural Identity—A Case of Cantonese Opera. *Sustainability, 13*(4), 1980. Q3. 10.3390/su13041980

Yi, Y. (2019). Cultural and creative product design from the perspective of the relationship between design and culture. *Art Education*, 9, 261–262.

Zhang, Y., & Li, J. (2022). Research on Hu bed design based on user research and hierarchical analysis. *Packaging Engineering*, 43(S1), 183–188. 10.19554/j.cnki.1001-3563.2022.S1.031

Zhao, Z. (2022). Analysis of Museum Cultural Creation from the Perspective of Cultural Industry. *Scientific Programming*, 2022, e8138574. 10.1155/2022/8138574

Zheng, Y. (2020). Design of Museum Cultural and Creative Product Development Based on Educational Function and Connotation: The Case of "Dinosaur Exhibition in China". *Natural Science Museum Research, 5* (2), 47-57+96+99. 10.19628/j.cnki.jnsmr.2020.02.007

Chapter 7
Cloud Computing Adoption at Quantity Surveying Firms:
Are We Ready?

Norhazren Izatie Mohd
https://orcid.org/0000-0003-3003-8205
Universiti Teknologi Malaysia, Malaysia

Nur Nilam Sari Md Yusoff
Universiti Teknologi Malaysia, Malaysia

Hamizah Liyana Tajul Ariffin
Universiti Teknologi Malaysia, Malaysia

Ahmad Faiz Azizi Ahmad Fauzi
https://orcid.org/0000-0003-1559-2275
Universiti Teknologi Malaysia, Malaysia

ABSTRACT

Despite notable technological progress, the construction industry has slowly embraced information and communication technologies (ICT). Cloud computing, a tool potentially enhancing information management productivity, operates through remote servers linked via the Internet. This allows for data storage and application accessibility from any device without requiring prior installation. This research addresses this technological gap by exploring how quantity surveying firms perceive the integration of cloud computing. The study aims to evaluate these firms' inclination to adopt cloud computing and identify the obstacles impeding its implementation. A quantitative approach and questionnaires were adopted to uncover quantity surveying firm's readiness for technological advancement. This study provides valuable insights into how these firms can reshape their operational and communicative practices in the context of IR 4.0, preparing them for the essential technological shifts that will shape future industry standards.

DOI: 10.4018/979-8-3693-1650-4.ch007

INTRODUCTION

The introduction and implementation of IT innovation remain an exciting topic of research. However, with progress in implementing IT innovation, companies can boost potency, improve service, cut back prices, and increase company worth. This is because when the implementation of IT innovation becomes a part of the enterprise, call manufacturers must perceive the facility and style components that build choices concerning the adoption and implementation.

However, within the competitive global business surroundings, decision-makers must create good choices and adopt new IT implementations for investment and institutional ways. Therefore, structure innovations that adopt this technology differ clearly in relation to the emergence of the latest IT technologies, such as cloud computing (Li et al., 2015). However, although cloud computing is said to supply varied advantages to companies, the implementation of cloud computing within companies has been far from expected. Consistent with Sallehudin et al. (2019), the varied firm is considering cloud computing quite warily by taking one step at a time or preferring to 'wait and see.' Therefore, a study on cloud computing implementation within quantity surveying firms must be dispensed to evaluate the perception and barrier factors that influence quantity surveying firms to implement this technology in the Malaysian construction industry. This is because cloud computing serves as a platform for integrating diverse information sources and tools used in quantity surveying. For example, Building Information Modelling (BIM) software, cost estimation tools, and project management systems can be hosted on cloud computing platforms, allowing seamless data exchange and workflow automation. Thus, this integration streamlines quantity surveying tasks, enhances data accuracy, and promotes better decision-making throughout the project lifecycle.

Following that, Kagermann et al. (2013) consider the 4th Industrial Revolution a trend towards automation and data exchange in manufacturing technology. Advancements in technologies such as the Internet of Things (IoT), 5G, cloud computing, data analysis, and robotics will change products, processes, and business models across all sectors, eventually creating new industrial patterns and a shift towards global value (Brussels, 2016). Cloud computing is one of the elements of the industrial revolution. Cloud computing is a trend for most companies worldwide, as it provides cost savings for maintenance and server resources of information and communication technology (ICT). For example, in the context of quantity surveying practices, cloud computing offers scalable and flexible computing resources over the Internet, enabling quantity surveyors to access data and software applications from anywhere with an Internet connection, thus facilitating collaboration among project stakeholders, enhancing data security, and improves the efficiency of quantity surveying processes. Therefore, this study is conducted to understand the acceptance of cloud computing within quantity surveying firms in Malaysia.

Furthermore, Sulaiman et al. (2017) argue that human communication has been in the fifth phase, based on the Internet and computing. Therefore, cloud computing is seen as a means of improving communication within the company. However, management and communication issues in the construction sector are complicated, challenging, and not currently well addressed by existing technologies and applications (Jiao et al., 2013). Similar to the quantity surveying practices, communication among quantity surveyors and project stakeholders is critical to ensure accurate and up-to-date information. This information includes project specifications, material costs, labour rates, and construction methodologies. Moreover, with the advent of digital technologies, vast amounts of data are generated throughout the project lifecycle and shall be communicated respectively. Thus, proper management and communication may enhance information transfer and significantly impact project outcomes, such as cost estimation, resource allocation, and risk management. Therefore, quantity surveying firms need to understand the acceptance factors of cloud computing in order to overcome them and embrace the utilization of cloud computing.

In conclusion, rapid economic growth requires companies to respond promptly to market changes and become more innovative to survive and develop within industrial competition. However, traditional IT infrastructure designs may impede the rapid Development of companies due to the high cost and existing technology infrastructure. Nevertheless, these restrictions can be mitigated with the Development of cloud computing. Additionally, cloud computing plays a pivotal role in transforming traditional quantity surveying practices and enabling more efficient and data-driven approaches to project management and cost control. Hence, this study aims to draw insights into quantity surveying firms.

Cloud Computing

Since 2007, cloud computing has gained widespread usage in the services and manufacturing industries, prompting numerous researchers to offer definitions based on different application perspectives. According to Microsoft (2018), cloud computing encompasses the provision of computing resources—servers, storage, databases, networks, and software—via the Internet, enabling rapid information access and flexible resource allocation. Cloud computing involves storing and processing data or information on a server for online sharing. As per Chandrasekaran (2014), the basic concept of cloud computing involves accessing data and applications over the Internet from remote locations rather than from local computer hard drives. Foster Ian et al. (2008) characterizes it as implementing a large-scale distributed computing model driven by economies of scale, establishing a virtual, remotely manageable infrastructure for computing power, storage, platforms, and on-demand services over the Internet.

According to the definition by the National Institute of Standard and Technology (NIST), cloud computing is a form of service opportunities for information and data to be present everywhere, providing facilities and appropriate network access to request to the location source computing designed, such as networks, servers, storage, applications, and services that can be provided quickly and launched with minimum interaction between service providers (Mell & Grance, 2011).

Characteristics of Cloud Computing

According to Mell and Grance (2011) and Shailja (2023), cloud computing has five essential characteristics: on-demand self-service, broad network access, resource pooling, rapid elasticity, and measured services (Mell and Grance, 2011; Shailja Tripathi, 2023).

- On-Demand Self-Service: This means users will automatically provide computing capability as needed without the restriction of interacting with humans. The capacity is obtainable to users over the network using mobile phones, tablets, laptops, and workstations.
- Broad Network Access: Cloud computing services should be accessed anywhere, at any time, and link all devices, such as laptops, mobile phones, displays, tablets, and other devices, via the Internet.
- Resource Pooling: Cloud computing services should be available to provide centralized and efficient resources. This is because a variety of users use cloud computing at the same time. Therefore, services would have to divide the load efficiently so that the system could be utilized at maximum.

- Rapid Elasticity: Cloud computing services should be able to raise and decrease the capacity or the load to fit the needs. For example, when a company's user number increases, the company can add the user to the cloud computing application and vice versa.
- Measured Service: Cloud computing services should be provided as measurable, as they will be used in the payment process. Cloud computing services are paid according to the requirements of the user's needs and should be measurable.

Type of Cloud Computing

Four development models are defined as services, such as public cloud, private cloud, hybrid cloud, and community cloud.

- Public cloud: Data kept in the cloud server that is found at a distant place elsewhere. It allows users to share and access data from any place and at any time. This implies that the public cloud promotes shared surroundings. Although a small amount of risk is associated with data security as business operations are done through the Internet, this offers exceptionally scalable surroundings.
- Private cloud: This is helpful for companies that do not need to share their confidential data with any third party. Data stored in an exceedingly private cloud is secured within the firewall settings and only enables the authorized parties to access data and not invite any third party.
- Hybrid cloud: A mix of each, offering users or business entities the advantage of both cloud environments. For example, suppose a commercial enterprise desires to share its services and merchandise with its clients worldwide but also wants to cover its confidential information. Hybrid cloud design would be best suited for such varieties of businesses.
- Community cloud: A cloud service model that provides a cloud computing solution to a restricted variety of people or companies that are ruled, managed, and secured ordinarily by all those taking part in the company or a third-party managed service provider.

Type of Cloud Computing Service Model

The types of service models that have emerged under cloud computing technology are Software-as-a-Service (SaaS), Platform-as-a-Service (PaaS) and Infra-structure-as-a-Service (IaaS).

- Software-as-a-Service (SaaS): Applications and software hosted on the server of cloud service suppliers and created on the market to customers over a network. SaaS is helpful for applications where the strain fluctuates, applications with demand for Internet and mobile access, short-term projects that need collaboration, and start-up businesses that need to quickly launch e-commerce sites without concern regarding server configurations and software updates.
- Platform-as-a-Service (PaaS): A service model that offers users a platform to create and run applications through a programming interface provided and supported by cloud service suppliers (Marston, 2011). PaaS is used once multiple developers are engaged in the Development or when external parties concerned are within the development method, companies that follow agile meth-

odology for software development, and the masses that need to customize applications. As PaaS reduces overhead costs, it is used by companies desiring to expand their capital investment.

- Infra-structure-as-a-Service (IaaS): Cloud service providers offer users a range of virtual infrastructures such as virtual servers, storage, and other fundamental computing resources that change them to deploy and run their operating system, applications, or transfer software or files into the cloud. IaaS may be used by companies that require complete control over their high-performing applications, start-ups, and small firms with time and energy constraints. In contrast, for a growing company that does not need hardware or software resources, applications see a volatile demand whenever scaling up, traffic spikes or valleys support down.

The Use of Cloud Computing Services

Integrating cloud-computing services has significantly transformed the landscape of business operations. This innovative technology offers versatile and scalable resources, comprising storage, processing power, and applications accessible over the Internet. Organizations are capitalizing on these services to boost efficiency, innovation, and cost-effectiveness, marking a fundamental shift in work dynamics across various industries.

Key applications of cloud computing services are delineated to underscore their significance. Cloud databases, for instance, provide robust and maintenance-free databases that relieve organizations of infrastructure management, leaving them responsible only for their data handling. Test and development environments benefit significantly from the cloud, allowing for cost reduction and rapid market deployment of firm applications. Developers can swiftly establish and dismantle these environments, scaling them based on project requirements, thus saving time and resources.

Another crucial application highlighted is big data analytics. Cloud computing facilitates access to vast data pools, enabling organizations to extract valuable insights for business enhancement. Retailers and suppliers leverage consumer-shopping patterns, while social networking platforms offer data patterns used for deriving meaningful insights.

Furthermore, the passage emphasizes cloud computing's role in file storage and sharing, enabling universal access to files and company data from any device with an Internet connection. This accessibility fosters seamless collaboration and remote work. The cloud is a reliable backup and recovery solution, ensuring data security and resilience against system failures or disasters. Its scalability allows for the swift expansion or reduction of databases based on organizational needs. Website hosting services in the cloud provide flexibility to transfer sites between servers as business requirements fluctuate. Additionally, cloud-based email services such as Hotmail and Gmail offer businesses mobility and accessibility without server capacity or location constraints.

In summary, cloud-computing services have redefined business operations. Their adaptability, cost-effectiveness, and scalability have positioned them as indispensable assets for modern enterprises. By enabling streamlined operations, enhancing data management, and facilitating seamless collaboration, these services are instrumental in meeting the ever-evolving technological demands of the industry. As a result, cloud computing has become a cornerstone for digital solutions, transforming how organizations leverage technology to drive productivity and innovation across diverse business sectors. The typical use of cloud computing is shown in Table 1.

Table 1. Typical use of cloud computing

	Test & Development	Big Data Analytics	File Storage & Sharing	Backup & Recovery	Cloud Database	Website Hosting	Email
(Karp, 2017)	/	/	/	/			
(Yvanovich, 2018)	/	/	/		/	/	
(Bone, 2017)			/	/			
(Rack, 2017)	/		/	/	/	/	/
(AlMabhouh &Alzaza, 2015)	/		/			/	/
(Ferkoun, 2014)	/	/	/	/			

The Perception of Quantity Surveying Firms on the Use of Cloud Computing

This perception of cloud computing adoption is structured around three pivotal facets: knowledge, cost considerations, and the firm's readiness. Understanding the technology's nuances, assessing its financial implications, and evaluating the organization's preparedness collectively influence the decision-making process regarding integrating cloud computing services. Knowledge ensures informed choices, cost evaluation helps budget allocation, and readiness determines the feasibility of incorporating cloud solutions into the firm's operational framework. These elements collectively shape the perspective and approach toward embracing cloud computing within an organizational setting.

1. Knowledge

In an increasingly competitive business environment, many companies pay significant attention towards adopting innovative technology in their management system to improve productivity, reduce risks, and produce quality products. The Development towards Industrial Revolution 4.0, also known as Industry 4.0 (IR 4.0), dramatically affects the operation and results of the company. Therefore, decision-makers and managers in the industry must invest to spur the introduction of automation in the existing manufacturing process (Moktadir, 2018).

Cloud computing is also one of the elements in the revolution of Industry 4.0 (Yusof, 2017). Cloud computing is one of the promising technologies that will help a company be more competitive in the market. Cloud computing allows employees to import data, email, documents, and other applications into the virtual platform through a connected server that can be accessed anywhere and at any time (Catalina, 2014). Cloud computing helps improve the workforce's productivity by allowing seamless real-time communication and collaboration. For example, Slack allows employees to view all communication in one place to avoid misunderstandings.

Cloud computing provides two significant benefits in information technology. The first is the efficiency of IT, where the power of modern computers is used more efficiently through source hardware and software that is highly scalable. Secondly, the rapidity of business-driven functions available on cloud computing has become a competitive tool to meet users' needs (Kim, 2009).

2. Cost

Cloud computing also helps increase the firm's work effectiveness with IT solutions that allow users to access computing services more flexibly at a low cost. This is because using computing does not need capital investment in hardware, software, IT support, or information (Alshamaila, 2013; Jain & Bhardwaj, 2010). Furthermore, in terms of the financial aspects of cloud computing, it also helps companies reduce costs because they have the option of a pay-per-use model (Al-Badi et al., 2017). In this context, IaaS tends to be the most cost-effective, as it provides the most flexibility and control over resources. Meanwhile, PaaS and SaaS often have higher per-user or per-application costs yet require less management overhead (Luxner, 2023). All in all, the cost of adopting cloud computing varies widely depending on factors such as the size of the companies, the scope of cloud deployment, the specific cloud services and applications used, and the existing IT infrastructure.

Additionally, cloud computing offers cost flexibility. Thus, companies can avoid significant upfront costs when purchasing and installing software applications and hardware systems. This is particularly useful to small companies that operate with a small budget. This allows them to rent software applications and storage space as needed only. In addition, cloud computing turns the cost from fixed to variable. Companies use and pay as they need, reducing the fixed overhead cost (Al-Badi et al., 2017). For example, instead of upfront investments in hardware and infrastructure, companies pay ongoing operational costs for cloud services. Nevertheless, while this can reduce upfront investment and provide greater flexibility, it is essential to consider the long-term implications of recurring expenses. All in all, companies shall carefully evaluate the cost implications and develop a comprehensive cost management strategy to optimize spending, maximize value, and mitigate risks associated with cloud adoption.

3. Readiness

The readiness of companies to represent the state of IT within the company for implementing innovations is also a perception aspect. The availability of these financial, human resources, and technical covers ought to be enough associated with an innovation to be implemented (Sallehudin et al., 2019). In the Malaysian public sector context, the company's readiness will cover all three aspects: adequate cost, the efficiency of human resources in the IT field for support or problem-solving, and the technology infrastructure or sufficient technical.

Cloud computing is hosted by the vendor and accessed via the Internet. At the same time, on-premise involves the installation of computers and the company's server. Therefore, with cloud computing, there is flexibility and customization from the seller. However, this means there is increased stability and continuous improvement. Some vendors often work closely with the company to make the necessary changes.

Meanwhile, hosting on-premises experiences a level of customization, which is excellent because it takes control of its solution. This leads to problems in terms of execution time and issues with software updates. However, the ability of a company to try cloud computing services on a limited basis paves the way for the company to decide to use it. Trialability is one of the factors suggested by studies before this to allow better acceptance of cloud computing (Alkhater et al., 2018). The company's readiness towards acceptance of cloud computing should be considered in terms of the needs and requirements of its customers, as the client should be satisfied with the services provided by the company. Therefore, cloud computing will not be impeded by customer service.

Barriers to the Use of Cloud Computing

There are a number of empirical studies investigating the barrier factors that influence the use of cloud computing in an organization. However, these studies were conducted solely on a specific number of factors and neglected other influential factors that may impact the intention to use cloud services, such as physical location. Thus, these factors examine their effect on an organization's decision to use cloud computing within the Quantity Surveying (QS) firm.

1. Technological Factors

In this study, the technological factors of cloud computing include quality of service, security, privacy, and trust. These factors determine the barriers that might stimulate the implementation of cloud computing by the QS Firm. The quality of services to end users has a significant impact on the use of online services. Cloud computing provides service to the end user via the Internet, anytime and anywhere. Therefore, if the quality of services is not satisfactory to the user, this will result in lower acceptance towards using cloud computing (Alkhater et al., 2018). Securities refer to the level of the procedure in the service cloud to secure the firm's or system's data from any threat or attack. Concerns about the security of data and documents are one of the main reasons many companies are not interested in using cloud computing (Lian et al., 2014).

In terms of privacy, the cloud service will be managed and stored by a third party, and the firm does not need to control this information entirely. This will increase the risk associated with data and stored information, possibly leading to privacy violations (Takabi et al., 2010). The privacy of data and information has become a concern for most companies that intend to store data and applications in cloud computing. Trust is another issue in cloud computing. The depth of transparency and management over data within the cloud might cause trust issues, which can concern enterprises. Trust is not simply a problem of trusting the cloud provider; there is conjointly a controversy of distrusting the technology and its capabilities to supply good service without interruption or loss of information. Though cloud suppliers guarantee customer service with Service-Level Agreements (SLA), unfortunately, this cannot save the organization's name if any data loss or breach occurs (Khan & Malluhi, 2010). During this study, this factor measures the willingness and confidence of organizations concerning storing their knowledge within the cloud and under third-party control.

2. Organsational Factors

Organizational factors signify the organizational elements described by this study as the element of the firm that determines its influence to use cloud computing. Commitment and support from the top management structure show favour for accepting and adopting technological innovation in organizations. This factor is related to the organization where top management can adopt cloud computing. Top management has an essential role in accepting new technology at the organizational level (Ramdani et al., 2009). Therefore, the support of top management can be a prohibitive factor in making changes and using cloud computing in the firm.

Organizational readiness represents the internal IT maturity for innovation implementation. The availability of financial, human, and technical readiness should be adequate for implementing the related innovation. For implementing cloud computing in the QS Firm, the company should be ready with IT staff to efficiently support and solve service problems during and after implementation. The company should also have a computer network infrastructure with a reliable Internet connection and faster download/upload speed. This is because access to cloud computing services is through the infrastructure of the computer network. Suppose the Internet connection is unreliable, or the speed of download/upload is slow. In that case, this would pose a problem or disruption of service and a failure to implement cloud computing (Sallehudin et al., 2019).

3. Environmental Factors

In this context, environmental factors represent the environmental aspects that might affect a QS Firm's intention to use cloud computing. Compliance issues may prevent the firm from moving to cloud technology. This is due to the lack of IT standards and the government and industry regulations in enforcing and supporting the QS Firm to use cloud computing (Marston, 2011; Nkhoma & Dang, 2013; Sun et al., 2024). For example, governments play a crucial role in establishing regulatory frameworks that govern the use of cloud computing, particularly concerning data privacy and security. By enacting appropriate legislation and regulations, governments can address concerns related to data protection, compliance, and liability in cloud environments, thereby fostering trust and confidence in cloud services among consumers, including quantity surveyors. Additionally, the location of data and resources in cloud services can be anywhere in the world. Cloud computing providers may store data in different countries and locations hidden from the end user. This will increase the firm's degree of concern regarding the use of cloud computing. In addition, there are still no international regulations or codes for protecting data stored with cloud technology. Therefore, physical location becomes an issue that can affect the decision of companies to use cloud computing (Jaeger, 2009; Marston, 2011).

Regarding external support, this factor is taken from earlier research conducted by IT experts (Alkhater et al., 2015). External support refers to support from cloud computing providers to meet the firm's needs. According to a study, the respondents agree that the organization requires external support from a cloud computing provider to meet the user's requirements. No outside support from a supplier will make the firm unable to meet its needs, which becomes an obstacle to using cloud computing. The state of the industry also influences the firm against IT innovation (Goode & Stevens, 2000). The type of industry also appeared as a factor in the research studies that have been made. From the perspective of experts, the organization should consider the nature of its business and requirements before using cloud services.

4. Human Factors

Culture is defined as the beliefs, values, habits, norms, and behaviour of a group in the community (Leung, 2005). This factor is taken from the studies by Alkhater et al. (2015). In this study, the culture of Saudi Arabia impacts the behaviour and attitude towards decision-making. This study measures the effectiveness of the cultural adoption of cloud technology through the behaviour and beliefs of the people against the use of cloud computing in the organization. The staff's capabilities, competencies, and knowledge will also impact the firm upon adopting an IT innovation, whether or not the staff has enough innovation, knowledge, and skills to adopt cloud computing. In this case, if the staff is skilled, the firm will have much confidence throughout the assimilation method.

METHODOLOGY

A quantitative approach was adopted to carry out this study. By employing a quantitative approach, the data can be collected from a large sample of firms, allowing for more reliable findings. Subsequent application of statistical analysis techniques can enable the identification of trends within the data, providing valuable insights into firms' readiness levels and the impediments they may encounter when adopting cloud computing. Overall, this approach aims to provide a comprehensive understanding of the perception surrounding cloud computing adoption in firms, offering actionable recommendations to enhance implementation processes.

Additionally, a set of questionnaires has been designed. Questionnaires are widely used to collect quantitative data, where they typically consist of structured questions designed to measure firms' perceptions and barriers related to cloud computing implementation. Hence, the questionnaire survey consists of three sections: firm details, firm perception, and barriers to implementation, using a 5-point Likert scale. Sampling involved random selection from construction workers by distributing questionnaires directly and via online surveys in collaboration with Quantity Surveying firms in Klang Valley.

Finally, descriptive analysis was utilized for data analysis. The mean range method was employed to determine the average response for each question, evaluating respondents' level of agreement and perception positions (Table 2).

Table 2. Level of agreement based on average mean

Mean Range	Level of Agreement
1.00 – 2.33	Low
2.34 – 3.67	Medium
3.68 – 5.00	High

The Perception of Quantity Surveying Firms Towards Utilisation of Cloud Computing

Table 3 displays 13 statements concerning respondents' views on utilizing cloud computing in Quantity Surveying firms, assessed across three aspects: knowledge, cost, and willingness. The analysis from the knowledge aspect reveals that respondents strongly believe that using computing clouds can notably enhance the firm's productivity. Table 3 substantiates this, displaying the highest mean range of 4.05 for the knowledge aspect statement. This top-rated agreement highlights the multitude of benefits associated with employing cloud computing.

Subsequently, respondents strongly endorse cloud computing as a suitable technology for quantity surveying, as demonstrated by their highest level of agreement with the statement. Table 3 reveals a mean range of 4.03, marking it the second highest among knowledge-related statements. Moreover, the analysis in Table 4 indicates unanimous agreement among respondents regarding cloud computing's pivotal role in Industry 4.0, garnering the highest level of agreement with a mean range of 3.98. This illustrates a comprehensive understanding among respondents of the technologies within Industrial Revolution 4.0. The mean range demonstrates respondents' confidence in the suitability of cloud computing for the operational structure of Quantity Surveying firms. Following the preceding analysis, respondents agreed that employing cloud computing aligns with most of the core business functions within Quantity Surveying firms, as indicated by a mean range of 3.70. The respondents' agreement level with this statement remains notably high.

Next, in terms of level of knowledge regarding the adoption of cloud computing. This indicates the comprehensive familiarity of all respondents with cloud computing. Additionally, the statement topped the respondents' perception regarding cost, which is that cloud computing can curtail additional capital expenses on hardware and software IT support, with a mean range of 3.55. The respondents' confidence in this statement demonstrates moderate agreement, as depicted in Table 3.

The cost aspect statement indicates a moderate agreement among respondent firms regarding their financial capability for cloud computing, with a mean range of 3.30 in Table 4, ranking as the least agreed-upon. However, in contrast, the willingness aspect statement regarding implementing cloud computing to meet customer needs secured the highest position in Table 3, displaying moderate agreement. The mean range of 3.58 in Table 3 supports this perception, indicating trust in using cloud computing for customer satisfaction. However, the statement about finding skilled consultants in cloud computing ranks last in willingness, with a mean range of 3.15, also expressing a moderate level of agreement by respondents.

Table 3. Level of agreement based on average mean

Construct	Items	Mean
Knowledge Aspect	Cloud computing is one of the elements of Industrial Revolution 4.0.	3.98
	The use of cloud computing will increase our firm's productivity.	4.05
	Cloud computing is an effective technology for our firm to adopt.	4.03
	Cloud computing is compatible with many aspects of our firm's core business.	3.70
	Cloud computing is suitable for our firm's working style.	3.93
Cost Aspect	Our firm has the financial resources to adopt cloud computing.	3.30
	Our firm has the financial resources to implement and maintain subscription fees for cloud computing services.	3.33
	Cloud computing enables our firm to reduce overhead costs.	3.23
	Cloud computing can prevent further capital expenditure on hardware, software, and IT support.	3.55
Willingness Aspect	Our firm can find skilled consultants in cloud computing.	3.15
	It is easier to use hosted applications in cloud computing than on-premise ones.	3.33
	A trial period on cloud computing will increase the chances of using it at our firm.	3.40
	Implementing cloud computing can meet customers' needs and requirements.	3.58

The Restraint Factors Against Utilisation of Cloud Computing by Quantity Surveying Firms

Table 4 outlines 14 inhibiting factors regarding the adoption of cloud computing within Quantity Surveying firms. These constraints are evaluated across four aspects: technology, organization, environment, and social. The primary technological constraint is data privacy and security concerns associated with storing data in cloud computing, ranking as the top restraint factor. Respondents show high agreement (average mean of 3.85) due to the vulnerability of data stored in cloud systems accessible via the Internet. The factor indicating third-party data management and the lack of firm control ranks lowest in the technological aspect, with a medium level of agreement (average mean value of 3.20).

Moving to the organizational aspect, the need for technology upgrades and human resources precede cloud computing use, securing the top position as a significant barrier, with an average mean of 3.78. Moreover, the hindrance caused by top management disinterest shows a medium level of agreement (mean value of 2.85).

In the environmental aspect, the absence of transparent government and industry rules to enforce cloud computing ranks as the top constraint, with an average mean of 3.73. Conversely, the construction industry's influence on firms' cloud computing adoption is the lowest in the environmental aspect, with an average mean of 3.20.

Regarding the social aspect, the top restraint factor is the need for staff to learn and be knowledgeable about cloud computing before adoption, achieving a high level of agreement (average mean of 3.85). This indicates the necessity for skill development and technology comprehension in the era of Industrial Revolution 4.0.

Table 4. List of restraint factors against the utilization of cloud computing by quantity surveying firms

Construct	Items	Mean
Technological Aspect	The firm's data and systems are vulnerable to cyber threats or attacks with cloud computing services.	3.58
	Data privacy and information are the primary concerns for storing in cloud computing.	3.85
	Third parties will manage and maintain these cloud computing services, and the firm does not fully control the data.	3.20
	There is no guarantee of agreement and quality of service by cloud computing suppliers.	3.63
	The firm's doubts about transparency and control over the data and information stored in cloud computing are alarming.	3.40
Organization Aspect	Top management is not interested in using cloud computing in the firm.	2.85
	Firms need to upgrade their technologies and provide human resources before using them.	3.78
	It is costly to build cloud computing in a firm.	3.53
Environment Aspect	The condition of the construction industry is restraining the firm's decision to adopt cloud computing.	3.20
	Firms need external support from cloud computing suppliers to meet the firm's needs.	3.65
	There are no clear rules by governments and industries to enforce cloud computing.	3.73
	The firm's data and information stored in cloud computing can be anywhere worldwide, and its location is hidden from users.	3.60
Social Aspect	Staff need to learn and have knowledge of cloud computing before using it.	3.95

CONCLUSION

In conclusion, this study underscores the pivotal role of cloud computing in the evolution of quantity surveying firms to enhance productivity amidst the challenges of the Industrial Revolution 4.0. The findings underscore the profound impact of cloud adoption, not only in mitigating capital expenses but also in addressing organizational constraints and environmental considerations. Nevertheless, the absence of a regulatory framework poses a significant hurdle, underscoring the imperative for governmental and industry intervention. Moreover, the study emphasizes the critical need for staff to acquire cloud computing competencies, highlighting the critical role of training and comprehensive understanding in navigating the demands of Industrial Revolution 4.0. Hence, as the firms chart their course in this digital era, embracing cloud computing emerges as not just a strategic imperative but also a stepping stone for driving innovation, sustainability, and competitiveness in Quantity Surveying firms. At the end of the study, it is essential to acknowledge its limitations. The future study shall underscore the need for a broader scope of respondents and a more robust discussion, potentially incorporating qualitative aspects among other stakeholders in the construction industry to complement these quantitative findings.

AUTHOR NOTE

This article is part of research titled "Cloud Computing Adoption Framework For Construction Firms," funded under the GUP Tier 2 Q.J130000.2652.16J72 by the Universiti Teknologi Malaysia.

Correspondence concerning this article should be addressed to Norhazren Izatie Mohd, Department of Quantity Surveying, Faculty of Built Environment Universiti Teknologi Malaysia, 81300 Johor Bahru, Johor Darul Takzim, Malaysia.

REFERENCES

Al-Badi, A., Tarhini, A., & Al-Kaaf, W. (2017). Financial Incentives for Adopting Cloud Computing in Higher Educational Institutions. *Asian Social Science*, 13(4), 162–174. 10.5539/ass.v13n4p162

Alkhater, N., Walters, R., & Wills, G. (2018). An empirical study of factors influencing cloud adoption among private sector organizations. *Telematics and Informatics*, 35(1), 38–54. 10.1016/j.tele.2017.09.017

Alkhater, N., Wills, G., & Walters, R. (2015, August). Factors affecting an organization's decision to adopt cloud services in Saudi Arabia. In 2015 3rd International Conference on Future Internet of Things and Cloud (pp. 553–557). IEEE, Rome, pp. 553–557.

Alshamaila, Y., Papagiannidis, S., & Li, F. (2013). Cloud computing adoption by SMEs in the north east of England: A multi-perspective framework. *Journal of Enterprise Information Management*, 26(3), 250–275. 10.1108/17410391311325225

Catalina, I. S. (2014). *Cloud computing - Impact on business*. [Master thesis, Aalborg University Copenhagen Denmark].

Chandrasekaran, K. (2014). *Essentials of cloud computing*. CRC Press. 10.1201/b17805

European Commission. (2016). *Digitizing European Industry—Reaping the Full Benefits of a Digital Single Market COM (2016) 180 Final*. EC.

Goode, S., & Stevens, K. (2000). An analysis of the business characteristics of adopters and non-adopters of World Wide Web technology. *Information Technology and Management*, 1(1), 129–154. 10.1023/A:1019112722593

Jaeger, P. T., Lin, J., Grimes, J. M., & Simmons, S. N. (2009). Where is the cloud? Geography, economics, environment, and jurisdiction in cloud computing. *First Monday*, 14(5). 10.5210/fm.v14i5.2456

Jain, L., & Bhardwaj. (2010). Enterprise cloud computing: Key considerations for adoption. *International Journal of Engineering and Information Technology*, 2(2), 113–117.

Jiao, Y., Wang, Y., Zhang, S., Li, Y., Yang, B., & Yuan, L. (2013). A cloud approach to unified lifecycle data management in architecture, engineering, construction and facilities management: Integrating BIMs and SNS. *Advanced Engineering Informatics*, 27(2), 173–188. 10.1016/j.aei.2012.11.006

Kagermann, K., Helbig, J., Hellinger, A., & Wahlster, W. (2013). *Recommendations for implementing the strategic initiative Industrie 4.0: Securing the future of German manufacturing industry; final report of the Industrie 4.0 Working Group*. Forschungsunion.

Khan, K., & Malluhi, Q. (2010). Establishing trust in cloud computing. *IT Professional*, 12(5), 20–27. 10.1109/MITP.2010.128

Kim, W. (2009). Cloud computing: Today and tomorrow. *Journal of Object Technology*, 8(1), 65–72. 10.5381/jot.2009.8.1.c4

Leung, K., Bhagat, R. S., Buchan, N. R., Erez, M., & Gibson, C. B. (2005). Culture and international business: Recent advances and their implications for future research. *Journal of International Business Studies*, 36(4), 357–378. 10.1057/palgrave.jibs.8400150

Li, M., Zhao, D., & Yu, Y. (2015). TOE drivers for cloud transformation: Direct or trust-mediated? *Asia Pacific Journal of Marketing and Logistics*, 27(2), 226–248. 10.1108/APJML-03-2014-0040

Lian, J. W., Yen, D. C., & Wang, Y. T. (2014). An exploratory study to understand the critical factors affecting the decision to adopt cloud computing in Taiwan hospital. *International Journal of Information Management*, 34(1), 28–36. 10.1016/j.ijinfomgt.2013.09.004

Luxner, T. (2023, May 18). Cloud computing stats: Flexera 2023 State of the cloud report. *Flexera Blog*. https://www.flexera.com/blog/cloud/cloud-computing-trends-flexera-2023-state-of-the-cloud-report/

Marston, S., Li, Z., Bandyopadhyay, S., Zhang, J., & Ghalsasi, A. (2011). Cloud computing—The business perspective. *Decision Support Systems*, 51(1), 176–189. 10.1016/j.dss.2010.12.006

Mell, P., & Grance, T. (2011). *The NIST Definition of Cloud Computing. Computer Security Division Information Technology Laboratory National Institute of Standards and Technology Gaithersburg*. National Institute of Standards and Technology.

Microsoft. (2018). *What is cloud computing? A beginner's guide*. Azure. https://azure.microsoft.com/en-us/resources/cloud-computing-dictionary/what-is-cloud-computing/

Moktadir, M. A., Ali, S. M., Kusi-Sarpong, S., & Shaikh, M. A. A. (2018). Assessing challenges for implementing Industry 4.0: Implications for process safety and environmental protection. *Process Safety and Environmental Protection*, 117, 730–741. 10.1016/j.psep.2018.04.020

Nkhoma, M. Z., Dang, D. P., & De Souza-Daw, A. (2013, May). Contributing factors of cloud computing adoption: A technology-organization-environment framework approach. In *Proceedings of the European Conference on Information Management & Evaluation* (Vol. 2, No. 1, pp. 180-188). 10.24924/ijise/2013.04/v1.iss1/30.41

Ramdani, B., Kawalek, P., & Lorenzo, O. (2009). Predicting SMEs' adoption enterprise systems. *Journal of Enterprise Information Management*, 22(1/2), 10–24. 10.1108/17410390910922796

Sallehudin, H. A. S. I. M. I., Razak, R. C., Ismail, M. O. H. A. M. M. A. D., Fadzil, A. M., & Baker, R. O. G. I. S. (2019). Cloud computing implementation in the public sector: Factors and impact. *Asia-Pasific Journal of Information Technology and Multimedia, 7*(2-2), 27-–42.

Shailja Tripathi, D. (2023). *Digital Transformation and Cloud Computing: System Dynamics Modeling Approach*. International Journal For Multidisciplinary Research.

Sulaiman. (2017). Pengaruh Teknologi Komunikasi Terhadap Perubahan Keorganisasian Di Jabatan Penyiaran Malaysia (Effect of Changes in Organizational Communication Technology in the Broadcasting Department of Malaysia). *e-Bangi, 12*(1).

Sun, Y., Wang, Z.-J., Deveci, M., & Chen, Z.-S. (2024). Optimal releasing strategy of enterprise software firms facing the competition from cloud providers. *Expert Systems with Applications*, 236, 121264. 10.1016/j.eswa.2023.121264

Takabi, H., Joshi, J. B., & Ahn, G. J. (2010). Security and privacy challenges in cloud computing environments. *IEEE Security and Privacy*, 8(6), 24–31. 10.1109/MSP.2010.186

Yusof, M. F. M. (2017). *Revolusi Perindustrian Keempat / Industri 4.0 / IR 4.0.*

Chapter 8
Comparative Analysis of Ecological Footprint in Five Cities of India:
A Pandemic Analysis

Neha Pranav Pranav Kolhe
https://orcid.org/0000-0001-6981-5863
Maulana Azad National Institute of Technology, India

Surabhi Mehrotra
https://orcid.org/0000-0002-6301-6603
Maulana Azad National Institute of Technology, India

Aditi Parihar
Maulana Azad National Institute of Technology, India

ABSTRACT

COVID-19, a new respiratory disease identified in December 2019, is one of the most recent risks to human health around the world. The purpose of this research is to examine the effects of COVID-19 on the ecological footprints of five cities across India: Delhi, Bhopal, Bareilly, Betul, and Razole, using the global footprint networks calculator tool. In this comparative analysis, six major factors were employed to examine EF which were examined through food, shelter, transportation, commodities, and services. Further, statistical test - T test and ANOVA analyse the difference ecological footprints in five cities. A total of 15 questions derived from six factors of land use type have been considered for the analysis using primary questionnaires collected via online google forms. The study presents a change in the ecological footprint through exploratory study of different cities with varying population sizes in the times of COVID-19. Results show significant difference in ecological footprint among the cities with maximum variation observed in class I cities like Delhi.

DOI: 10.4018/979-8-3693-1650-4.ch008

INTRODUCTION

From the emergence of human civilization, there has been a need for food, shelter and home. Human beings with a span of time began to exploit nature for their own benefit. Technological advancements and emerging technologies are frequently used to help harmonize both urban and rural environments (Kudryavtseva et al., 2021). To fulfil the demand of rapid urbanization, the global climate is deteriorating and thus the quality of life in urban areas (Mehrotra et al., 2020). Globalization along with subsequent growth of economies has significantly led to the rise in carbon emissions (Shahbaz et al., 2018). Globally, the current pace of urbanisation in megacities leads to the evolvement of natural ecosystems as environment roles change (Sharma et al., 2020). Given the complex consequences of excessive human use of natural resources plus following land use changes, ecosystem services remain particularly vulnerable to a variety of influences (Nasiri et al., 2020). The present worldwide environmental issue has sparked interest in learning how individuals could become more conscious of the environmental effects of human actions (Villarroel et al., 2017). Infectious diseases have already altered the architecture, design, and urban planning of our cities (Megahed & Ghoneim, 2020). The ongoing coronavirus disease, known as COVID-19 2019, has been declared a pandemic that has resulted in mayhem on people, their behavior, lifestyle and society at large (Echegaray, 2021; Yang et al., 2020). To promote social distancing and prevent infection of coronavirus disease, national and local governments worldwide have announced emergencies, encouraged ed safer-at-home directives, and mandated business closures (Mofijur et al., 2021; Weill et al., 2020). This isolated billions of people around the world and less local and global traffic, influencing nature in a variety of ways (Corlett et al., 2020). During this period of crisis, limited human contact with nature has proven to be a protection for nature and the environment (Lokhandwala & Gautam, 2020). Lowered energy demand as well as corresponding greenhouse gas (GHG) and additional pollutants emissions result from fewer vehicles on the road and congestion (Khursheed et al., 2020). Following this, there was a substantial depletion in greenhouse gas emissions on account of decreased transportation, manufacturing, schooling, as well as several other activities; however, when represented that this was insufficient to minimize pollution in the air for all contaminants (Eroğlu, 2020a).

It is well established that the virus has no direct influences on the surroundings or the energy sector individually (Eroğlu, 2020). However, the pandemic has impacted humans and their interaction with the ecological processes in various ways. Human impact on ecology was first measured by the Ecological Footprint (EF) - a tool to calculate human demand for natural resources, showing, the aggregate of nature required to sustain humans or an economy (Bazan, 1997; M. Wackernagel et al., 1999). It helped to understand issues like resource use limits, and the international dealing out of the natural resources of the planet (Borucke et al., 2013). It included strategies for ensuring the sustainability in the long run of environmental resources utilised worldwide the spread of the virus has unquestionably affected the Ecological Footprint. Carbon Footprint (CF) - another tool that evaluates the number of greenhouses especially carbon dioxide that are released by human activities in the atmosphere of the area, locally and globally. The sector-wise studies of Food, Shelter, Mobility, Goods, and Services have been conducted but a holistic study of the Ecological Footprint and Carbon Footprint pertaining to COVID-19 still lacking.

The Global Footprint Network by Wackernagel has risen to prominence in carrying out this procedure via the enhancement of the National Footprint Accounts including the current Footprint Standards method because it keeps updating the data. The primary objectives of this research are to use a structured approach for calculating Ecological Footprint along with Carbon Footprint, to look at the effects of the pandemic on society. Secondly, the study aims to examine changes in the lifestyle and resource consumption of people before and after COVID-19, and their collective impact. The calculations of Ecological Footprint and Carbon Footprint are crucial for understanding our impact on Earth, and also for analyzing the changes required in lifestyle, home, travelling patterns, etc. The study would explore the variation in 5 different classes of Indian cities. Further, the t-test was used to determine whether there is a significant difference between the five cities in order to determine whether there is a difference between the before and after EF values for each city. In all cities, the tests revealed a significant difference (p 0.05). However, the greatest change in EF and CF values was recorded in Delhi.

RELATED WORK

The Ecological Footprint is a concept which calculates the aggregate natural resources required for people and the economy to sustain (Meena & Yadav, 2019). This accounting tool makes it possible to estimate a given human population's or economy's resource usage and waste integration demands in order of a corresponding productive land area (Bazan, 1997). The Ecological Footprint is a land-based representation of the society's demands on natural resources (Rees & Wackernagel, 2012). The Ecological Footprint represents human demand, and biocapacity, represents the supply of the ecosystem, the pair is measured in global hectares, which are hectares comparable to the overall productivity of each and every bio-productive hectare on Earth. It is viewed as a straightforward operational measure for tracking progress toward (un)sustainability, i.e., the preservation (loss) of natural resources. An EF may be calculated for individuals, events, or areas, varying in size from one city to the entire globe (Bergh & Verbruggen, 1999). The EF methodology builds on previous environmental impact evaluation methods such as accounting for primary production, resource and 'Emergy' accounting, carrying capacity appraisal, and lifespan consideration, while being a unique type of SD indicator (Mathis Wackernagel & Yount, 2000). Emergy assessment was used to assess carrying capacity in order to improve problem diagnosis and make sustainability more understandable, thereby assisting in the formulation of public policies (Nakajima & Ortega, 2016).

Ecological Footprint analysis has been widely used at national and local levels in many countries (Barrett et al., 2005). It depicts natural resource requirements with regard to a theoretically similar land/sea (biosphere) region, along with the scale of the footprint (additionally known as 'appropriated carrying capacity') to a community along with a set period of time (usually one year) that is determined by the population's way of life (Hunter & Shaw, 2007). The Ecological Footprint's strength comes from the fact that it clusters and transforms usually composite resource use trends into a single number (Costanza, 2000).

Long-term environmental damage and disruption a significant problem that poses a direct threat to human life and non-human life on Earth (Jorgenson, 2003). Every individual has an effect on the planet, no matter how small or large, because humans consume natural products and services. As per M. Wackernagel et al. (1999), the following are two basic facts that are used to calculate Ecological Footprint:

i. The majority of people's consumption and most of the waste they produce can be tracked.
ii. It is possible to quantify the biologically productive areas required so that the processing of this utilization and the integration of waste.

Six underlying assumptions underpin Ecological Footprint accounting as per Wackernagel et al (2002):

i. People can monitor the majority of the resources they use and the waste they produce.

ii. Biologically productive areas essential to sustain the majority of such resources and waste flows could be determined. The calculation excludes flows of resources and waste that cannot be calculated, resulting in a systemic underestimation of mankind's true Ecological Footprint.
iii. Various sorts of regions can be transformed into global hectares, hectares with the planet's estimated bio productivity, by weighting every region in ratio to its bio productivity.
iv. Since a common global hectare shows a common usage and all global hectares in any given year show a similar quantity of bio productivity, a collective measure of Ecological Footprint or biocapacity can be calculated by adding them together.
v. When both are measured in global hectares, human demand, as measured by the Ecological Footprint, is comparable straightaway to nature's availability, and biocapacity.
vi. If pressure on an ecosystem exceeds the ecosystem's regenerative ability, the area required can exceed the area provided. Overshoot is a condition in which the Ecological Footprint exceeds the available biocapacity.

Six key categories for ecologically beneficial areas are distinguished in the study of Ecological Footprint: cropland, pasture, forest, water area, and built-up along energy land (Zhao et al., 2005). It was designed as a planning instrument to compare the utilization of various populations using a single-point indicator representing suitable carrying capacity, so as to compare these populations with regard to sustainability (Lenzen & Murray, 2001). Nevertheless, much knowledge about the impacts on regional ecosystems is neglected in the current transformation to world-average productivity (Lenzen & Murray, 2001). The Ecological Footprint measures how people use natural resources by contrasting resource utilization and waste production to the planet's regenerative potential (M. Wackernagel et al., 1999). To achieve a sustainable, just, and efficient economy, monitoring the distribution of natural resource usage would be important (White, 2007).

Coronaviruses are RNA viruses that infect both people and animals, affecting the respiratory, digestive, and central nervous systems (Cui et al., 2019). The most contemporary outbreak was identified in Wuhan, China, and has since then extended across China and several other nations (Huang et al., 2020). The COVID-19 disease has been humanity's most serious threat in a century (Jeyanathan, et al.). COVID-19 is characterized by viral pneumonia symptoms including fever, fatigue, dry cough, and lymphopenia, according to the literature (Cucinotta & Vanelli, 2020). Many elderly patients who become critically ill have proof of underlying illnesses such as heart failure, liver disease, kidney disease, or malignant tumors (Cucinotta & Vanelli, 2020). As successful preventive measures to stop the SARS-CoV-2 virus from spreading, handwashing, using hand sanitizers, and routine disinfection of high-touch surfaces, along with wearing face masks, are all encouraged (Kumar & Morawska, 2019). We live in a society which is highly integrated in terms of public, commodities, and food, and also in compact societies (Watkins, 2020). The death rate continues to increase, and many countries have been forced to implement social distancing and lockdown measures (Yuki et al., 2020).

The coronavirus outbreak has forced many people to reconsider their lifestyles and goals, resulting in significant changes in how they go about their daily lives (Workie et al., 2020; Shamshiripour et al., 2020). The virus's new condition can make it difficult to maintain a balanced and varied diet, as well as daily exercise. For example, reduced accessibility to everyday grocery shopping can cause a reduction in fresh food intake, particularly fruit, vegetables, and fish, which ultimately advocate extremely processed foods like pre-packaged meals, unhealthy snacks treat, and fully prepared cereals, which are high in calories, sugars, and salt (Di Renzo et al., 2020). There may be a connection between the intake of extremely palatable foods, for example, ultra-processed foods, and a disruption in innate and adaptive immune systems' temporal coordination (Renzo et al., 2020). According to studies, people in the UAE saw a negative shift in lifestyle, unbalanced eating habits, a decline in regular exercise, and facing problems both emotionally and psychologically because of the COVID-19 pandemic (Ismail et al., 2020).

Long-established habits like travelling to work and in-store shopping have been substituted by work-from-home and online shopping. Many of these changes have been in the works for some time, but the pandemic has intensified them dramatically (Shamshiripour et al., 2020). COVID-19 fear was also found to be linked to non-negligible changes in the likelihood of staying at home and reducing trip frequency. Since shopping for essential goods was necessary for survival it was done on a regular basis, despite the potential risk. As a result, it's possible that people would choose to defend themselves by wearing masks, physically separating themselves from others in shops, and so on, rather than reducing their shopping frequency (Parady et al., 2020).

Several contaminants, the majority of which come from road traffic, have a significant effect on air quality (Marino et al., 2016). Public transportation vehicles are enclosed spaces that allow infectious diseases to spread from person to person (Shen et al., 2020). As a result, several countries have recorded several outbreaks of cases of respiratory virus infections in public transportation vehicles, as well as SARS-CoV-2 (Shen et al., 2020). Unprecedented limits on public gatherings and movement have been followed by frightening, and sometimes clashing political announcements. As public transportation has been related to a variety of health issues, government officials recommended that its usage have to be limited. This has resulted in an observable change toward the use of privately owned vehicles and the implementation of mobility by micro-level initiatives such as pedestrianization, bike network expansion, and road closures (Gutiérrez et al., 2020). This COVID-19 pandemic has wreaked havoc on all modes of transportation, both public and private, on a national and international scale (Wielechowski et al., 2020). The rise of something like the coronavirus pandemic, in addition to many other problems, further devastated major cities, prompting many implications for the future of sustainability (Bhattacharjee & Sattar, 2021). The current work is therefore needed to explore the cumulative impacts of COVID-19 on people's behavior in urban areas.

STUDY AREA

India is a diverse country in terms of cultural, social and economic value. People of different cities live a different type of lifestyle. The preferences for the use of resources by people change based on their availability. The lifestyle of a person living in a metropolitan city contrasts to the of one living in the village. The Ecological Footprint analysis has been carried out at 5 urban settlements shown in Fig. 1, as per URDPFI Guidelines (Town and Country Planning Organisation; Ministry of Urban Development; Government of India, 2014) based on population (refer to Table 1). By taking these 5 distinct classes of urban settlements we aimed to incorporate the socio-economic and cultural variations and analyse the Ecological Footprint.

Figure 1. Location of cities on the map of India

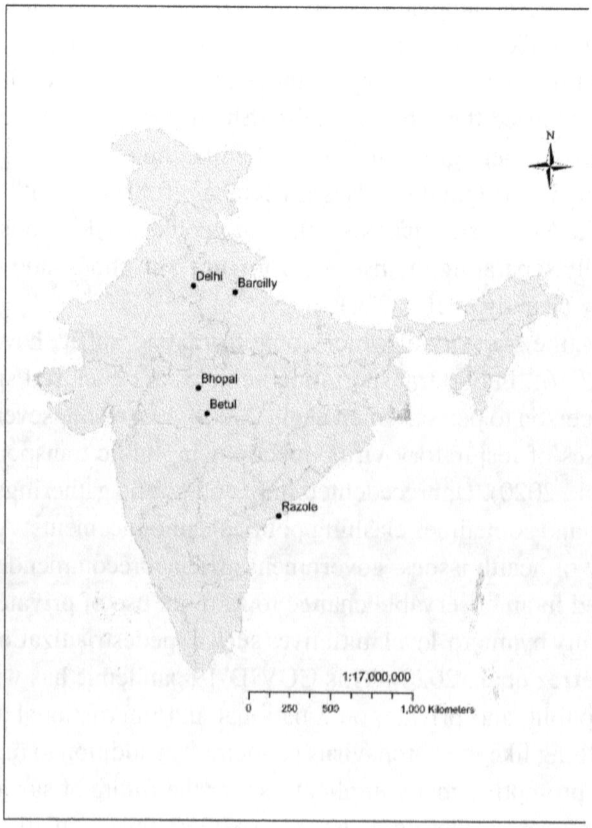

Table 1. Cities under the study with details (source: census 2011)

Particulars	Razole	Betul	Bareilly	Bhopal	Delhi
State	Andhra Pradesh	Madhya Pradesh	Uttar Pradesh	Madhya Pradesh	Union Territory
Location	16.4743° N, 81.8402° E	21.92° N, 77.9° E	28°10′ N, 78°23′ E	23.2599° N, 77.4126° E	28.7041° N, 77.1025° E
Total Population	13,597	1,03,330	9,03,668	17,98,218	1,10,34,555
Male Population	6,585	52,823	476,927	936,168	5,882,117
Female Population	7,012	50,507	426,741	862,050	5,152,438
Native Language	Telegu	Hindi	Hindi	Hindi	Hindi
Administrative Body	Razole Gram Panchayat	Betul Municipal Corporation	Bareilly Municipal corporation	Bhopal Municipal Corporation	Delhi Municipal Corporation
Class of Urban Settlement	Small Town	Medium Town	Large city	Metropolitan city	Megapolis

METHODOLOGY

It is deductive research, where the inferences are derived from primary data for sufficient evidence to evaluate the conclusion, and the rejection or corroboration of the conclusion is all part of deductive research (Baker, 2001). The primary data is collected through telephonic interviews due to the pandemic situation. The reliable survey questionnaire was taken into consideration. Over time, reliable questionnaires provide consistent findings from several samples and different researchers (Boynton & Greenhalgh, 2004). A total of 15 questions were elicited from 6 parameters of land use type considered. Types of land use in Ecological Footprint are mainly cropland, grazing land, fishing grounds, forest area, built-up land and carbon land.

The data was collected in 2 phases - before COVID-19 and during COVID-19. The set of questions was kept the same in order to gain insights into the change in the scenario. A question or series of questions must be asked in a simple, understandable, and acceptable manner so that the respondents can easily compose, express, and communicate their responses (Stone, 1993). Eventually, 50 samples from each city were collected making total samples as 250 of the 5 different cities. The interviews were transcribed, and a thematic analysis was conducted. This involved the categorization of 6 main parameters for analysis (see Fig. 2), that were important in understanding the Ecological Footprint of the area. The interviews gave an in-depth understanding of the participant's lifestyle and needs from nature.

Figure 2. Research methodology

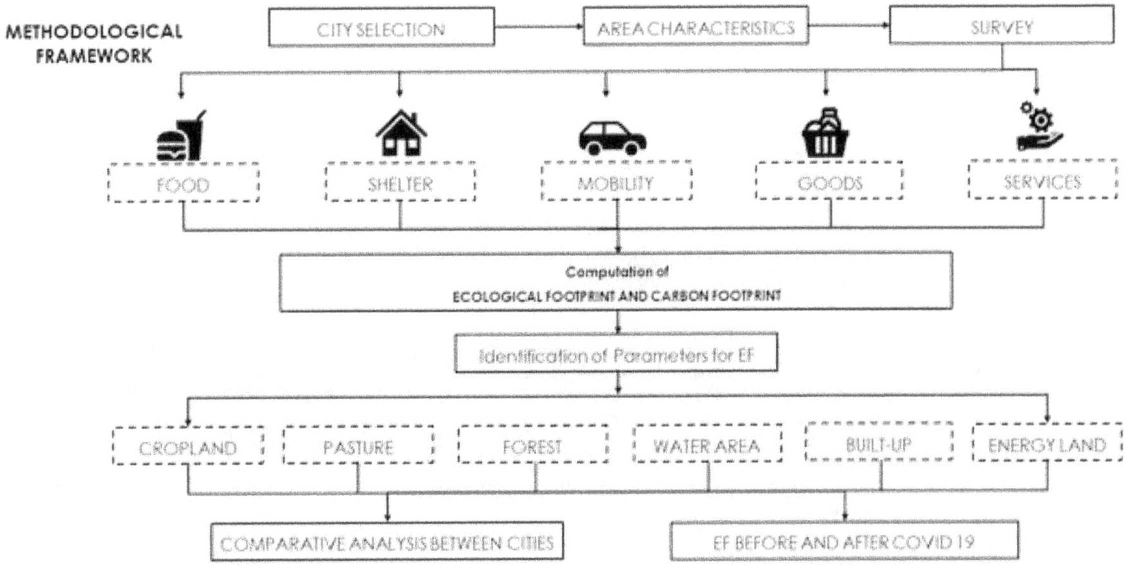

By taking into use the Ecological Footprint calculator by M. Wackernagel et al., (1999), each sample's Ecological and Carbon Footprint was measured. The details were extracted from Global Footprint Network. It released Ecological Footprint Explorer in April 2017 as an open data platform to calculate Ecological Footprints and explore national footprints. A t-test has been used when comparing the means of two groups i.e., pairwise comparison for per COVID-19 and during COVID-19 study for each city individually. Moreover, to compare all 5 cities together, and to do multiple pairwise comparisons, the ANOVA test has been used.

RESULTS

Earths Needed

The Ecological Footprint, a tool for calculating how much resource is used along with the planet's regenerative potential (or "biocapacity") a particular action demands (Lin et al., 2018). It measures the earth's sustainability. The earths needed shows that if we continue to live the same lifestyle and use resources in the same way then how many earths we will require. It is impossible to have other earths and hence it acts as an alarm for us to use our resources judicially to save for future needs.

Figure 3. Graph showing earths needed by the study areas

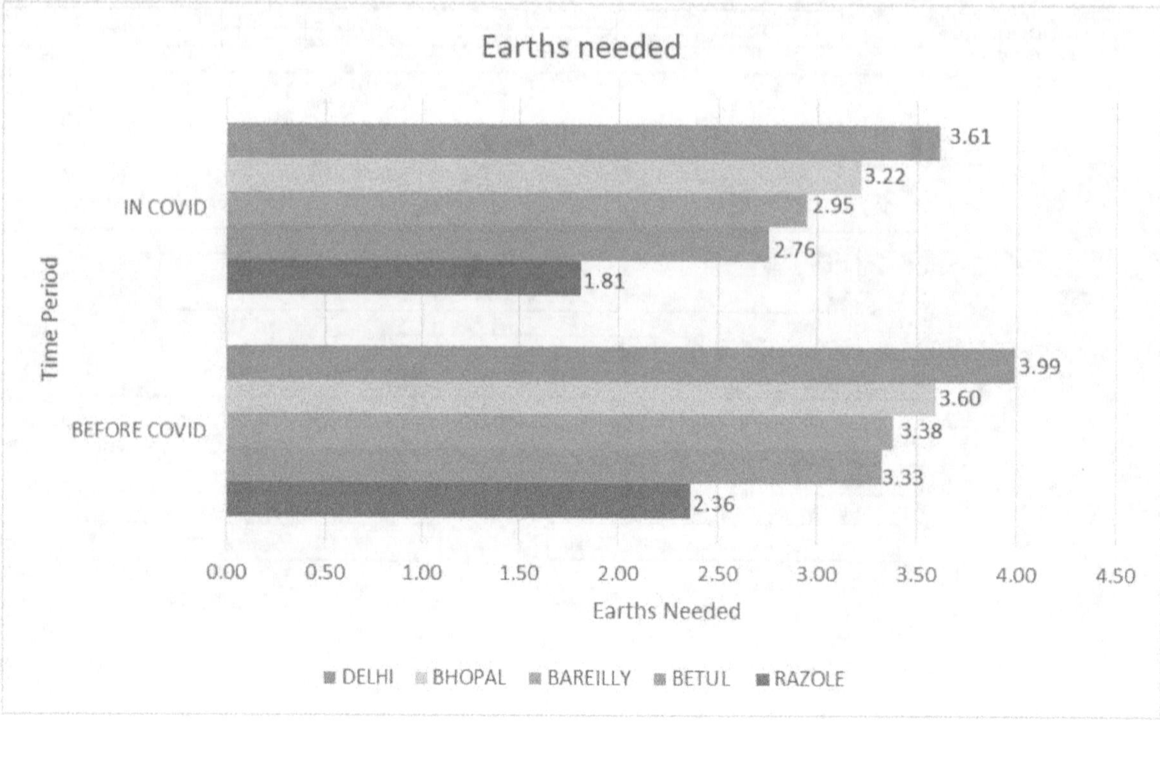

It was observed that there has been a decline in the earths needed in each class of city during the times of COVID-19 as people have been using resources wisely. As of now, if we live the lifestyle same as we lived during the COVID-19 pandemic, people of Razole will need 1.81 earths, Betul will need 2.76 earths, Bareilly will need 2.95 earths, Bhopal will need 3.22 earths and Delhi will need 3.61 earths as shown in Figure 3.

Ecological Footprint

It shows reflection of what humans expect from the environment, or the amount of natural resources needed to sustain humans or an economy (Meena & Yadav, 2019). It keeps track of demands using an environmental accounting system. The distinction between an individual's biologically productive area and the biologically productive area accessible of a country or around planet (biocapacity). In a nutshell, this is a calculation of human effects on the nature and environment along with the human economy's reliance on natural resources. The EF has been widely praised as a useful heuristic and pedagogic tool for communicating current total human resource usage in a way that almost all can understand (Costanza, 2000). This approach measures a specific ecological condition for sustainability by contrasting human influence with the earth's limited bio productive area (Holmberg et al., 1999).

Figure 4. Graph showing ecological footprint by the study areas

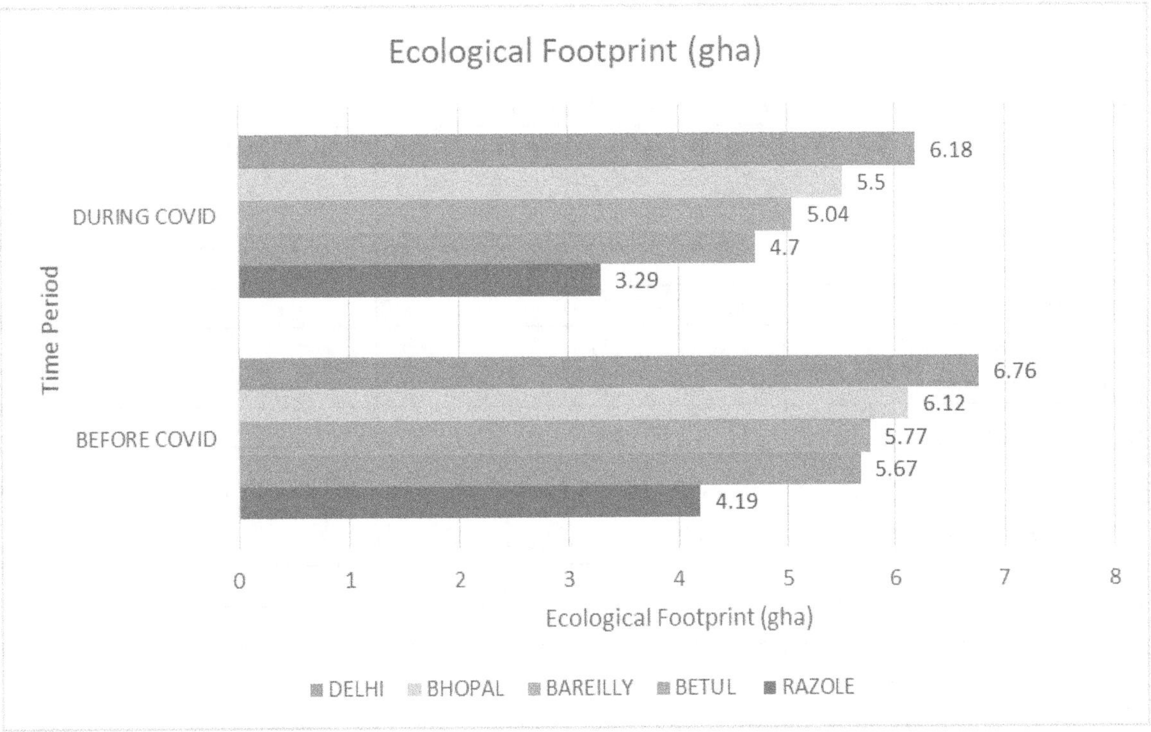

It can be clearly noted that the Ecological Footprint of cities with large population is more as compared to that with less population (refer Fig. 4). In the times of COVID-19 the Ecological Footprint of all the types of cities have reduced significantly. The Ecological Footprint of Delhi, before COVID-19 was 6.76 gha while in COVID-19 it reduced to 6.18 gha.

Stratification and analysis of covariance are two techniques that can be applied. Data that are organically coupled form useful strata and frequently allow for more exact comparisons than data from unrelated subjects (Hsu & Lachenbruch, 2014). This results in the paired t test. Further to understand the change of all the 5 cities t test has been applied to the collected data. This provides a validation regarding the hypothesis generated.

Table 2. T-Test: Paired two sample for means, Delhi, India

	During COVID-19	Before COVID-19
Mean	6.75	6.17
Variance	3.69	3.57
t Stat	3.16	
P(T<=t) one-tail	0.001	
t Critical one-tail	1.69	
P(T<=t) two-tail	0.003	
t Critical two-tail	2.045	

Table 3. T-Test: Paired two sample for means, Bhopal, India

	During COVID-19	Before COVID-19
Mean	6.11	0.79
Variance	2.58	0.039173
t Stat	22.11	
P(T<=t) one-tail	0.05	
t Critical one-tail	1.68	
P(T<=t) two-tail	0.05	
t Critical two-tail	2.02	

Table 4. T-Test: Paired two sample for means, Bareilly, India

	During COVID-19	Before COVID-19
Mean	5.77	5.04
Variance	3.50	2.77
t Stat	4.0	
P(T<=t) one-tail	0.05	
t Critical one-tail	1.71	
P(T<=t) two-tail	0.05	
t Critical two-tail	2.06	

Table 5. t-Test: Paired two sample for means, Betul, India

	During COVID-19	Before COVID-19
Mean	5.67	4.7
Variance	1.35	0.62
t Stat	7.68	
P(T<=t) one-tail	0.05	
t Critical one-tail	1.71	
P(T<=t) two-tail	0.05	
t Critical two-tail	2.06	

Table 6. t-Test: Paired two sample for means, Razole, India

	During COVID-19	Before COVID-19
Mean	38.9	63.4
Variance	452.1	422.93
t Stat	-2.12	
P(T<=t) one-tail	0.03	
t Critical one-tail	1.83	
P(T<=t) two-tail	0.05	
t Critical two-tail	2.26	

The analysis showed that the P(T<=t) value for all the 5 cities that is Delhi (0.003594031) (Table 2), Bhopal (0.05) (Table 3), (Bareilly (0.0005) (Table 4), Betul (0.05) (Table 5) and Razole (0.06) (Table 6) is less than 0.05 which shows that the test results are considerable for analysis. Moreover, it can also be observed from the tables that the mean and variance of all the cities differ before COVID-19 and during COVID-19.

Carbon Footprint

The land area required to soak all of the CO2 released by humans in the course of its lifetime is referred to as the Carbon Footprint (Pandey et al., 2011). Many nations are now under enormous pressure to cut CO2 emissions (Bosupeng, 2017). This 'Carbon Footprint,' which is generally calculated in tonnes of CO2 equivalents, is a component of comprehensive Ecological Footprint study (Kitzes & Wackernagel, 2009). In several developing countries, transportation is among the major contributors to total greenhouse gas emissions. Since food items often commute long distances to finally reach to the consumers, one would look forward to transportation as the primary source of greenhouse gas releasing from the food related industry (Wakeland et al., 2012).

Figure 5. Graph showing carbon footprint by the study areas

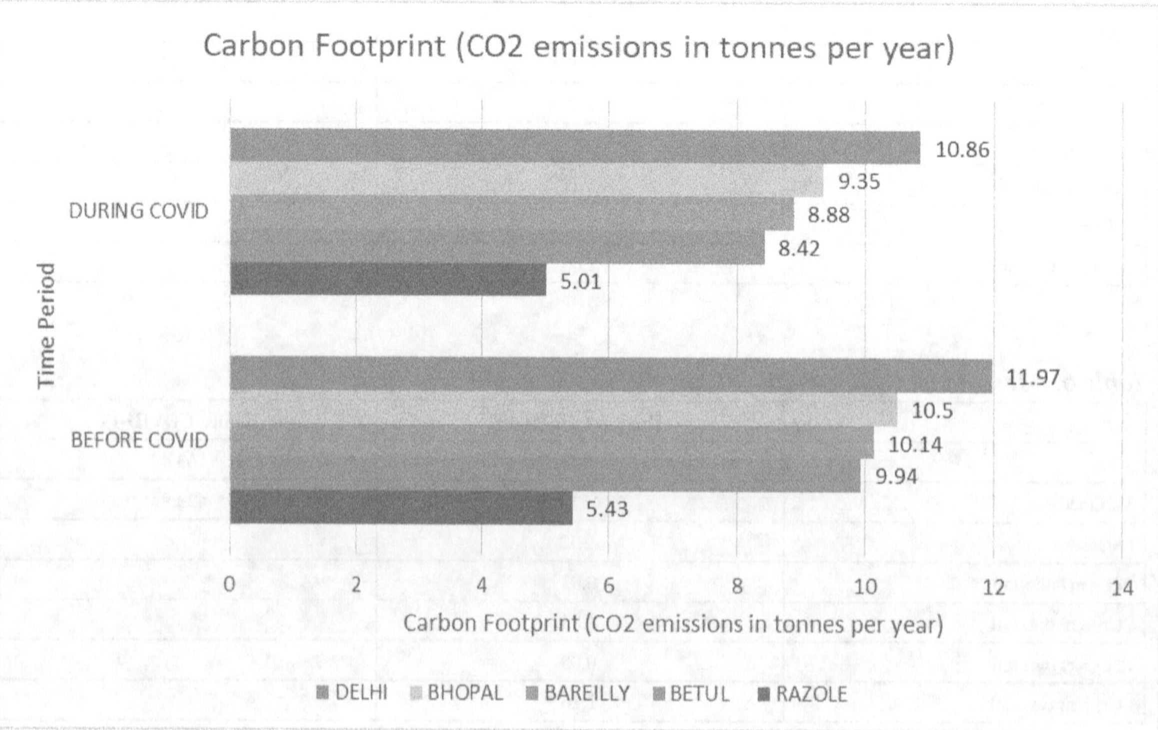

The Carbon Footprint holds a prime component while calculating Ecological Footprint. The Carbon Footprint of people has decreased during COVID-19 due to various reasons like subsequent lockdowns, work from home, etc. The Carbon Footprint of Delhi reduced from 11.97 CO2 Emissions in tonnes per year to 10.86 CO2 Emissions in tonnes per year similarly in Bhopal it fell from 10.50 CO2 Emissions in tonnes per year to 9.35 CO2 Emissions in tonnes per year (refer Fig. 5).

Comparison of EF and CF for All the Cities

As observed while analysing the cities individually that more of the metropolitan cities are affected due to the spread of virus and subsequent change in lifestyle. Moreover, the higher number of cases were observed in large urban areas. The same hypothesis can be seen when ANOVA test was applied to the result for analysing the major change in different cities.

Table 7. Summary of ANOVA test

Groups	Sum	Average	Variance
Razole	47.6	3.4	1.73
Betul	63.2	4.514	0.61
Bareilly	78.6	5.61	4.05
Bhopal	142.4	10.17	2.62
Delhi	85.7	6.12	4.32

Table 8. ANOVA test for ecological footprint after COVID-19

Source of Variation	SS	df	MS	F	P-value	F crit
Between Groups	371.35	4	92.84	34.80	0.05	2.52
Within Groups	173.43	65	2.67			
Total	544.78	69				

The ANOVA test allowed for a comparison of all five cities at the same time in order to establish whether there was a relationship between them. The F statistic (also known as the F-ratio) produced by the ANOVA formula allowed for the examination of 5 cities to assess the variability between groups and within groups. The major variance (4.32489) can be observed in Delhi which is a metropolitan city with a population of millions (refer Table 7). The results also show an f critical value of 2.41304 which indicates that there is a difference between the EF of the cities after COVID-19 (refer Table 8) as the result of the ANOVA's F-ratio statistic is not close to 1.

DISCUSSIONS

Results show major changes have been observed to account for the 6 parameters influencing the Ecological Footprint of a city. The t-test was used to understand the difference between the before and after EF values for each city. The tests showed significant differences in before and during COVID-19 ($p < 0.05$) in all cities. To understand whether there exists a significant difference amongst the 5 cities, having different classes, one-way ANOVA was employed. Each parameter is discussed in detail below:

Eating Habits

Due to COVID-19 people prefer to eat food that is prepared at home rather than eating outside food due to safety constraints. The myth that people can suffer from coronavirus by eating animal-based products like meat, although do not have any scientific proof has altered the eating habits of people and they tend to avoid it as of now.

The food sector's long-term viability is being challenged as the world's population grows and buying power rises, as environmental externalities, primarily greenhouse gas emissions, result from consumption (Pais et al., 2020). Increased dietary Carbon Footprint before COVID-19 was also linked to higher recorded intakes of beef, butter, minced beef, meat, and seafood, as well as lower intakes of potatoes, rice, and pasta during the study period than before COVID-19 (Hjorth et al., 2020). Reduced meat and dairy consumption is proposed to be a feasible mitigation alternative recommended by the IPCC (Frenette et al., 2017). When eating out, meat is often preferred over vegetarian dishes, a choice that many people have momentarily abandoned (Attwood & Hajat, 2020). As a result of COVID-19, people's eating patterns have changed. People are choosing not to consume animal-based products for a variety of reasons, the most common of which is a fear of infection. SARS-CoV-2, a novel beta-coronavirus, is thought to have originated from bats in Wuhan in 2019 (Kaur & Gupta, 2020). Attwood and Hajat (2020) suggested that, if rapid improvements in eating habits due to COVID-19 are maintained in the long run, a significant chance to arise as a result of the ongoing pandemic, there could be a shift away from meat consumption, which could have long-term benefits on earth resources.

Household Size

Owing to the national lockdown, the number of members of a household has also increased. In many households, the children studying in other cities have returned back to their homes and thus this has increased the household size. In contemporary society, the family and its members in the house are by far the most important socioeconomic entities (Bongaarts, 2001). Household size, or the number of family members living under the same roof increased suddenly, as many people across India returned back that included students, travellers, pilgrims, and migrant workers as a result of the unexpected lockdown (Srivastava, 2020). This has impacted the household-level Ecological Footprint in most of the studied cities.

Usage of Public Transport

Observations from the survey indicate that people from different cities do not prefer to use public transport like the bus or metro. The major reason behind it is fear of the spreading of coronavirus due to a lack of social distancing. The COVID-19 pandemic has had a detrimental influence on how people work and move in cities, as well as the community as a whole. Following the halt in regular daily life due to the pandemic, many people preferred to use private vehicles over public transport due to concerns about sanitation and social distancing (Shakibaei et al., 2021). Movement restrictions, such as the closing of non-essential facilities and stay-at-home orders, played a role in the drop-in operation, although to a minor degree (Brinkman & Mangum, 2020). Owing to an increase in working remotely, e-learning, and a drop in the number of public activities and events, travel demand could decline as a result of social distancing. People are more likely to participate in activities at home with their families or close friends leads to less pollution during rush hour. However, the dramatic drop in demand for public transit as a result of COVID-19 infection raises many concerns about the potential viability of urban mobility in future (Tirachini & Cats, 2020).

Flying Pattern

The responses of the people are taken during the time of the COVID-19 pandemic and so there has been a change in lifestyle of people. With the forced lockdowns travelling by airflight was reduced to only extremely necessary trips. For resuming air travel, risk evaluation was needed, which entails determining the traveller's baseline risk, identifying factors that could change that threat, measures to reduce that threat, and taking into account the traveller's perceived risk and tolerance (Khatib et al., 2020). COVID-19 had negative impacts on the overall travel and tourism and travel industry (Chua et al., 2020).

Carpooling Behaviour

Most people were not inclined towards carpooling due to the increased risk of getting the coronavirus and their comfort level. The results from the study by Chen, Y., Jiao, J., Bai, S., & Lindquist, J. (2020) indicate higher rates of COVID-19-positive cases can be linked to clinical density, urban green density, average distance travelled, sex ratio, and travelling (walking, carpooling, and public transportation). The major reason for carpooling is to reduce the cost of travel, as well as environmental concerns, however, with the necessity to keep a safe distance from others, carpooling could be declining in importance (Molina et al., 2020).

DISCUSSION AND RECOMMENDATIONS

Evaluation

The EF and CF studies can be narrowly divided into those that cover emissions from person, product, institutional, city, and country-level activities. Typically, national emissions inventories are based on demand view from industrial and household practices, only tracking emissions produced within the territory. Territorial emissions are a prevalent theme in political processes like negotiations on climate change, but these factors are disconnected from the factors that drive the manufacture of products that contribute to these emissions (West et al., 2016). In comparison, consumption-based emissions are directly related to the goods and services that put the consumers' usage to a close. They are known for climate change and thus are especially appropriate for appealing to people with it. Approaches from the top down to lowering carbon pollutants have proven ineffective, in part due to the unavailability of international decisions (van Aalst et al., 2008). Government-led carbon-lowering practices (for example, the UK's Act on CO2 campaign) incline towards concentrating towards educating people regarding changes in climate knowledge, as it is vital because surveys have proven that people have many misunderstandings with reference to the causes (Lorenzoni et al., 2007). Individual EF as well as CF calculators may be a successful way to engage people at the neighbourhood level and assist in highlighting people's contributions to pollution, which is critical because the general public often disregards their contribution.

The use of EF and subsequently CF calculator will encourage individuals, in particular, to see their ability to make a difference to their footprint. It's vital to note that awareness is just a dimension of actions. Stern (2000) identifies four factors that affect environmentally compelling behaviour: (2) forces of circumstances (socioeconomic, political, or organisational); (3) exclusive resources along with capabilities (example, accomplishment and awareness); and (4) routine. The tool provides consumers with options to minimise impact by including a series of pledges including various conditions of behaviour or buying choices that can be overlooked or chosen based on their financial situation (West et al., 2016).

Research has shown that human's ability to alter behaviours diversifies confiding on the part of their way of living it affects. People are more likely to recycle and save resources at home, but changes in travel habits are met with resistance (Lorenzoni et al., 2007). The individual commitment to lifestyle advancement would help mitigate to enhance the public's inclination for mitigation of climate change

(Whitmarsh, 2009) and implementing the behaviour change. The psychological process by which this happens is ambiguous yet effective (Lokhorst et al., 2013).

Tool Development Recommendations

The method, using a consumer questionnaire to refine postcode-level expenditure estimates, offers better geographic descriptions of footprints in the neighbourhood. This approach helps researchers to assess whether individual households and neighbourhoods have increased or decreased emissions than predicted, and it is more beneficial to prioritise behavioural legislative changes.

Secondly, EF and CF calculator tools can be future-proofed (minimizing the effects of future events). Designers who design tools should upgrade the tools, as well as consider the trade-offs between precision and accuracy when designing tools. Both commodity rates and the carbon efficiency with which they are updated, as well as measurement methods, can become obsolete over time.

Thirdly, other possible measures, such as land use, jobs, and natural resources in the provision of the supply chain of products and services, may be used instead of "carbon" as the key factor in Carbon Footprint. One choice is to present a panel of indicators that allows the public to reduce negative influences on things which are important to them (West et al., 2016). Some people are more bothered with fair exchange and equity, while the remaining find environmental messages as motivating. In the most "sustainable" choices, the effects of consumption choices may be incompatible with the set of indicators (Talukder et al., 2021).

User Engagement Recommendations

It is necessary to recognize the demand for continuous advancement of the tool when designing tools along with the results of advancement efforts that must be tracked as well as appraised. interrogating, common questions such as who is the one involved in using the tool and how have they heard about it can be extremely helpful in determining the merits of various promotional approaches. Individuals are not only self-interested: attitudes are rooted in social contexts, and social norms can influence individual acts. This sense of citizenship could be provided by calculators which can enable people to equate their daily lifestyle and some of its implications with remaining. It can also cause behaviour to change by generating an anticipation of competition within a community of users. Given that several people are driven to turn towards eco-friendly behaviour for inferences other than environmental protection, such as a desire to minimise energy costs, economic encouragement may be a crucial driver of advancement (Seyfang, 2010). It should be considered when designing engagement initiatives whether foot printing tools function as self-contained "products" that could promote broader involvement, which is important both in order to facilitate the usage of the instrument and for activating change.

The study has been conducted with certain limitations. Such as the consumption by a person is not only related to the mentioned: food, shelter, mobility, goods and services. Land appropriation is explicitly assigned to final consumption groups when measuring Ecological Footprint. Due to COVID-19 and subsequent lockdowns, the study has been conducted through a telephonic survey so human errors are possible. There have been a lot of studies on Ecological Footprint but specifically, the changes in Ecological footprint in an area due to COVID-19 are not done yet. Therefore, this has been analysed through an exploratory study and further continuation of this work will be an add-on to the research.

CONCLUSION

This paper has successfully analysed the change in the Ecological Footprint through an exploratory study of different cities with varying population sizes during the times of COVID-19. The key focus of the research was to analyse how Ecological Footprint varies based on the size of cities and what impact COVID-19 has on the Ecological Footprint of cities. To do so, surveys in 5 different urban settlements based on population were carried out. The samples were analysed, and sorted and the Ecological Footprint was calculated. Later, the t-test was applied to samples of each city. The results showed that cities with less population have a low Ecological Footprint as compared to cities with more population which has been statistically shown by applying ANOVA test. It also showed the change in Ecological Footprint during the time of COVID-19 when most of the activities were put on hold. It is suggested that raising awareness about our resources and environmental conservation is very important today and so. The consequent lockdowns, social distancing norms and various other safety measures had an overall positive effect on the nature and environment (as observed the Ecological Footprint was significantly reduced). The main question lies why do we need a pandemic to help us act judiciously for the conservation of our resources? It is high time for every individual to act cautiously towards our resources. United Nations Development Programme specifies that the Sustainable Development Goals (SDGs) have fairly provoked social, financial, and ecological protection so as to transform the world as well as have given 17 SDGs to be accomplished by countries by 2030. This is equally important to reach the goal of carbon neutrality, as developing nations like India, will benefit us in many environmental improvement aspects too. The findings prove the impact we have on the environment and resources which are scarce to us, was reduced by change in our daily lifestyle as seen during the pandemic. The current work is a pertinent realization for humans to implement changes in our lifestyles and shows ways to reduce Ecological Footprint for sustainable development.

REFERENCES

Al Nasiri, N., Al-Awadhi, T., Hereher, M., Ahsan, R., & AlRubkhi, A. G. (2020). Changing Urban Ecology a Challenge for Coastal Urban Resilience: A Study on Muscat. *Environment & Urbanization Asia*, 11(1), 10–28. 10.1177/0975425320906307

Attwood, S., & Hajat, C. (2020). How will the COVID-19 pandemic shape the future of meat consumption? *Public Health Nutrition*, 23(17), 3116–3120. 10.1017/S136898002000316X32782062

Barrett, J., Birch, R., Cherrett, N., & Wiedmann, T. (2005). Exploring the application of the ecological footprint to sustainable consumption policy. *Journal of Environmental Policy and Planning*, 7(4), 303–316. 10.1080/15239080500441095

Bazan, G. (1997). Our Ecological Footprint: Reducing human impact on the earth. *Electronic Green Journal*, 1(7). Advance online publication. 10.5070/G31710273

Bergh, J. C. J. M. Van Den, & Verbruggen, H. (1999). *Spatial Sustainability, Trade and Indicators: An Evaluation of the 'Ecological Footprint.'* EconPapers.

Bhattacharjee, S., & Sattar, S. (2021). Pandemic And Urban Sustainability: Analyzing The Covid-19 Scenario In Mumbai, India. Geography, Environment. *Sustainability*, 14(1), 17–24. 10.24057/2071-9388-2020-189

Bongaarts, J. (2001). Household size and composition in the developing world in the 1990s. *Population Studies*, 55(3), 263–279. 10.1080/00324720127697

Borucke, M., Moore, D., Cranston, G., Gracey, K., Iha, K., Larson, J., Lazarus, E., Morales, J. C., Wackernagel, M., & Galli, A. (2013). Accounting for demand and supply of the biosphere's regenerative capacity: The National Footprint Accounts' underlying methodology and framework. *Ecological Indicators*, 24, 518–533. 10.1016/j.ecolind.2012.08.005

Bosupeng, M. (2017). Is China's target of a 40-45% reduction in carbon dioxide emission plausible? *Environmental & Socio-Economic Studies*, 5(1), 46–50. 10.1515/environ-2017-0005

Brinkman, J., & Mangum, K. (2020). *The Geography of Travel Behavior in the Early Phase of the Covid-19 Pandemic*. SSRN. https://papers.ssrn.com/abstract=3717188

Chen, Y., Jiao, J., Bai, S., & Lindquist, J. (2020). Modeling the spatial factors of COVID-19 in New York City. https://ssrn.com/abstract=360671910.2139/ssrn.3606719

Chua, B. L., Al-Ansi, A., Lee, M. J., & Han, H. (2020). Tourists' outbound travel behavior in the aftermath of the COVID-19: Role of corporate social responsibility, response effort, and health prevention. *Journal of Sustainable Tourism*. 10.1080/09669582.2020.1849236

Corlett, R. T., Primack, R. B., Devictor, V., Maas, B., Goswami, V. R., Bates, A. E., Koh, L. P., Regan, T. J., Loyola, R., Pakeman, R. J., Cumming, G. S., Pidgeon, A., Johns, D., & Roth, R. (2020). Impacts of the coronavirus pandemic on biodiversity conservation. *Biological Conservation*, 246, 108571. 10.1016/j.biocon.2020.10857132292203

Costanza, R. (2000). The dynamics of the ecological footprint concept. *Ecological Economics*, 32(3), 341–345. 10.1016/S0921-8009(99)00150-0

Cucinotta, D., & Vanelli, M. (2020). WHO declares COVID-19 a pandemic. In *Acta Biomedica, 91*(1). 10.23750/abm.v91i1.9397

Cui, J., Li, F., & Shi, Z. L. (2019). Origin and evolution of pathogenic coronaviruses. *Nature Reviews. Microbiology*, 17(3), 181–192. 10.1038/s41579-018-0118-930531947

Di Renzo, L., Gualtieri, P., Cinelli, G., Bigioni, G., Soldati, L., Attinà, A., Bianco, F. F., Caparello, G., Camodeca, V., Carrano, E., Ferraro, S., Giannattasio, S., Leggeri, C., Rampello, T., Lo Presti, L., Tarsitano, M. G., & De Lorenzo, A. (2020). Psychological aspects and eating habits during covid-19 home confinement: Results of ehlc-covid-19 italian online survey. *Nutrients*, 12(7), 1–14. 10.3390/nu1207215232707724

Di Renzo, L., Gualtieri, P., Pivari, F., Soldati, L., Attinà, A., Cinelli, G., Cinelli, G., Leggeri, C., Caparello, G., Barrea, L., Scerbo, F., Esposito, E., & De Lorenzo, A. (2020). Eating habits and lifestyle changes during COVID-19 lockdown: An Italian survey. *Journal of Translational Medicine*, 18(1), 1–15. 10.1186/s12967-020-02399-532513197

Echegaray, F. (2021). What POST-COVID-19 lifestyles may look like? Identifying scenarios and their implications for sustainability. *Sustainable Production and Consumption*, 27, 567–574. 10.1016/j.spc.2021.01.02534722841

Eroğlu, H. (2020). Effects of Covid-19 outbreak on environment and renewable energy sector. *Environment, Development and Sustainability*, 0123456789. 10.1007/s10668-020-00837-432837274

Frenette, E., Bahn, O., & Vaillancourt, K. (2017). Meat, Dairy and Climate Change: Assessing the Long-Term Mitigation Potential of Alternative Agri-Food Consumption Patterns in Canada. *Environmental Modeling and Assessment*, 22(1), 1–16. 10.1007/s10666-016-9522-6

Gutiérrez, A., Miravet, D., & Domènech, A. (2020). COVID-19 and urban public transport services: Emerging challenges and research agenda. *Cities & Health*, 00(00), 1–4. 10.1080/23748834.2020.1804291

Hjorth, T., Huseinovic, E., Hallström, E., Strid, A., Johansson, I., Lindahl, B., Sonesson, U., & Winkvist, A. (2020). Changes in dietary carbon footprint over ten years relative to individual characteristics and food intake in the Västerbotten Intervention Programme. *Scientific Reports*, 10(1), 1–14. 10.1038/s41598-019-56924-831913331

Holmberg, J., Lundqvist, U., Robèrt, K. H., & Wackernagel, M. (1999). The ecological footprint from a systems perspective of sustainability. *International Journal of Sustainable Development and World Ecology*, 6(1), 17–33. 10.1080/13504509.1999.9728469

Hsu, H., & Lachenbruch, P. A. (2014). Paired t Test. Wiley StatsRef. *Statistics Reference Online*, 1–3. 10.1002/9781118445112.stat05929

Huang, C., Wang, Y., Li, X., Ren, L., Zhao, J., Hu, Y., Zhang, L., Fan, G., Xu, J., Gu, X., Cheng, Z., Yu, T., Xia, J., Wei, Y., Wu, W., Xie, X., Yin, W., Li, H., Liu, M., & Cao, B. (2020). Clinical features of patients infected with 2019 novel coronavirus in Wuhan, China. *Lancet*, 395(10223), 497–506. 10.1016/S0140-6736(20)30183-531986264

Hunter, C., & Shaw, J. (2007). The ecological footprint as a key indicator of sustainable tourism. *Tourism Management*, 28(1), 46–57. 10.1016/j.tourman.2005.07.016

Ismail, L. C., Osaili, T. M., Mohamad, M. N., Al Marzouqi, A., Jarrar, A. H., Jamous, D. O. A., Magriplis, E., Ali, H. I., Al Sabbah, H., Hasan, H., Almarzooqi, L. M. R., Stojanovska, L., Hashim, M., Obaid, R. R. S., Saleh, S. T., & Dhaheri, A. S. A. (2020). Eating habits and lifestyle during covid-19 lockdown in the united arab emirates: A cross-sectional study. *Nutrients*, 12(11), 1–20. 10.3390/nu1211331433137947

Jorgenson, A. K. (2003). Consumption and Environmental Degradation: A Cross-National Analysis of the Ecological Footprint. In *Social Problems, 50*(3). University of California Press. 10.1525/sp.2003.50.3.374

Kaur, S. P., & Gupta, V. (2020). COVID-19 Vaccine: A comprehensive status report. In *Virus Research (Vol. 288)*. Elsevier B.V. 10.1016/j.virusres.2020.198114

Khatib, A. N., Carvalho, A. M., Primavesi, R., To, K., & Poirier, V. (2020). Navigating the risks of flying during COVID-19: A review for safe air travel. *Journal of Travel Medicine*, 27(8), 1–9. 10.1093/jtm/taaa21233184655

Khursheed, A., Alam, S., Tyagi, V. K., Nagpure, A. S., Khan, A. A., Gaur, R. Z., Singh, S., Bhattacharya, P., Mukherjee, S., & Kumar, M. (2020). Future liasing of the lockdown during COVID-19 pandemic: The dawn is expected at hand from the darkest hour. *Groundwater for Sustainable Development*, 11(336), 100433. 10.1016/j.gsd.2020.10043332835071

Kitzes, J., & Wackernagel, M. (2009). Answers to common questions in Ecological Footprint accounting. *Ecological Indicators*, 9(4), 812–817. 10.1016/j.ecolind.2008.09.014

Kudryavtseva, O. V., Malikova, O. I., & Egorov, E. G. (2021). Sustainable Urban Development And Ecological Externalities: Russian Case. Geography, Environment. *Sustainability*, 14(1), 81–90. 10.24057/2071-9388-2020-151

Kumar, P., & Morawska, L. (2019). Could fighting airborne transmission be the next line of defence against COVID-19 spread? *City and Environment Interactions*, 4, 100033. 10.1016/j.cacint.2020.10003334235420

Lenzen, M., & Murray, S. A. (2001). A modified ecological footprint method and its application to Australia. *Ecological Economics*, 37(2), 229–255. 10.1016/S0921-8009(00)00275-5

Lin, D., Hanscom, L., Murthy, A., Galli, A., Evans, M., Neill, E., Mancini, M. S., Martindill, J., Medouar, F. Z., Huang, S., & Wackernagel, M. (2018). Ecological footprint accounting for countries: Updates and results of the national footprint accounts, 2012-2018. *Resources*, 7(3), 58. 10.3390/resources7030058

Lokhandwala, S., & Gautam, P. (2020). Indirect impact of COVID-19 on environment: A brief study in Indian context. *Environmental Research*, 188(June), 109807. 10.1016/j.envres.2020.10980732574854

Lokhorst, A. M., Werner, C., Staats, H., van Dijk, E., & Gale, J. L. (2013). Commitment and Behavior Change: A Meta-Analysis and Critical Review of Commitment-Making Strategies in Environmental Research. *Environment and Behavior*, 45(1), 3–34. 10.1177/0013916511411477

Lorenzoni, I., Nicholson-Cole, S., & Whitmarsh, L. (2007). Barriers perceived to engaging with climate change among the UK public and their policy implications. *Global Environmental Change*, 17(3–4), 445–459. 10.1016/j.gloenvcha.2007.01.004

Marino, C., Nucara, A., Pietrafesa, M., & Pudano, A. (2016). The Assessment of Road Traffic Air Pollution by Means of an Average Emission Parameter. *Environmental Modeling and Assessment*, 21(1), 53–69. 10.1007/s10666-015-9489-8

Meena, A. K., & Yadav, T. K. (2019). What is Ecological Footprint and Why Is It Important? *Agrobios, 18*(June), 25–26. http://www.mpict.org/ict_education_defined_importance.html

Megahed, N. A., & Ghoneim, E. M. (2020). Antivirus-built environment: Lessons learned from Covid-19 pandemic. *Sustainable Cities and Society*, 61, 102350. 10.1016/j.scs.2020.10235032834930

Mehrotra, S., Bardhan, R., & Ramamritham, K. (2020, June). Diurnal thermal diversity in heterogeneous built area: Mumbai, India. *Urban Climate*, 32, 100627. 10.1016/j.uclim.2020.100627

Mofijur, M., Fattah, I. M. R., Alam, M. A., Islam, A. B. M. S., Ong, H. C., Rahman, S. M. A., Najafi, G., Ahmed, S. F., Uddin, M. A., & Mahlia, T. M. I. (2021). Impact of COVID-19 on the social, economic, environmental and energy domains: Lessons learnt from a global pandemic. *Sustainable Production and Consumption, 26*(September 2020), 343–359. 10.1016/j.spc.2020.10.016

Molina, J. A., Ignacio Giménez-Nadal, J., & Velilla, J. (2020). Sustainable commuting: Results from a social approach and international evidence on carpooling. *Sustainability (Basel)*, 12(22), 1–12. 10.3390/su12229587

Nakajima, E. S., & Ortega, E. (2016). Carrying capacity using emergy and a new calculation of the ecological footprint. *Ecological Indicators*, 60, 1200–1207. 10.1016/j.ecolind.2015.08.054

Pais, D. F., Marques, A. C., & Fuinhas, J. A. (2020). Reducing Meat Consumption to Mitigate Climate Change and Promote Health: But Is It Good for the Economy? *Environmental Modeling and Assessment*, 25(6), 793–807. 10.1007/s10666-020-09710-0

Pandey, D., Agrawal, M., & Pandey, J. S. (2011). Carbon footprint: Current methods of estimation. *Environmental Monitoring and Assessment*, 178(1–4), 135–160. 10.1007/s10661-010-1678-y20848311

Parady, G., Taniguchi, A., & Takami, K. (2020). Travel behavior changes during the COVID-19 pandemic in Japan: Analyzing the effects of risk perception and social influence on going-out self-restriction. *Transportation Research Interdisciplinary Perspectives*, 7, 100181. 10.1016/j.trip.2020.10018135782787

Rees, W., & Wackernagel, M. (2012). Urban ecological footprints: Why cities cannot be sustainable and why they are a key to sustainability. *The Urban Sociology Reader, 9255*(96), 157–165. 10.4324/9780203103333-26

Seyfang, G. (2010). Community action for sustainable housing: Building a low-carbon future. *Energy Policy*, 38(12), 7624–7633. 10.1016/j.enpol.2009.10.027

Shahbaz, M., Shahzad, S. J. H., & Mahalik, M. K. (2018). Is Globalization Detrimental to CO2 Emissions in Japan? New Threshold Analysis. *Environmental Modeling and Assessment*, 23(5), 557–568. 10.1007/s10666-017-9584-0

Shakibaei, S., de Jong, G. C., Alpkökin, P., & Rashidi, T. H. (2021). Impact of the COVID-19 pandemic on travel behavior in Istanbul: A panel data analysis. *Sustainable Cities and Society*, 65, 102619. 10.1016/j.scs.2020.10261933251093

Shamshiripour, A., Rahimi, E., Shabanpour, R., & Mohammadian, A. (2020). How is COVID-19 reshaping activity-travel behavior? Evidence from a comprehensive survey in Chicago. *Transportation Research Interdisciplinary Perspectives*, 7, 100216. 10.1016/j.trip.2020.10021634173469

Sharma, S., Nahid, S., Sharma, M., Sannigrahi, S., Anees, M. M., Sharma, R., Shekhar, R., Basu, A. S., Pilla, F., Basu, B., & Joshi, P. K. (2020). A long-term and comprehensive assessment of urbanization-induced impacts on ecosystem services in the capital city of India. *City and Environment Interactions*, 7, 100047. 10.1016/j.cacint.2020.100047

Shen, J., Duan, H., Zhang, B., Wang, J., Ji, J. S., Wang, J., Pan, L., Wang, X., Zhao, K., Ying, B., Tang, S., Zhang, J., Liang, C., Sun, H., Lv, Y., Li, Y., Li, T., Li, L., Liu, H., & Shi, X. (2020). Prevention and control of COVID-19 in public transportation: Experience from China. *Environmental Pollution*, 266, 115291. Advance online publication. 10.1016/j.envpol.2020.11529132829124

Srivastava, R. (2020). Growing Precarity, Circular Migration, and the Lockdown in India. *The Indian Journal of Labour Economics : the Quarterly Journal of the Indian Society of Labour Economics*, 63(0123456789), 79–86. 10.1007/s41027-020-00260-332929315

Stern, P. C. (2000). Toward a Coherent Theory of Environmentally Significant Behavior. *The Journal of Social Issues*, 56(3), 407–424. 10.1111/0022-4537.00175

Stone, D. H. (1993). Design a questionnaire. *British Medical Journal*, 307(6914), 1264–1266. 10.1136/bmj.307.6914.12648281062

Talukder, B., vanLoon, G. W., Hipel, K. W., & Orbinski, J. (2021). COVID-19's implications on agri-food systems and human health in Bangladesh. *Current Research in Environmental Sustainability*, 3, 100033. 10.1016/j.crsust.2021.10003334977606

Tirachini, A., & Cats, O. (2020). COVID-19 and public transportation: Current assessment, prospects, and research needs. *Journal of Public Transportation*, 22(1), 1–34. 10.5038/2375-0901.22.1.136118518

Town and Country Planning Organisation. (2014). *Ministry of Urban Development*. Government of India.

van Aalst, M. K., Cannon, T., & Burton, I. (2008). Community level adaptation to climate change: The potential role of participatory community risk assessment. *Global Environmental Change*, 18(1), 165–179. 10.1016/j.gloenvcha.2007.06.002

Villarroel, J. D., Antón, Á., Zuazagoitia, D., & Nuño, T. (2017). Young children's environmental judgement and its relationship with their understanding of the concept of living things. *Environmental & Socio-Economic Studies*, 5(1), 1–10. 10.1515/environ-2017-0001

Wackernagel, M. Schulz, N. B., Deumling, D., Linares, A. C., Jenkins, M., Kapos, V., Monfreda, C., Loh, J., Myers, N., Norgaard, R., & Randers, J. (2002). *Tracking the ecological overshoot of the human economy*. PNAS. www.pnas.org

Wackernagel, M., Lewan, L., & Hansson, C. B. (1999). Evaluating the use of natural capital with the ecological footprint - Applications in Sweden and subregions. *Ambio*, 28(7), 604–612.

Wackernagel, Mathis, & Yount, J. D. (2000). *Footprints For Sustainability: The Next Steps Division; WCED-World Commission on Environment and Development.* WWF-World Wide Fund for Nature (originally World Wildlife Fund). http://www.rprogress.org

Wakeland, W., Cholette, S., & Venkat, K. (2012). Food transportation issues and reducing carbon footprint. In *Food Engineering Series* (pp. 211–236). Springer. 10.1007/978-1-4614-1587-9_9

Watkins, J. (2020). Preventing a covid-19 pandemic. In *The BMJ* (Vol. 368). BMJ Publishing Group. 10.1136/bmj.m810

Weill, J. A., Stigler, M., Deschenes, O., & Springborn, M. R. (2020). Social distancing responses to COVID-19 emergency declarations strongly differentiated by income. *Proceedings of the National Academy of Sciences of the United States of America*, 117(33), 19658–19660. 10.1073/pnas.200941211732727905

West, S. E., Owen, A., Axelsson, K., & West, C. D. (2016). Evaluating the Use of a Carbon Footprint Calculator: Communicating Impacts of Consumption at Household Level and Exploring Mitigation Options. *Journal of Industrial Ecology*, 20(3), 396–409. 10.1111/jiec.12372

White, T. J. (2007). Sharing resources: The global distribution of the Ecological Footprint. *Ecological Economics*, 64(2), 402–410. 10.1016/j.ecolecon.2007.07.024

Whitmarsh, L. (2009). Behavioural responses to climate change: Asymmetry of intentions and impacts. *Journal of Environmental Psychology*, 29(1), 13–23. 10.1016/j.jenvp.2008.05.003

Wielechowski, M., Czech, K., & Grz da, Ł. (2020). Decline in mobility: Public transport in Poland in the time of the COVID-19 pandemic. *Economies*, 8(4), 1–24. 10.3390/economies8040078

Workie, E., Mackolil, J., Nyika, J., & Ramadas, S. (2020). Deciphering the impact of COVID-19 pandemic on food security, agriculture, and livelihoods: A review of the evidence from developing countries. *Current Research in Environmental Sustainability*, 2, 100014. 10.1016/j.crsust.2020.10001434977605

Yang, L., Liu, S., Liu, J., Zhang, Z., Wan, X., Huang, B., Chen, Y., & Zhang, Y. (2020). COVID-19: Immunopathogenesis and Immunotherapeutics. *Signal Transduction and Targeted Therapy*, 5(1), 1–8. 10.1038/s41392-020-00243-232712629

Yuki, K., Fujiogi, M., & Koutsogiannaki, S. (2020). COVID-19 pathophysiology: A review. In *Clinical Immunology (Vol. 215).* Academic Press Inc. 10.1016/j.clim.2020.108427

Zhao, S., Li, Z., & Li, W. (2005). A modified method of ecological footprint calculation and its application. *Ecological Modelling*, 185(1), 65–75. 10.1016/j.ecolmodel.2004.11.016

APPENDIX A

Survey Questionnaire

The survey questionnaire included 15 questions as per Global Footprint (Www.Footprintnetwork.Org.) are as follows:

A.1. How often do you consume items derived from animals? (Not at all/ Occasionally/ Often/ Very Often)

A.2. What percentage of your food is raw, unprocessed, or grown nearby? (in percentage)

A.3. Which of the following housing types better related to your house? (Multi-story building/ Duplex houses in a row/ a single building with 2-4 houses/ Condominium/ Freestanding with no running water/ Self-supporting, running water)

A.4. What materials are used to build your home?

A.5. What is the number of people residing in your house? (number)

A.6. What is the square footage of your home? (Tiny (0-20 sq m)/ Small (21-50 sq m)/ Medium (51-140 sq m)/ Large (141- 460 sq m)/ Huge (461 sq m and more))

A.7. Do you have electricity in your home? (Yes/No)

A.8. What percentage of the electricity in your home is produced from renewable sources? (percentage)

A.9. What is the energy efficiency of your home? (Very inefficient/ Below average/ Average/ Above average/ Efficiency-centred design)

A.10. How much waste do you produce in comparison to your neighbours? (Much less/ Less/ Same/ More/ Much more)

A.11. How much distance do you ride each week by car or motorcycle? (Car/ Two- wheeler) (in km)

A.12. What is the average fuel efficiency of the most frequently used mode of transportation? (Car/ Two-wheeler) (in litres/100 km)

A.13. How often do you carpool when you travel by car? (Never/ Infrequently/ Occasionally/ Often/ Always)

A.14. On a weekly basis, how far do you travel by public transportation? (in km)

A.15. How many hours a year do you fly? (in hrs).

Webpages

Global Footprint Network. https://www.footprintcalculator.org/ (accessed 25 March, 2021)

Chapter 9
Consequences of the COVID-19 Pandemic on Declining Cities, Urban Vacant Land Uses, Green Infrastructure, and Their Impact on Climate Change Hazards

José G. Vargas-Hernandez
Budapest Centre for Long-Term Sustainability, Hungary

Omar C. Vargas-Gonzàlez
https://orcid.org/0000-0002-6089-956X
Tecnològico Nacional de Mèxico, Ciudad Guzmàn, Mexico

Omar A. Guirette-Barbosa
https://orcid.org/0000-0003-1336-9475
Universidad Politécnica de Zacatecas, Mexico

Selene Castañeda-Burciaga
https://orcid.org/0000-0002-2436-308X
Universidad Politécnica de Zacatecas, Mexico

ABSTRACT

This chapter aims to analyze the implications between the declining and shrinking cities, urban vacant lands, urban land uses, green infrastructures, urban green areas, and their impact on climate hazards change. The analysis departs from the basic assumption that urban vacant land sites and spaces have a negative connotation, but supported by the appropriate policies and programs of incentives, can turn around and develop the essential green infrastructure to enable the mitigation of climate change hazards, economic growth, and socio ecological development. The method used is the analytical-descriptive base on the theoretical and empirical literature review. It is concluded that the land uses of vacant land sites more vacationed towards urban green innovation infrastructure and forest areas contribute to mitigate the climate change hazards.

DOI: 10.4018/979-8-3693-1650-4.ch009

INTRODUCTION

Urban processes linked to shrinking cities feature characteristics including economic decline associated to urban land uses, and other socio-economic dimensions that may lead to the adaptive capacity building which in turn increases the urban green infrastructure and moderates the impact of climate hazards change. The vacant land is used as green infrastructure to improve the health of urban ecosystems.

City regions experiencing economic growth and increases in urban population are subject to the adversity of losing green infrastructure resources leading towards a disequilibrium in adaptive capacity to climate change hazards and the quality-of-life goals.

The objective of this analysis is to find the relationships of the implications between the declining and shrinking cities, urban vacant lands, urban land uses, green infrastructures, urban green areas, and their impact on climate hazards change. First, it begins analyzing the processes of declining and shrinking cities to establish the relevance of the major urban vacant land sites and the land uses derived. The land uses of vacant land sites more vacationed towards urban green innovation infrastructure and forest areas are the ones that lead to mitigate the climate change hazards.

RESEARCH METHODOLOGY

A bibliometric review was used, through which a quantitative analysis of publications related to declining cities, urban vacant land uses, green infrastructure and its impact on climate change risks was carried out. The bibliometric indicators taken as indicators were the quality of the publications, the contexts analyzed in the documents, the relevance of the topic and the temporality.

To carry out this bibliometric review, literary and scientific sources on declining cities, urban vacant land uses, green infrastructure and its impact on climate change risks were used as a basis. Eighty-five documents were used for the research. This was done in two stages. In the first stage, the search criteria to be used in the different databases were defined, which included temporality, date of consultation, type of document and journal, topic and key words. The second stage consisted of classifying and analyzing the documents based on the identification criteria; these were grouped to identify the connections between the documents, selecting the most relevant ones.

DECLINING AND SHRINKING CITIES

City-regions and urban areas develop, expand or decline, remaining sensitive to environmental and climate change agendas guided by spatial planning regulations. Cities that are experiencing shrinkage and decline facing different challenges. Around the globe, cities are either growing and expanding or shrinking and declining because some dynamic processes such as deindustrialization, globalization, demographic change (Martinez-Fernandez, Audirac, Fol, & Cunningham-Sabot, 2012; McKinsey and Company, 2016; Sassen, 2001; Hall, 1993). Besides the complexity of these challenges, there are the global warming and climate change affecting urban settings (Ruth and Coelho, 2007).

The economic growth processes of cities that benefited from the development of some comparative and competitive advantages in the past, have led to the corresponding deindustrialization processes while unplugged from the global value chain and resulting in urban population and economic decline

(Hall, 1993; Pike et al., 2016; Martinez-Fernandez et al., 2012: 220). Patterns of economic decline and depopulating in communities caused by contextual conditions result in vacant land with implications in institutional, fiscal, and programmatic responses that may bring some potential benefits.

Urban areas in large cities of the developed world regions experience population decline (McKinsey and Company, 2016). Economic and urban decline is linked to limited regulation over the land use in the different locations and forms of new urban development supported by deficient urban spatial planning, small changes in green infrastructure cover and lack of strategies aimed to the mitigation of socioeconomic inequality and environmental degradation. All these factors have an impact on the related implications for adaptive capacity to cope the risks from urban weather and climate change.

The concept of shrinking cities is widely accepted as a global phenomenon which dates to the urban sociology theories of cyclical growth and decline (Martinez-Fernandez et al., 2012). Shrinking cities are places characterized by economic and population decline that harbor vacant properties and land parcels (Dietersdorfer et al., 2012: 9; Burkholder, 2012; Schilling & Logan, 2008).

The urban expansion, declining and shrinkage processes are projected to continue influencing the global landscape of city spaces, urban land uses and green infrastructure. Declining urban land uses, derelict and disused land experiencing declines in surface area, may replace informal open green spaces and can be considered prospects for urban green infrastructure.

URBAN VACANT LAND

Many cities are dominated by abandoned and vacant land. Cities that have experience of land abandonment, use vacant land for creating open green spaces and sites for productive purposes (Alexander, 2005; U.S. Department of Housing and Urban Development, 2009; Wachter, S.; Scruggs, G.; Voith, R.; Huang, 2010).

The term vacant land has a negative connotation considered as a problem (Jakle, Wilson, 1992; Coleman, 1982). Vacant land in physical blight conveys negative images that contribute to disinvestment, reducing reinvestment and development, and increases urban vacancy rates (Neukrug, 2009). Vacant land deters and discourage investments in the community imposing the costs of maintaining vacant properties. Urban vacant land, empty urban spaces and abandoned buildings have become a wasteland undervalued as a normal part of the urban fabric despite the lack of public policy and economic investment for renewal.

Most cities do not have adequate programs of economic incentives to support maintenance and improvements of abandoned buildings and vacant land. Vacant land and abandoned building are not maintained and have a negative impact on the quality of life and property values of the community.

Urban vacant lands are structural assets with economic and socioecological value relative to other urban land uses. Urban vacant lands are ecological habitats for a wide range of biodiversity of the wildlife of plants and animals and provide ecological corridors and green network systems, wetland and storm water management, regulation services to disperse pollutants (Kamvasinou, 2011).

The emergence of open vacant land spaces provides some benefits and opportunities associated to secure adaptive capacity with new land uses and green infrastructure. Revitalization of urban vacant land reuse of urban wasted land are critical to urban land use in urban spaces. The capacity of urban vacant land to design, planning and management promotes the public socio ecological value of urban landscape as natural resource and community asset to improve the quality of life in the surrounding neighborhood. Vacant urban land covers a significant spatial territory of the urban landscape that provides green infrastructure for ecosystem services used to enhance health and better quality of life of residents (Pagano, Bowman, 2000; Kim, Miller, Nowak, 2015).

Urban vacant land is a valuable resource that creates a variety of urban open spaces opportunities available to city residents. Urban vacant land is a valuable ecological resource as green infrastructure used to enhance the urban socio ecosystems and promote the health and quality of life of residents (Kim et al, 2015). Vacant land is a valuable urban landscape and source of ecosystem services that benefits the urban communities and provides transformation opportunities for community development (Burkholder, 2012; Little, 2008).

Urban vacant land has a public value in terms of the potential to enhance socioecological benefits if they are redeveloped and integrated with other urban green infrastructure (Kim, 2016). Planning for urban vacant land leads to better appreciation of public value and to design strategies to optimize the utilization of these spaces in landscape design, urban environmental regeneration, and renewal. Urban vacant land has potential as a valuable socio ecological resource to provide socio ecosystem services in the form of public open and urban green spaces, etc. Urban vacant land has relevant functions as a strategic use and reuse in the urban landscape provided by the ecological opportunities and benefits in creating a more prosperous, livable, healthy, and sustainable cities.

From the economics, the vacant space should be subject to development, contribute to socio ecological benefits (Németh, Langhorst, 2014; Kim, Miller, and Nowak, 2016Kim et al., 2015). Urban vacant land benefits provided as green infrastructure are not well recognized since it has both economic and political problems.

There are different types of urban vacant land sites which can be identified and explored by a wide range of typologies (Blakeman, Brown, Fitzpatrick, Shaw, Williamson, 2008). These are different types of urban vacant land habitats, among other vacant lots and areas alongside parking lots and roads, abandoned buildings in industrial areas, wastelands, old towns centers, railway sites, etc., where can be sustained diversity of urban plant communities and animal populations depending on the environmental conditions, natural habitats and management practices (Robinson, Lundholm, 2012; Darlington, 1969; Jehlík, 1986; Klimeš, 1987; Sukopp, Blume, Kunick, 1979; Brandes, 1995; Hope, Gries, Zhu, Fagan, Redman, Grimm, Kinzig, 2003; Burkholder, 2012).

The different types of urban vacant land and lot have potential values and uses to contribute to urban socio ecology and landscape as integrated green infrastructure, despite the minimal considerations for design, planning and management. A typology of vacant land is a planning and designing tool to plan for better utilization of the spaces and open creative approaches to envisioning urban space and landscape design in enhancing the ecological and social benefits.

A typology of urban vacant land has implications for policy design, planning and development to utilize for an advantage seeking to identify and catalog post-industrial land, derelict, unattended land with vegetation, natural land, and transportation-related land. This typology categorizes the vacant land in terms of potential uses, as natural sites, unattended sites, post-industrial sites, derelict sites, and transportation-related sites. The derelict sites were previously developed bur wasted, abandoned or unused sites. The unattended vacant land with vegetation that are empty and inactive.

A typology on urban vacant land may support urban planning policies to design different policies, regulations, and programs in advancing redevelopment to support design, maintenance, and implementation of site green infrastructure for the different types of vacant urban land and abandoned buildings. Urban vacant residential, industrial, and commercial sites with historical relevant background can be developed, preserved, and enhanced in their ecosystem values.

Vacant land is an alternative to public spaces because it has easy access to accommodate social groups, including marginalized and young people hanging around, provides playgrounds accommodating a wide range of activities for creative uses and in the social dimension, families enjoying contact with nature, matching the loose and open spaces as a resting area and extend choice for other activities. Public maintenance of the different types of vacant land should be supported by local authorities by enforcing public and private maintenance requirements, registration fees, procuring green infrastructure credits for socio ecological and environmental benefits.

Urban vacant land has the potential to contribute to socio ecology an environmental system besides the economic best viable use by promoting green infrastructure and open urban green spaces when it is managed properly. Urban vacant land offers creative landscape design alternatives for alternative open spaces and opportunities in built-up city environment. Caring for the vacant land increases people's engagement and promotes neighborhood identity.

Properly managed, urban vacant land provides beneficial socioecological services that have a positive impact to the surrounded communities by creating and developing infrastructure services, community and wildlife gardens, public plantings, and recreational areas (Bonham, Spilka, Rastorfer, 2002; Bonham and Spilka,1995). Socio ecological values and benefits provided by urban vacant land as a resource, improve urban policies through urban planning, development, and management aimed to enhance environmental quality, human health, and quality of life for residents in the community.

In highly vacant urban land areas, the real market is weak, although space planning and development for vacant urban land with an adequate socioecological systems knowledge can be used for green infrastructure and agroecology to provide ecosystem services, adaptive capacity to climate change, healthy fresh food, and water, among other benefits.

Clusters of urban vacant lots with overlapping urban areas of low concentration of green spaces and ecological value, requires a socio-ecological functions associated to ecosystem services. Vacant land can be reused by its socioecological value and the financial value is an obstacle for redevelopment. Land vacancy rates in developed urban areas tend to increase due to de-industrilization processes, declining city population, urban sprawl, and preferences for new residential choices (Kremer et al., 2013).

The transformation of vacant land includes projects that may be components of urban community development comprehensive plans that may include greenways, waterfront parks and pathways development (City of Pittsburgh. Greening Resources. 2012; The Greening of Detroit, 2012). The transformation of different types of vacant land and abandoned buildings into green infrastructure can have the goal to promote community stabilization, economic growth, socio ecological well-being, and quality of life.

City authorities and private market must be able to stablish a redevelopment program of vacant land, empty spaces, and abandoned buildings and to reuse them. City authorities and private market must be able to stablish a redevelopment program of vacant land, empty spaces, and abandoned buildings and to reuse them.

Investment policies, incentive programs and regulatory strategies on rehabilitation on vacant land and abandoned buildings such as enforcement codes on maintenance requirements and increasing ownership rates, may encourage owners to improve their vacant properties and neighborhood stability. Local authorities can maintain enforcement codes if the owners fail to improve their vacant land and abandoned buildings (Accordino, Johnson, 2000).

A positive incentive program and graded property tax system on the requirement of vacant land, empty spaces and abandoned buildings as primary assets for productive speculation, restructuring is promoted encouraging redevelopment (Kim, Miller, Nowak, 2015; Bowman, Pagano, 2004).

URBAN LAND USES

Use of vacant land as green infrastructure to perform the function to provide ecosystem services and socio ecological values and benefits for the welfare of communities. The socio ecosystem values of urban green vacant sites can be developed and protected for a variety of types of land uses and revitalized those that are wasted, underused, and undervalued compared to other types. Transformation of expanding uncontrolled urban land uses leads to loss of natural forest and vegetation followed by open spaces, a decline of ecosystem services provision and undergoing environmental and climate changes (Sackey, 2008; and Weng, 2001).

Urban vacant land sites are a valuable ecological resource despite that has fewer impervious ground cover than other urban land uses. To assess the value of the urban vacant land, it can be classified in categories based on their physical, biological, and socio ecological characteristics to support urban planning and developing.

Uses of urban land spaces have contrasting implications for cities, which in turn is a driver committed to adapt green infrastructure to changing climate, this in turn can influence and enhance the capacity of urban land uses of green spaces and infrastructure to provide some functions to adapt to climate change, such as for example rainwater infiltration. Land use changes focusing on green infrastructure may contribute to develop the capacity building for urban spaces to adapt to global warming, climate change and extreme weather hazards.

Transdisciplinary approaches can be applied to address vacant urban land uses and urban design and planning to implement strategies aimed to improve the urban green areas and urban green infrastructure which may mitigate the hazards of climate change. Urban land use planning must be based on a comprehensive benchmark for ecosystem services productivity considering the benefits of other land uses and management of vacant land. Urban land uses planning and development, urban forestry, and green spaces can help to mitigate the problems associated with climate change hazards. Vacant land policies and programs may alter the land use structure and nature to achieve higher redevelopment potential utility (Blakeman, Brown, Fitzpatrick, Shaw, Williamson, 2008).

Uses of different types of vacant land should be supported by local authorities, communities, land trusts and greening organizations to maintain green programs and crops initiative aimed enhance the socio ecosystem. Social organizations may manage special programs for the improvement and maintenance aimed to different types of vacant land using project funding from tax levies, foundations, endowments from private donations and personal funds.

GREEN INFRASTRUCTURE AND FOREST STRUCTURE

Urban vacant land sites are a component of green infrastructure systems that improve the health of the socio ecosystems with an enduring value for the community. Urban vacant land is a vital source of green infrastructure and creates opportunities for urban green open spaces and creative landscapes.

Vacant parcels contain unimproved green areas and forests. Cultivated and vacant land uses, and expansion of urban built-up spaces are the result of the decrease in vegetation, wetlands, and water. As the result of urban expansion, the environmental functions of vegetation changes when the urban land uses changes altering windspeeds, evaporation of water, storage and exchange of heat and influencing local climate changes in urban areas with their consequences in human comfort and quality of life.

Greening vacant parcels can be transformed into public green spaces to support economic development, social and environmental wellbeing of local neighborhoods by encouraging reinvestment and decreasing gun violence (Branas, Cheney, MacDonald, Tam, Jackson, Ten Have, 2011). Some cities have begun greening their urban vacant land and public open spaces supported by management tools incorporating the urban stormwater management strategy to reduce the rainfall drains.

A city can utilize the greening of vacant lots as integral urban planning strategy for an urban green infrastructure system with benefits for the socio ecologically underserved regions and communities while utilizing green infrastructure. Greening of vacant land projects may be developed on a lot-by-lot base and through site aggregation as for example, next to major roadways, urban development of open public spaces, community gardens, etc.

Urban farming in vacant lots may not be the optimal application to improve urban agroecology and mitigate climate change. Traditional low tech farming hold more advantages that high tech farming in terms of climate change and resource use in urban vacant land sites. Conventional and traditional production methods of tomatoes have potentially lower environmental burdens than greenhouse production in relation of climate change in urban vacant lots. The use of solar panels has greater benefits than urban benefits to mitigate climate change. Planning and implementation capacities to acquire the ownership of different types of vacant land until construction though multiple strategies can be completed by local authorities and local market.

Urban vacant land providing ecosystem services from urban forest structure and green infrastructure may differ from those provided among the different types of vacant land with different land uses. Urban vacant land can be used for urban forests including park systems, street trees, natural resources, and vegetation to create a more sustainable, healthy, and livable city.

Some of the functional benefits of urban vacant land used for forest structures are the structural value of trees, the value of air pollution removal, carbon sequestration and storage, energy savings, avoided runoff, etc. (Nowak, Crane, Dwyer, 2002). Forest infrastructures in urban vacant land sites help to counteract the climate change by removing carbon dioxide through photosynthesis and alter the energy consumption by reducing the emissions from the fossil-fuels burned by power plants (Abdollahi, Ning, Appeaning, 2000).

Vacant land can be a strategically valuable ecological resource that can be utilized as green urban forests infrastructure through for storm water with an impact on the runoff for socioecological services and environmental benefits. Forest structure on current urban vacant land help cities to manage urban storm water to prevent floods and filter the polluted water running off leading to recharge ground water systems.

Urban vacant land infrastructure is a cost-effective way of reducing an expensive infrastructure for storm water management system. The assessment of urban vacant land forest structure functions as urban green infrastructure to provide ecosystem services is required for an effective and efficient urban forest management.

Vacant land vegetation is cost-effective way of reducing storm water management infrastructure able to mitigate urban runoff (Crauderueff, Margolis, Tanikawa, 2012). Urban vacant land cover vegetation and bare soil is a major driver to regulating services and climate regulation, provisioning socio ecosystem services among others food production, air pollution removal, water supply and supporting services, such as nutrient cycling and soil building, etc. (Millennium Ecosystem Assessment, 2005).

The urban forest structures provide economic values through the ecosystem services from urban vacant land and other land uses. Well-maintained urban green vacant land vegetative structures affect both the land raising property values increasing tax receipts and the improvement of the quality of life in the community.

Urban green and forest structures have a structural value in providing substantially greater environmental and socio-ecosystem services, including among others the climate change adaptation, decreased urban heat island effect, energy saving, air pollution removal, carbon sequestration and storage, rainfall interception, and other benefits (Rosenthal, Crauderueff, Carter, 2008). Determining the characteristics of the different forest structures of urban vacant land is vital for a detailed evaluation an accurate estimate of the ecosystem services productivity and value of the urban green infrastructure.

Forest infrastructure on urban vacant land reduce the energy consumption by evaporative cooling in nearby buildings and contributes to regulate climate changes.

In forested areas, vacant lands may produce ecosystem benefits in similar or greater amount of ecosystem services to other forested vacant lands. The plantable space available on urban vacant land sites are a valuable ecological resource when become high relative to other urban land uses and planting on this unused land increases, have considerable potential of being utilized strategically for reducing surface runoff. These urban forests become ecological resources utilized as urban green storm water infrastructure, intercepting precipitation, and promoting infiltration and storage in the soil to reach the vegetation root systems.

Green infrastructure projects on urban vacant land have an impact on urban design properties focusing more on functions of the socio ecosystem services considering their impact on sustainability, social responsibility, and health in socio ecologically underserved communities. Many cities are greening urban vacant land sites as a strategy of storm water management infrastructure and potentially becoming a useful component.

Urban infrastructure as an urban green innovation can contribute in the long term to capacity building aimed to adapt to global warming and climate change in urban areas given their severity, increased risks, and related impacts. Urban green landscapes have an influence either to intensify or to moderate climate change impacts.

Urban vacant land requires the design and implementation of policies, planning and management of urban open green spaces with the inclusion of urban forests, street trees, and any other vegetation as a valuable natural and ecological resources to provide ecosystem services. Urban forest structures as well as green roofs and other green innovation areas have a duality of providing benefits but also some catastrophic events, disease spreading and other negative effects critical for urban planning, design, and development. This can be prevented by implementing and integrated approach for urban forest structure management of vacant land sites as part of the urban green infrastructure, aimed to achieve an optimum balance and maximize benefits and advantages (Baró, Chaparro, Gómez-Baggethun, Langemeyer, Nowak, Terradas, 2014).

Spatial planning policies may enhance urban green infrastructure which be interrupted and lost due to changes driven by pressures in economic development, population growth and urban expansion (Lennon & Scott, 2014; Angel, Sheppard, & Civco, 2005). The design and implementation of policy mechanisms aimed to greening urban vacant lots improve the environmental benefits to the community such as the promotion of urban agriculture to grow food crops for self-consumption and sale Hodgson, K.; Campbell, M.C.; Bailkey, M. (2011)Ackerman, K. (2012). Lawson, L.J. (2005).

Urban planning policies have an impact on green infrastructure of vacant lots based on their physical, biological, and sociological characteristics, which may be of help to identify and catalog on an inventory many different types of vacant land spaces. An inventory to identify vacant lots, open green spaces, green infrastructure and underutilized green areas and parks with agricultural potential is a step to determine vacant land to be committed to urban agroecology and forest infrastructure. However, it is necessary a site assessment before a negotiation process of potential conflicting interests of the stakeholders.

Planning the greening of vacant lots as a cost effective and potential for converting to a natural asset for the benefit of the public health and support the climate change adaptation of the urban community and promotes the quality of life. The use of vacant land forest must be encouraged to preserve it where it would be beneficial and healthier for the community, and the owners must be supported by a structure of tax credits and incentives.

SOCIO-ECOSYSTEM SERVICES

Vacant land contributes to ecosystem services that results on benefits for local communities (Schilling, and Mallach, 2012). Ecosystem services are the benefits obtained from ecosystems as provisioning services, regulating services such as climate regulation, supporting services, recreation, and education (Millennium Ecosystem Assessment, 2003). Urban vacant land has potential value in socio ecosystem benefits based on actively greening vacant lots using urban planning and development management tools (Lynch, 1995). Vacant land supports the ecosystem services obtaining benefits depending on the natural systems, environmental conditions, and management practices. Vacant land has species diversity in the ecosystem relative to other land uses providing natural sustainability and resilience to recover from natural hazards (Shah, 2015).

Urban vacant land provides ecosystem services than other urban land uses, such as biocontrol in gardens and lawns, soil food web, rainwater retention, habitat provision services, carbon capture and climate regulation, among others (Yadav, Duckworth, Grewal, 2012; Grewal, Cheng, Masih, Wolboldt, Huda, Knight, Grewal, 2011; Shuster, Dadio, Drohan, Losco, Shaffer, 2014; Robinson, Lundholm, 2012).

To assess the ecosystem services productivity as the benefits from the forest structures of urban vacant land uses, must be considered that may have different results and outcomes depending on the type and extent of residential, industrial, commercial, etc., characteristics of the surrounding urban areas.

CLIMATE CHANGE HAZARDS

Rapid urbanization processes have increased the socioecological and environmental threats resulting in climate change and its consequences in alterations of biochemical cycles, ozone depletion, chemical pollutants, flooding, modifications of the water balance, etc. (Fohrer et al., 2001; Rosa et al., 2004). The new context of urban areas requires to build and maintain an adaptive capacity as the ability to anticipate and influence the urbanization features of their landscapes aimed to moderate and manage climate related risks and climate change impacts (Sun, Li, Fu, Li, & Tang, 2013). Risk is a function of climate hazards and the vulnerability of the urban area. In urban areas, climate change patterns of vulnerability and risk are uneven (Carter et al., 2015).

It is broadly accepted that climate change is occurring assuming that the different potential of land use has effects on the production of temperature and precipitation (IPCC, 2013). The surface temperatures of land uses exchange moisture and energy with the environment with rapid variations depending on time and space and having an impact on the climate change. These land surface temperatures have an impact on heat and radiation balance in global climate changes (Wubet, 2003; Ifatimehin, 2007 and Weng, 2001).

The long-term relationship between urban green infrastructure and urban land use change has implications on the adaptive capacity of urban spaces to enhance the spatial context of climate change impacts on related biophysical processes (Blanco, McCarney, Parnell, Schmidt, & Seto, 2011). Temporal and spatial changes in the land uses, vacant land and built-up spaces increase the land surface temperature and encourage environmental hazards associated with climate change.

Climate change is an alarming current trend in urban dynamics in developing countries challenging the energy consumption in transport-land uses (Grazi, van den Bergh, 2008). The exploration of oil sands and shale oils is a relevant option to fossil fuel scarcity and not necessarily climate change risks and therefore creates environmental damage (Rogner, 1997).

The capacity of urban spaces to address climate-related risks is affected by changes on land uses. A gradual urban decline in urban land uses is reflected in the challenges of the declining economic activities related to the slowing of development activities, other socioeconomic factors, and the raising of a set of environmental issues of urban adaptive capacity in the context of climate change. This could increase the risks linked to climate change hazards such as increase in the level of evapotranspiration of land cover, heatwaves, and flooding in urbanized spaces lacking green infrastructure. etc.

A systematic conceptual approach of the relationship between spatial planning, structure and transport and climate change relates to the spatial transportation planning influencing climate change control (Greene and Schafer, 2003, IEA, 2001b). Climate change analysis is dominated by models of local spatial organization of urban dynamics using aggregate approaches of urban planning and development, aimed to design climate policy and the implications in urban sustainability (Newman and Kenworthy, 1999).). Climate change hazards are projected to emerge more strongly and become more severe with higher influence on growth and development of urban areas.

The urban-local dimension of climate policy identifies planning options in direct and indirect transport policies and non-transport policies. Spatial local planning and climate policy contributes to effective and efficient mitigation of greenhouse gas emissions. Urban climate hazard change impacts are projected to intensify (IPCC, 2014). Urban green spaces provide several among which is the climate change adaptation (Rosenthal, Crauderueff, Carter, 2008). There are several applications based on scenario approaches to analyze changes in the urban climate as an impact of land use change although there is a gap in the research of the relationship between urban land use and climate change adaptation (Storch and Downes, 2011; Tong, Sun, Ranatunga, He, and Yang (2012

An adaptive capacity is related to a set of enabling conditions supporting the adaptation to climate change impacts (EEA 2012: 63). There are many elements of adaptive capacity that are detrimental to climate change, including, among others, the lack of governance structures, social and urban community cohesion. The loss of adaptive capacity linked to green infrastructure functions increases the level of risks associated with climate change hazards.

Adaptive capacity is the capacity for preventing and managing the impacts of climate change (Swart et al. 2012: 11). Adaptive capacity interrelates a wide range of issues like the awareness of climate change, urban land use and urban green infrastructure coverage, institutional factors such as strategies and governance (Haddad, 2005; Smit & Wandel, 2006; Swart et al., 2012; Yohe & Tol, 2002). Urban green infrastructure enhances adaptive capacity to respond to global warming and climate change impacts (Swart et al. 2012)

The adaptive capacity that may exist within cities in the context of the urban green land use areas and infrastructure spaces and networks has an important role in the adaptation to climate change increasingly recognized by research communities and policy making (European Commission, 2013; Gill, Handley, Ennos, & Pauleit, 2007). The adaptive capacity support activities to address the gap by reducing the urban climate risks. An urban spatial organization analysis of climate change policy for effective and efficient mitigation of greenhouse gas emissions allows assessing mechanisms of spatial urban policy, planning and development in relation to climate change policies.

Climate change adaptation in declining and expanding cities requires to be supported by strategic planning of green infrastructure Cities must develop urban strategies to grow and expand by enhancing urban green infrastructure to moderate climate change risks. Designing and implementing climate change adaptation strategies, support decision making to respond to the implications of shrinking and declining urban spaces, land uses, and enhance the potential value of adaptive capacity to foster urban green innovation.

RESULTS

The dizzying changes faced by today's society, whether of an economic or health nature (COVID-19), affect the development of cities, causing them to expand or contract. This results in the generation of vacant land, which represents a possibility for the creation of green spaces. Now, as part of the bibliometric study, it was possible to recognize through the literature review, different relevant success cases that allow us to glimpse the uses of urban vacant land, green infrastructure and its impact on the risks of climate change.

With respect to the above, the adaptation of abandoned land into new green areas provides a better image to cities, in addition to the fact that these recreational places encourage community participation, thus improving the quality of life of the inhabitants, as described by Bonham, Spilka, Rastorfer, (2002); Bonham and Spilka, (1995). Similarly, as described by Branas, Cheney, MacDonald, Tam, Jackson, Ten Have, (2011), the greening of vacant plots of land, transformed into public green spaces help social and environmental well-being, thereby reducing gun violence.

It should be noted that urban brownfield forestry infrastructure contributes to climate change mitigation by removing carbon dioxide, as noted by Abdollahi, Ning and Appeaning (2000) and Nowak, Crane and Dwyer (2002). Likewise, brownfield vegetation reduces stormwater management infrastructure, as described by Crauderueff, Margolis and Tanikawa (2012). According to Schilling and Mallach; Yadav, Duckworth and Grewal (2012); Grewal, Cheng, Masih, Wolboldt, Huda, Knight and Grewal (2011); Shuster, Dadio, Drohan, Losco and Shaffer (2014); Robinson and Lundholm (2012), therefore, brownfields contribute to systemic services that, in turn, have a beneficial impact on the community.

Therefore, based on the methodology used in this study, it is possible to identify the significance of brownfields rehabilitation, as well as its impact on climate change abatement.

DISCUSSION

The changes in land surface temperatures of the land uses lead to an increase in surface temperature of urban vacant land, urban built-up spaces, and stream. Poor urban planning of uncontrolled land uses, and inadequate environmental management leads to socioecological and environmental problems that alter urban climate (UNCHS, 1996; Zhau *et al*, 2006; Kalnay and Cai, 2003). The consequences of urban climate change calls for immediate action to avert negative environmental implications.

Revitalization of vacant land can be improved through the design and implementation of strategies addressing tax incentive structures, low rates to infill development of land forest structures and rehabilitation abatement to increase the value.

Adaptive capacity is needed to design and implement urban planning related strategies to adapt and deliver benefits linked to climate hazard change including land use banking and the use of vacant sites to create urban green spaces where develop green infrastructure projects (Schilling & Logan, 2008). Urban areas must maintain and enhance the adaptive capacity to the changing climate to determine the vulnerability to the severity of climate hazards and related risks. High adaptive capacity in urban areas tends to moderate the vulnerability and severity of climate-related risks.

Urban planning may incorporate vacant lots in its urban management strategy to promote cost effective greening vacant lots, as the case of stablishing the use of vacant lots in the stormwater management strategy to reduce the drains of the rainfalls, protecting the supply of drinking water, improving the safety of waterways, and mitigating the impact of hazards such as flooding.

REFERENCES

Abdollahi, K. K., Ning, Z. H., & Appeaning, A. (2000). Gulf Coast Regional Climate Change Council. In *Global Climate Change & the Urban Forest*. Franklin Press.

Accordino, J., & Johnson, G. T. (2000). Addressing the vacant and abandoned property problem. *Journal of Urban Affairs*, 22(3), 301–315. 10.1111/0735-2166.00058

Ackerman, K. (2012). *The Potential for Urban Agriculture in New York City: Growing Capacity, Food Security, & Green Infrastructure*. Urban Design Lab at the Columbia Earth Institute.

Alexander, F. S. (2005). *Land Bank Authorities: A Guide for the Creation and Operation of Local Land Banks*. Local Initiatives Support Corporation. http://www.nhc.org/media/files/Land_Bank_Authorities -Alexander.pdf

Angel, S., Sheppard, S., & Civco, D. (2005). *The dynamics of global urban expansion*. Transport and Urban Development Department, World Bank.

Baró, F., Chaparro, L., Gómez-Baggethun, E., Langemeyer, J., Nowak, D. J., & Terradas, J. (2014). *Contribution of ecosystem services to air quality and climate change mitigation policies: The case of urban forests in Barcelona*, Spain. *Ambio*, 43(4), 466–479. 10.1007/s13280-014-0507-x24740618

Blakeman, C., Brown, B., Fitzpatrick, B. T., Shaw, I., & Williamson, A. (2008). City-Wide Brownfield Redevelopment Plan*; Report Prepared for the City of Roanoke*. City Council: Roanoke, VA.

Blakeman, C., Brown, B., Fitzpatrick, B. T., Shaw, I., & Williamson, A. (2008). *City-Wide Brownfield Redevelopment Plan*. City Council.

Blanco, H., McCarney, P., Parnell, S., Schmidt, M., & Seto, K. (2011). The role of urban land in climate change. In Rozensweig, C., Solecki, W., & Hammer, S. (Eds.), *First Urban Climate Change Research Network (UCCRN) assessment report on climate change in cities*. Cambridge University Press. 10.1017/CBO9780511783142.014

Bonham, J. B.Jr, & Spilka, G. (1995). *Urban Vacant Land: Issues and Recommendations*. Pennsylvania Horticultural Society.

Bonham, J. B., Spilka, G., & Rastorfer, D. (2002). *Old Cities/Green Cities: Planning Advisory Service Report Number 506/507*. American Planning Association: Chicago, IL, USA.

Bowman, A. O. M., & Pagano, M. A. (2004). *Terra Incognita: Vacant Land and Urban Strategies*. Georgetown University Press.

Branas, C. C., Cheney, R. A., MacDonald, J. M., Tam, V. W., Jackson, T. D., & Ten Have, T. R. (2011). A difference in differences analysis of health, safety, and greening vacant urban space. *American Journal of Epidemiology*, 174(11), 1296–1306. 10.1093/aje/kwr27322079788

Brandes, D. (1995). The flora of old town centers in Europe. In Sukoop, H., Numata, M., & Huber, A. (Eds.), *Urban Ecology as the Basis for Urban Planning*. SPB Academic Publishing.

Burkholder, S. (2012). The new ecology of vacancy: Rethinking land use in shrinking cities. *Sustainability (Basel)*, 4(6), 1154–1172. 10.3390/su4061154

Carter, J. G., Cavan, G., Connelly, A., Guy, S., Handley, J., & Kazmierczak, A. (2015). Climate change and the city: Building capacity for urban adaptation. *Progress in Planning*, 95, 1–66. 10.1016/j.progress.2013.08.001

City of Pittsburgh. (2012). *Greening Resources*. https://www.pittsburghpa.gov/neighborhoodinitiatives/greenup/resources.htm

Coleman, A. (1982). Dead space in the dying inner city. *The International Journal of Environmental Studies*, 19(2), 103–107. 10.1080/00207238208709976

Crauderueff, R., Margolis, S., & Tanikawa, S. (2012). *Greening Vacant Lots: Planning and Implementation Strategies*. The Nature Conservancy.

Darlington, A. (1969). *Ecology of Refuse Tips*. Heinemann Educational.

Dietersdorfer, L., Efremova, V., Fernandez Agueda, B., Fleschurz, R., Mangialardi, G., Piscitelli, C., & Wolff, M. (2012). *Urban shrinkage and chances for adaptation to climate change*. Final report EU-COST action: TU0803.

European Commission. (2013). *Building a green infrastructure for Europe*. European Commission.

European Environment Agency (EEA). (2012). *Urban adaptation to climate change in Europe: Challenges and opportunities for cities together with supportive national and European policies*. Office for Official Publications of the European Union.

Fohrer, N., Haverkamp, S., Eckhardt, K., & Frede, H. G. (2001). Hydrologic Response to Land Use Changes on the Watershed Scale. *Hydrology. Oceans and Atmosphere*, 26(7-8), 577–582.

Gill, S., Handley, J., Ennos, A., & Pauleit, S. (2007). Adapting cities for climate change: The role of the green infrastructure. *Built Environment*, 33(1), 115–133. 10.2148/benv.33.1.115

Grazi, F., & van den Bergh, J. C. J. M. (2008). Spatial organization, transport, and climate change: Comparing instruments of spatial planning and policy. *Ecological Economics*, 67(4), 630–639. https://www.sciencedirect.com/science/article/pii/S0921800908000426. 10.1016/j.ecolecon.2008.01.014

Greene, D. L., & Schafer, A. (2003). *Reducing greenhouse gas emissions from U.S transportation*. Pew Center on Global Climate Change.

Grewal, S. S., Cheng, Z., Masih, S., Wolboldt, M., Huda, N., Knight, A., & Grewal, P. S. (2011). An assessment of soil nematode food webs and nutrient pools in community gardens and vacant lands in two post-industrial American cities. *Urban Ecosystems*, 14(2), 181–194. 10.1007/s11252-010-0146-3

Haddad, B. (2005). Ranking the adaptive capacity of nations to climate change when socio-political goals are explicit. *Global Environmental Change*, 15(2), 165–176. 10.1016/j.gloenvcha.2004.10.002

Hall, P. (1993). Forces shaping urban Europe. *Urban Studies (Edinburgh, Scotland)*, 30(6), 883–898. 10.1080/00420989320080831

Hodgson, K., Campbell, M. C., & Bailkey, M. (2011). *Urban Agriculture: Growing Healthy, Sustainable Places*. American Planning Association.

Hope, D., Gries, C., Zhu, W., Fagan, W. F., Redman, C. L., Grimm, N. B., & Kinzig, A. (2003). Socio-economics drive urban plant diversity. *Proceedings of the National Academy of Sciences of the United States of America*, 100(15), 8788–8792. 10.1073/pnas.153755710012847293

IEA. (2006). *Energy technology perspectives: Scenario and strategies to 2050, International Energy Agency*. IEA.

Ifatimehin, O. O. (2007). An Assessment of Urban Heat Island of Lokoja Town and Surroundings Using Landsat ETM Data. *FUTY Journal of the Environment*, 2(1), 100–108.

IPCC. (2001). *Climate Change 2001: The Scientific Basis* (Albritton, D. L., Eds.). Cambridge University Press.

IPCC. (2014). *Climate change 2014: Impacts, adaptation, and vulnerability. Part A: Global and sectoral aspects. Intergovernmental Panel on Climate Change. Contribution of working group II to the fifth assessment report of the intergovernmental panel on climate change.* Cambridge: Cambridge University Press.

Jakle, J. A., & Wilson, D. (1992). *Derelict Landscapes: The Wasting of America's Built Environment*. Rowman & Littlefield.

Jehlík, V. (1986) *The Vegetation of Railways in Northern Bohemia*. Academia: Praha, Czech Republic.

Kalnay, E & Cai, M. (2003). Impact of Urbanization and Land Use change on Climate. *Nature*, *423*, 528-531. , http://dx.doi.org/10.1038/nature01675

Kim, G. (2016). The Public Value of Urban Vacant Land: Social Responses and Ecological Value. *Sustainability (Basel)*, 8(5), 486. 10.3390/su8050486

Kim, G., Miller, P. A., & Nowak, D. J. (2015). Assessing urban vacant land ecosystem services: Urban vacant land as green infrastructure in the City of Roanoke, Virginia. *Urban Forestry & Urban Greening*, 14, 519–526. 10.1016/j.ufug.2015.05.003

Kim, G., Miller, P. A., & Nowak, D. J. (2016). The value of green infrastructure on vacant and residential land in Roanoke, Virginia. *Sustainability (Basel)*, 8(4), 296. 10.3390/su8040296

Klimeš, L. (1987). Succession in road bank vegetation. *Folia Geobotanica et Phytotaxonomica*, 22(4), 435–440. 10.1007/BF02853242

Kremer, P., Hamstead, Z. A., & McPhearson, T. A. (2013). Social-ecological assessment of vacant lots in New York City. *Landscape and Urban Planning*, 2013(120), 218–233. 10.1016/j.landurbplan.2013.05.003

Lawson, L. J. (2005). *City Bountiful: A Century of Community Gardening in America*. University of California Press. 10.1525/9780520931558

Lennon, M., & Scott, M. (2014). Delivering ecosystems services via spatial planning: Reviewing the possibilities and implications of a green infrastructure approach. *The Town Planning Review*, 85(5), 563–587. 10.3828/tpr.2014.35

Little, G. M. (2008). *629 Main Street—A Revisioning of an Urban Void: An Exploration into Challenging Cultural Perceptions of an Urban Vacant Lot*. The University of Manitoba.

Lynch, K. (1995). The openness of open space. In Banerjee, T., & Southworth, M. (Eds.), *City sense and city design: Writings and projects of Kevin Lynch* (pp. 396–412). The MIT Press.

Martinez-Fernandez, C., Audirac, I., Fol, S., & Cunningham-Sabot, E. (2012). Shrinking cities: Urban challenges of globalization. *International Journal of Urban and Regional Research*, 36(2), 213–225. 10.1111/j.1468-2427.2011.01092.x22518881

McKinsey and Company. (2016). *Urban world: Meeting the demographic challenge*. McKinsey and Company.

Millennium Ecosystem Assessment. (2003). *Ecosystems and Human Well-Being: Synthesis. World Health*. Island Press.

Németh, J., & Langhorst, J. (2014). Rethinking urban transformation: Temporary uses for vacant land. *Cities (London, England)*, 40, 143–150. 10.1016/j.cities.2013.04.007

Neukrug, H. M. (2009). *Clean Water, Green City: Blending Interests of Land and Water*. Office of Watersheds.

Newman, P., & Kenworthy, J. (1999). *Sustainability and Cities: Overcoming Automobile Dependence*. Island Press.

Nowak, D. J., Crane, D. E., & Dwyer, J. F. (2002). Compensatory value of urban trees in the United States. *Arboriculture & Urban Forestry*, 28(4), 194–199. 10.48044/jauf.2002.028

Pagano, M. A., & Bowman, A. O. M. (2000). *Vacant Land in Cities: An Urban Resource*. Brookings Institution, Center on Urban and Metropolitan Policy.

Pike, A., MacKinnon, D., Coombes, M., Champion, T., Bradley, D., Cumbers, A., & Wymer, C. (2016). *Uneven growth: Tackling city decline*. Joseph Rowntree Foundation.

Robinson, S. L., & Lundholm, J. T. (2012). Ecosystem services provided by urban spontaneous vegetation. *Urban Ecosystems*, 15(3), 545–557. 10.1007/s11252-012-0225-8

Rogner, H. H. (1997). An assessment of world hydrocarbon resources. *Annual Review of Energy and the Environment*, 22(1), 217–262. 10.1146/annurev.energy.22.1.217

Rosa, E. A., York, R., & Dietz, T. (2004). Tracking the Antropogenic Drivers of Ecological Impacts. *Ambio*, 33(8), 509–512. 10.1579/0044-7447-33.8.50915666682

Rosenthal, J. K., Crauderueff, R., & Carter, M. (2008). *Urban Heat Island Mitigation Can Improve New York City's Environment*. Sustainable South Bronx.

Ruth, M., & Coelho, D. (2007). Understanding and managing the complexity of urban systems under climate change. *Climate Policy*, 7(4), 317–336. 10.1080/14693062.2007.9685659

Sackey, A. N. A. (2008). Environmental Impacts of Rapid Urbanization of Ethiopia's Cities in M.P. van Dijk and J. Fransen (eds) *Managing Ethiopian Cities in an Era of Rapid Urbanization*. Eburon Delft.

Sassen, S. (2001). The global city: New York, *London, Tokyo*. Princeton, NJ: Princeton University Press. 10.1515/9781400847488

Schilling, J., & Logan, J. (2008). Greening the rust belt: A green infrastructure model for right sizing America's shrinking cities. *Journal of the American Planning Association*, 74(4), 451–466. 10.1080/01944360802354956

Schilling, J. M., & Mallach, A. (2012). *Cities in Transition: A Guide for Practicing Planners*. American Planning Association.

Shah, A. (2015) *Why Is Biodiversity Important? Who Cares?* Global Issues. https://www.globalissues.org/article/170/why-is-biodiversity-important-who-cares

Shuster, W. D., Dadio, S., Drohan, P., Losco, R., & Shaffer, J. (2014). Residential demolition and its impact on vacant lot hydrology: Implications for the management of stormwater and sewer system overflows. *Landscape and Urban Planning*, 125, 48–59. 10.1016/j.landurbplan.2014.02.003

Smit, B., & Wandel, J. (2006). Adaptation, adaptive capacity, and vulnerability. *Global Environmental Change*, 16(3), 282–292. 10.1016/j.gloenvcha.2006.03.008

Storch, H., & Downes, N. (2011). A scenario-based approach to assess Ho Chi Minh City's urban development strategies against the impact of climate change. *Cities (London, England)*, 28(6), 517–526. 10.1016/j.cities.2011.07.002

Sukopp, H., Blume, H. P., & Kunick, W. (1979). The soil, flora, and vegetation of Berlin's waste lands. *In Nature in Cities: The Natural Environment in the Design and Development of Urban Green Space*; Wiley. *New Jersy, NY, USA*, 1979, 115–132.

Sun, Z., Li, X., Fu, W., Li, Y., & Tang, D. (2013). Long-term effects of land use/land cover change on surface runoff in urban areas of Beijing, China. *Journal of Applied Remote Sensing*, 8(1), 084596. 10.1117/1.JRS.8.084596

Swart, R., Fons, J., Geertsema, W., van Hove, B., Gregor, M., Havranek, M., & Peltonen, L. (2012). *Urban vulnerability indicators: A joint report*. ETC-CCA & ETC-SIA. http://cca.eionet.europa.eu/docs/TP_3-2012,

The Greening of Detroit. (2012). Urban Agriculture and Open Space. The Greening of Detroit. http://detroitagriculture.met/open-space-stories/

Tong, S., Sun, Y., Ranatunga, T., He, J., & Yang, Y. (2012). Predicting plausible impacts of sets of climate and land use change scenarios on water resources. *Applied Geography (Sevenoaks, England)*, 32(2), 477–489. 10.1016/j.apgeog.2011.06.014

UNCHS. (1996). *An Urbanizing World. World Global Report on Human Settlements*. Oxford University Press, New York.

U.S. Department of Housing and Urban Development. (2009). *Revitalizing Foreclosed Properties with Land Banks*. Diane Publishing.

Wachter, S., Scruggs, G., Voith, R., & Huang, L. (2010). *Redevelopment Authority of the City of Philadelphia Land Use and Policy Study*. Penn Institute for Urban Research, Econsult Corporation.

Weng, Q. (2001). A Remote Sensing-GIS Evaluation of Urban Expansion and Its Impact on Surface Temperature in the Zhujiang Delta, China. *International Journal of Remote Sensing*, 22(10), 1999–2014.

Wubet, M. T. (2003). *Estimation of Absolute Surface Temperature by Satellite Remote Sensing*, [Unpublished M. Sc Thesis, International Institute for Geoinformation Science and Earth Observation, Netherlands].

Yadav, P., Duckworth, K., & Grewal, P. S. (2012). Habitat structure influences below ground biocontrol services: A comparison between urban gardens and vacant lands. *Landscape and Urban Planning*, 104(2), 238–244. 10.1016/j.landurbplan.2011.10.018

Yohe, G., & Tol, R. (2002). Indicators for social and economic coping capacity: 'Moving towards a working definition of adaptive capacity. *Global Environmental Change*, 12(1), 25–40. 10.1016/S0959-3780(01)00026-7

Zhau, S., Da, L., Tang, Z., Fang, H., Song, K., & Fang, J. (2006). Ecological Consequences of Rapid Urban Expansion: Shanghai, China *Frontiers in Ecological Environ. The Ecological Society of America*, 4(7), 341–346.

Chapter 10
Integration of Building Information Modelling (BIM) Into the Construction Project Scheduling

Lam Tatt Soon
Taylor's University, Malaysia

Jasmine Yi Xuan Liew
Taylor's University, Malaysia

Tze Shawn Lim
https://orcid.org/0000-0002-6291-5287
Taylor's University, Malaysia

Kenn Jhun Kam
https://orcid.org/0000-0003-4927-4611
Taylor's University, Malaysia

Boon Tik Leong
Taylor's University, Malaysia

ABSTRACT

Construction management assumes a crucial role in overseeing a project's time, cost, and quality, particularly in large-scale projects. Building information modelling (BIM) is expected to enhance construction management by providing improved insights. To refine construction management further, 4D BIM can be integrated into projects by incorporating time or schedule-related information into the model, thereby establishing a more structured workflow. The aim of this research is to examine the application of 4D BIM in construction management, showcasing the value of BIM brings to construction projects. Additionally, the research also establishes a 4D BIM scheduling process for residential buildings and develops a 4D project scheduling workflow within a BIM environment. The research methodology includes desk analysis based on a case study of a residential building in Malaysia. This research serves

DOI: 10.4018/979-8-3693-1650-4.ch010

as a guide for developers and contractors in simulating and accessing construction sequence scheduling within the context of 4D BIM implementation.

INTRODUCTION

Construction management is a recognized professional service that provides effective oversight of project scheduling, coordination, cost, quality, safety, scope, and function on behalf of the project's key stakeholders from the early stages until completion. Construction management is compatible with various project procurement methods, including the traditional method, Design and Build, Design-bid-Build, and Public Private Partnerships (PPP). Therefore, construction schedules need to be meticulously planned to determine project duration and can be compared with measured performance (Gledson & Greenwood, 2016). However, preparing Gantt charts can be challenging without the aid of advanced software, especially when updates occur frequently. Construction professionals may find it confusing, as adding more activities or making changes to the schedule affects the entire chart. Traditional Gantt charts fail to incorporate other critical factors, such as the level of difficulty and resources needed for each task. Hence, the adoption of Building Information Modelling (BIM) can enhance the construction management process.

BIM is a model-based practice that enables users to efficiently manage building data. The 4-dimensional (4D) BIM was introduced as the next generation of construction management for developing schedules, detailed planning, and data analysis. The 4D BIM phase extends the progress of the 3D BIM model by incorporating the fourth dimension, time (scheduling), into the 3D model of the project (Daga, 2022). Construction professionals can adopt BIM scheduling by linking specific start and end dates to every activity involved in the construction project. Therefore, Bentley Synchro 4D will be analysed to increase awareness among construction professionals and promote good software development practices. BIM adoption implies 'the successful implementation whereby an organization, following a readiness phase, crosses the 'Point of Adoption' into one of the BIM capability stages, namely, modelling, collaboration, and integration (Succar & Kassem 2015). Although BIM is recognized as advanced technology in Malaysia, its full benefits have not been realized. The adoption of 4D BIM needs to be managed effectively to improve construction scheduling and planning. To fully leverage the capabilities of 4D BIM, it is proposed to conduct research on BIM implementation steps in the project lifecycle (Wang, 2015). The Malaysian Construction Management will determine the implementation of 4D BIM in future construction projects for the country's economy.

RESEARCH PROBLEMS

As Malaysia gradually embraces digitalization, 4D BIM becomes a convenient application for construction management, especially for developers, contractors, and project managers aiming to boost sustainable growth by spurring innovation. This technique has a high capacity to enhance project management and delivery, regardless of the project's size. However, the complexity of projects increases when managers face challenges involving trade-offs among various resources, such as time, cost, quality, safety, and the environment. Hence, BIM was introduced to Malaysian Construction Management, and its usage has increased over the years. Large Malaysian construction companies have recognized the value of BIM to the extent that they have started implementing it in both the design and construction phases for construction planning (Franco et al., 2015).

Malaysian construction management regularly faces unrealistic expectations from clients and stakeholders. The construction industry largely operates based on reputation, as people tend to work with those they know and trust. Despite offering great advantages, there are still skill gaps among teams causing project delays. To mitigate potential pitfalls, 4D BIM software can be adopted. For 4D BIM, differing priorities affect users' choices. Some prioritize performance, while others prioritize extensibility. Though it is impossible to establish a single standard that satisfies both needs, the adoption of 4D BIM scheduling faces limitations and complications. Construction professionals appear to have limited knowledge regarding the utilization of 4D BIM scheduling and planning (Kassem et al., 2012). According to Migillinskasa et al. (2013), the lack of knowledge in 4D software is a limitation to data exchanges and effective hardware, leading to potential issues. Zhang and Li (2010) argued that 4D software requires high hardware that weakens 3D outputs, resulting in the poor use of information. Without further knowledge, construction professionals might not recognize the best 4D software to implement BIM scheduling in construction management, regardless of their project requirements. Some 4D software does not allow new tasks or even complex discrete functions for programs, although it might not be compulsory for animations, but those not supporting management functions (Tamer, 2015).

There is a common misinterpretation that, before creating a 4D project, the 3D model must be fully completed by the architects and engineers. Alternatively, that is not the case for 4D BIM; BIM scheduling and planning can proceed even with a rough figure from the 3D model. However, unstructured information in 3D BIM models affects the outcome of BIM scheduling, as the 4D software fails to interpret or understand what is represented. Unstructured information includes datasets that are not stored in a preset data model, as it might be human-generated or even in a non-textual format. 3D model data are usually obtained from the as-built structural components within the 3D software sensing range and view (Kim et al., 2013). Thus, this research shall conduct a desk analysis and case study to gain in-depth knowledge and understanding of 4D BIM software; specifically, Bentley Synchro 4D. The case study will be carried out by developing 4D scheduling models in Synchro software, analysing and differentiating the construction management workflow of BIM. This will be constructed for planning and scheduling in Malaysian Construction Management to further understand their specific requirements.

RESEARCH OBJECTIVES

1. To determine a 4D BIM scheduling process for residential buildings.
2. To develop a 4D project scheduling workflow within a BIM environment.

LITERATURE REVIEW

Malaysian Construction Management

Construction management is a well-known service that involves the planning, budgeting, coordination, monitoring, control, and reporting of key business processes, such as procurement, production, marketing, accounts, and financial requirements. Its objective is to achieve financial success and profitability for an organization engaged in construction facilities (Harris et al.,2021). Typically, clients, contracting companies, consulting firms, or stakeholders undertake a series of projects, ensuring a safe budget, on-time completion, social responsibility, and adherence to quality and expected standards, impacting innovation, aesthetics, and the environment (Harris et al., 2021). To maintain project schedules on time and within budget, precise and structured scheduling must be employed for effective construction management (Stonemark, 2022). However, Malaysian construction management faces numerous challenges, particularly concerning time management issues leading to project delays. Poor construction management can result in cost and time overruns, litigation, disputes, and even project abandonment. Causes of these challenges include poor software installation, inadequate time management, subpar site work, and more (Westland, 2006).

Building Information Modelling (BIM) Scheduling

Building Information Modelling (BIM) is an emerging technology gaining traction in Malaysian construction management. BIM has demonstrated its clear capabilities and benefits within the construction industry in Malaysia, representing a significant leap toward project success. Furthermore, recognizing and understanding the value of BIM enables high-performance possibilities in construction management (Al-Ashmori et al., 2020). To facilitate multiparty collaboration, synchronous and asynchronous data control for the 3D model is necessary (Zhou et al., 2012). The construction management team can adopt BIM for a clearer visualization of the project, easy identification of potential issues, and effective tracking of site work progress. BIM scheduling provides time-based data that, when used with various information model components, allows pinpointing a specific work area or item, sometimes with a precise design and deployment lead time (Hadavi & Tavakolan, 2018). Users employing BIM scheduling can integrate a conventional project schedule from tools like Primavera or Astra with a 3D model to visualize the timeframe for building. When visualizing a project plan, BIM scheduling proves valuable in achieving optimal project results, as visuals have the ability to easily convey crucial information to users (VIATechnik, 2022).

BIM scheduling broadens the perspective for construction planning, encompassing building, designing, maintaining, or creating a standard information model that contains all the necessary data for the project's life cycle. The coordination between planners and clients during the planning phase is significantly enhanced by 4D scheduling. This enables project stakeholders to monitor project development and identify problems early on (Hadavi & Tavakolan, 2018). BIM scheduling software facilitates the construction of virtual projects and the simulation of various scenarios, providing all stakeholders with a better understanding of how to detect project hazards earlier and minimize such risks by taking appropriate action (Charlesraj & Dinesh, 2020). The key connecting tool between 3D modelling at the design level and planning and scheduling at the construction level in BIM is 4D modelling. By providing a graphical presentation and creating opportunities to identify unforeseen issues and incompatibility conflicts, 4D modelling helps in better understanding the consequences of BIM scheduling. The timetable is a key component for all team members and one of the major determinants of a project's success. The construction management team member responsible for creating the schedule at the start of a project (and regularly before that) is a specialized engineer who plays a significant role in the construction process. This schedule typically reflects experiences with construction timing, material lead time, weather, crew, and equipment concerns. Therefore, new engineers or planners need to be trained with this new technology and its applications.

Building Information Modelling (BIM) Model Development

Initiating with 3D BIM modelling is imperative for any project adhering to the BIM methodology. This process involves collecting both graphical and non-graphical data to create 3D models, and then sharing that data in a Common Data Environment (CDE). 3D BIM modelling includes both high Levels of Development (LOD) and high Levels of Information (LOI) (Sam, 2022). To identify and address any problems before commencing building or construction work, and to avoid unnecessary rework expenditures, project partners must submit accurate information.

Furthermore, when performed correctly, 3D BIM serves as the building block of a BIM project, expediting decision-making from the initial concept design stage through the project's lifespan (Sam, 2022). Used throughout the Architecture, Engineering, and Construction (AEC) industry, 3D BIM can significantly enhance project performance and delivery while providing a unified strategy that encourages cooperation and development. During the design phase, the 3D BIM model offers all sectors a more accurate depiction of the design. Improved communication between various design disciplines can lead to ongoing improvement, significantly reducing design errors and omissions (Ghaffarianhoseini, 2017). To keep all parties involved in the design process informed of the financial consequences as the design develops, cost estimates can be generated from the building model to ensure the most updated design and model without the worry of time consumption (Ghaffarianhoseini, 2017).

Constraints Generating BIM Schedules

Despite the advantages that 4D BIM offers to construction management, a few technical and non-technical obstacles are still impeding the widespread use of BIM scheduling (Hadavi & Tavakolan, 2018). Consideration must be given to technology precedence restrictions and resource constraints when construction scheduling requires specifics at the activity level to develop a useful, detailed schedule. One of the key characteristics of BIM is interoperability, enabling effective information sharing across stakeholders using various applications (Tommasi & Achille, 2017). While there are many software programs on the market (Abanda et al. 2015), none of them can currently handle all BIM application domains. This complicates information exchange between sectors using different tools (Steel et al., 2012), preventing users from sharing sufficient information with one another. Consequently, this software limitation makes it unlikely that individual concurrent planning and 4D modeling efforts would result in collaborative teamwork (Zhou et al., 2012) .

Proper utilization of BIM necessitates significant standardization of BIM procedures and software usage. Every aspect of corporate processes, from vendor component specifications to compliance software, should be streamlined and integrated into a unified system. Additionally, the substantial cost of training at various levels poses a challenge, as organizations may lack the necessary expertise to embrace BIM scheduling. Consequently, constraints on recruiting 4D BIM scheduling experts may persist. To fully appreciate the advantages of BIM adoption, the industry may face significant challenges related to software, training, and expertise (Lymath, 2014). It has also been noted that BIM scheduling has significant hardware requirements and that poor utilization of this data could result from weak 3D outputs (Zhang & Li, 2010).

Analysis of 4D BIM Scheduling

In order to identify existing clashes and design flaws, 4D BIM enables various project teams to be involved early in the design and construction phases of the project. 4D BIM scheduling allows the fusion of BIM and traditional project management scheduling methods. Traditional scheduling utilizes 2D schedules and bar charts, whereas BIM scheduling links the construction schedule to the 3D model for better visualization of the project. Bar charts are particularly useful for conveying a project's present status and planned actions. Traditional scheduling lacks visualization, leaving clients and stakeholders uncertain about their requirements. Moreover, clients and stakeholders might struggle to arrive at a decision, leading to longer resolution times as multiple points of view and individual obstinacy among participants could impede a meeting from reaching a decision (Chand, 2014).

In contrast, BIM scheduling meetings facilitate discussions by helping clients and other participants visualize the issue through 4D models, highlighting the problem. With traditional scheduling, the project management approach is complex, as construction professionals are fragmented and very hierarchical during the design, production, and construction phases, leading to conflicts of interest among the parties involved and undermining collaboration (Zigurat Global Institute of Technology, 2021). Nevertheless, in BIM scheduling, all construction parties focus on one 4D BIM model, allowing information to be shared among them, improving communication, and enabling informed decisions during the project lifecycle.

Researchers have incorporated the notion of measuring the performance of 4D BIM into construction processes, integrating the time dimension into BIM models Huhnt et al., (2010). Unfortunately, these 4D models lack comprehensive information and constraints at the process level. There is no linear relationship between the resources used at each hour and job progress, resulting in information overloaded. A constraint occurring when schedulers fail to extract relevant information from drawings and specifications of the construction design. The computer processing of 3D and 4D models is hampered by excessive details. Additionally, the Level of Development (LOD) of 4D BIM lacks the required degree of information and connectivity at the six-week look-ahead level, often being limited to the master planning or phase planning level (Soman, 2019). A lack of necessary LOD or insufficient detail in the 3D model can lead to inaccurate or inconsistent project scope scheduling (Singh, 2021). Information constraints, such as geometric inconsistencies from linking 3D models to construction schedules and missing data from drawings due to late design, affect the accuracy of BIM scheduling and can result in project delays. The scheduling process suffers from a lack of coordination and synchronization between the design and construction phases, leading to incomplete information (Büchmann & Andersson, 2010).

Scheduling constraints, which are restrictions placed on a project schedule, can impact the start or finish date of an activity. Fixed deadlines imposed on a work item might serve as a type of scheduling constraint. For example, rather than leaving dates to be decided by the planning process, a constraint date can be used to limit the commencement of an activity to occur no earlier than a given date. If the construction schedule defies physics and generates inaccurate values during extraction, it can delay the project, requiring the construction management team to reschedule the timeline. Once building plans or 2D models are accessible, software like Revit can integrate them into 3D models. After creating a 3D model, the next step is to generate construction schedule data from scheduling tools. Every construction project must have an accurate schedule of all events and activities involved to proceed with further planning (Alzarrad et al., 2021). Subsequently, the Industry Foundation Classes (IFC) format should be used to export the 3D Revit file (rvt.). Finally, the prepared 3D model and construction schedule will be imported into the 4D BIM tool to generate BIM scheduling.

RESEARCH METHODOLOGY

The research approach for this study is qualitative research, utilizing a case study and desk analysis to focus on data collection and analysis. This research process does not rely on a body of pre-existing theory but rather on abstractions, concepts, and hypothesis testing. The chosen case study is a double-storey semi-detached house located in Selangor, Malaysia. This residence includes 5 bathrooms, 4 bedrooms, a living room, a dining area, a family area, a dry and wet kitchen, a utility room, a store, and a laundry area. For the desk analysis, BIM models created based on this existing residential house will be used for further analysis and observation, accompanied by framework development. The constructed BIM models specifically focus on the architectural and structural elements of the project. The elements modelled in the BIM model are listed below in Table 1, created using Autodesk Revit 2022. Further analysis of BIM scheduling will be conducted using Synchro 4D Pro.

Table 1. The elements modelled in the BIM model

Architectural Works	Structural Works
Wall & Wall Finishes	Beam
Floor & Floor Finishes	Slab
Staircase	Column
Ceiling	Column Stump
Roof	Pad Footing
Doors & Windows	

RESEARCH FINDINGS

BIM has been implemented into this project as a baseline to study experimentally on the model of the building by creating an architectural and structural model, as well as to deliver precise data for monitoring the building site. This research prioritises on the analysis of 4D BIM scheduling in construction management and scheduling-related issues. The architectural 3D model created using Autodesk Revit at LOD 200-300 based on architectural design drawings that are shown in the tender drawings. The 4D BIM model and construction schedule generated using Synchro 4D Pro after importing the 3D models from Autodesk Revit for the analysis of BIM scheduling.

Development of BIM Scheduling Process

To construct an architectural model in Autodesk Revit, initiate the process by creating a new project and selecting the 'Architectural Template'. Establish levels and grids according to the building drawings, and adjust the Grid Bubble visibility by clicking the square node offset at the end of the gridline. Levels can be drawn by selecting the desired insert spot, and renaming levels is accomplished through a double-click on the existing ones. Proceeding with the setup of levels and grids, the creation of the 3D model begins with walls and wall finishes. Revit's default parameters facilitate the snapping of walls into nearly horizontal or vertical orientations automatically. Unlike previous elements, floors, staircases, railings, and other components are created in 'Sketch Mode', where modifications can be confirmed with a 'tick' or discarded with a 'cross'.

Following the assembly of walls and floors, doors and windows are inserted using the 'Door' and 'Window' options from the 'Architecture' tab. Staircases are created by selecting the 'Stair' option on the 'Architecture' tab, and adjustments to calculation rules, maximum rise height, minimum tread depth, and minimum run width can be made through the 'Edit Type' option. The automatic ceiling feature is enabled by dragging the pointer over a closed area, and the ceiling finishes are automatically inserted once placed. However, ceilings are not visible in floor plans because they are created in plan view, and their pattern is only observable from the top. Roofs, generated by footprint, are defined by the building's boundaries. Overhangs can be added by inserting the measured length from the exterior walls to the extended roof. Modelling the roof in this research study proves challenging due to the complexity of the design and the need for a deeper understanding of the modelling concept. Further analysis is conducted to refine the approach to modelling the roof. Upon completing the architectural model, the next step involves exporting the model to a CAD drawing for the development of the structural model.

To create a structural model in Autodesk Revit, start by generating a structural template. Once the template is in place, proceed to draw levels and grids, ensuring they align with the architectural template to maintain measurement accuracy. It is crucial that the levels and grids match those of the architectural model. For the structural model, include a foundation level to accommodate work below the lowest floor level. Import the architectural CAD drawings into the structural template, aligning them with the structural model. Begin by inserting columns; column stumps are positioned in the foundation level by lengthening the column and adjusting the offset to extend below ground level. Follow this with beams, slabs, and pad footings. The foundation can snap at each column stump, given the prior lengthening of the columns. Synchro 4D Pro requires a BIM model to generate a 4D BIM schedule for further analysis. Consequently, link and bind both the architectural and structural models in Revit to create a comprehensive BIM model of the project. After linking and binding, select the BIM model, ungroup the elements, and edit individual elements as needed. Subsequently, export the BIM model to Industry Foundation Classes (IFC) format for BIM scheduling in Synchro 4D Pro. The IFC format precisely describes every element of the construction project. Alternatively, Synchro 4D Pro also accommodates DWFx format, depending on the preferences of the construction parties. To establish BIM scheduling in Synchro 4D Pro, import the BIM model in IFC format. Manually input schedule data into Microsoft Excel and import it into Synchro 4D Pro. Figure 7 illustrates the imported Excel file in Synchro 4D Pro.

To develop a structural model in Autodesk Revit, initiate the process by creating a structural template. Subsequently, draw levels and grids using the architectural template as a reference, ensuring their alignment with the architectural model's corresponding levels and grids to maintain measurement accuracy. Introduce a foundation level in the structural model to accommodate work below the lowest floor level. Import the architectural CAD drawings into the structural template, aligning them with the structural model. After aligning the architectural CAD drawing and structural model, proceed to insert columns. Column stumps are positioned in the foundation level by lengthening the columns and adjusting the offset to extend below ground level. This is followed by the addition of beams, slabs, and pad footings. The foundation can easily snap at each column stump, given the prior lengthening of the columns. To facilitate further analysis and generate a 4D BIM schedule in Synchro 4D Pro, a BIM model is necessary. Thus, both the architectural and structural models in Revit must be linked and bound together to create a comprehensive BIM model for the entire project. After linking and binding, select the BIM model, ungroup the elements, and edit individual elements within the group. Subsequently, export the

BIM model to Industry Foundation Classes (IFC) format for BIM scheduling in Synchro 4D Pro. The IFC format ensures a comprehensive description of every element within the construction project. Additionally, Synchro 4D Pro also supports DWFx format, catering to the preferences of the construction parties involved.

To create BIM scheduling in Synchro 4D Pro, initiate the process by importing the BIM model in IFC format. Manually input schedule data into Microsoft Excel, and subsequently import this data into Synchro 4D Pro. After importing both the BIM model and construction schedule, the BIM model needs to be subdivided to separate design phases and avoid interference. Start by selecting the objects to be subdivided under the '3D' tab, choosing 'Subdivide.' The 3D Subdivision bounding box will appear, including only the selected objects, with nearby objects appearing as greyed-out references. Subdivision can be applied to multiple objects by clicking the 'Subdivide Multiple Objects' box. To create a time-based 4D BIM model, connect the "components" from the 3D model to the construction activities derived from the construction schedule details. Link the schedule to the BIM model by assigning resources to tasks. Select objects manually, such as first-floor columns, and drag the highlighted box into the task schedule. The disappearance of highlighted objects indicates a successful link between the BIM model and the schedule. Review linked tasks by adjusting the 'Focus Time.' Auto matching is available in Synchro 4D Pro, automatically assigning resources to tasks based on data in the model and schedule. The Work Breakdown Structure (WBS) is automatically organized when creating the construction schedule.

To create a critical path, adjust reschedule options under 'Plan' > 'Reschedule.' Insert an additional column for the critical path in the task schedule, and click 'Compute Critical Path' to generate it. Growth simulation can be applied to show construction progress swiftly. Define the appearance profile and link it to the task properties. 4D BIM scheduling enables project animation after completing the schedule and BIM model. Navigate the 3D window to set up the starting camera angle, define the starting view, and adjust the camera angle in the Animation Editor. Add camera keyframes to update the 3D view as needed, and use the '4D Review' > 'Animation Editor' and 'Animation' to navigate the model. The focus time marker can be dragged to preview different angles, and camera keyframes are added to save specific views during the animation.

Analysis of Construction Management in 4D BIM Scheduling

In this research analysis, the construction period has been adjusted to exclude weekends, mirroring the real-life construction practices in Malaysian construction management. The working hours are set by default from 9:00 a.m. to 5:00 p.m. An average construction duration of 203 days (approximately 6-7 months) was applied to the research case study project. This duration accounts for various factors such as the size and complexity of the unit, material availability, weather conditions, and the efficiency of the construction crew.

Benefits of 4D BIM Scheduling

Based on the data collected in this research study, it is evident that 4D BIM scheduling brings multiple benefits to Malaysian construction management. Synchro 4D Pro offers a range of tools and parameters that streamline construction processes, enabling quicker and more efficient procedures. The construction management in Malaysia can utilize Synchro 4D Pro to closely monitor each task, identify and mitigate design risks, and present 4D visualizations for conveying project information in a digital environment.

The coordination capabilities of 4D BIM scheduling help prevent clashes, and the ability to construct virtual projects and model various scenarios provides stakeholders with a better understanding of identifying potential hazards early on, enabling risk reduction (Charlesraj & Dinesh, 2020).

The data analysis from this research also indicates an improvement in project performance with the implementation of 4D BIM scheduling. Malaysian construction management can leverage 4D BIM scheduling to monitor, assess, and update construction progress effectively. For instance, the linking of task schedules in Synchro 4D Pro, whether done manually or through auto-matching resources, facilitates rapid updates to the project's construction timeline and allows easy tracking of changes in construction sequences. Furthermore, the Gantt chart can be easily kept up-to-date by examining the 3D view in Synchro 4D Pro after linking the schedules, providing enhanced visualization of the project. This stands in contrast to traditional scheduling methods, which lack instructiveness as they do not illustrate the interconnectedness of different activities, leading to a deficiency in visualisation.

Due to the dynamic nature of the construction process, traditional project scheduling methods like the Critical Path Method (CPM) and the program evaluation and review technique often fall short in accommodating the necessary time considerations (Abbasi et al., 2020). Manual scheduling becomes particularly challenging in adjusting designs frequently, making it difficult for construction management to grasp the project schedule accurately. However, the application of 4D BIM scheduling in Synchro 4D Pro addresses this challenge by automatically calculating the critical path, a departure from traditional approaches. Synchro 4D Pro possesses the capability to automatically calculate and display the Critical Path (least float or longest path) through the Critical Path dialog. This feature ensures efficient decision-making and aids in delivering the project on time.

Constraints of 4D BIM Scheduling

During the data analysis process, certain limitations were identified. While these limitations are beyond the researcher's control, acknowledging them is crucial for contextualizing study results, assessing the reliability of the scientific work, and establishing the credibility of published research. In the analysis of 4D BIM scheduling, several limitations emerged, including information, schedule, and software constraints. Despite the evident advantages of 4D BIM scheduling, it becomes apparent that its performance fails to encompass the knowledge and constraints at the construction process level. Researchers evaluating the performance of 4D BIM have integrated this concept into construction processes (Huhnt et al., 2010). Unfortunately, 4D BIM models inadequately represent information and constraints at the construction steps because there is no direct correlation between the resources used at each hour and the progress of the task, requiring the breakdown of positions into sub-tasks to identify causal links among essential elements, as mentioned by (Giretti et al.,2012).

A significant limitation observed during the analysis of 4D BIM scheduling pertains to the constraints of the 4D BIM software. The analysis indicates that only at the data level can plans be updated or synchronized in Synchro 4D Pro. This software limitation hinders effective communication among construction parties, reducing the likelihood of collaborative teamwork in simultaneous planning, scheduling, and 4D modelling efforts. This limitation aligns with findings from Zhou et al. (2012), who noted that while Synchro 4D has the capability to integrate all plan data for individual planners to conduct 4D modelling and 3D planning, Synchro 4D Pro only allows data level plans, limiting users from gaining comprehensive insights from other parties.

Fig. 1 summarizes the research findings, outlining a workflow for the adoption of BIM 4D scheduling in construction management. The depicted research workflow serves as a guide for data collection and analysis, providing researchers with a systematic approach to minimize inconsistencies, errors, and biases in their research. The workflow illustrates that 4D BIM planning and scheduling can be effectively executed using Autodesk Revit, Microsoft Excel, and Synchro 4D Pro. Autodesk Revit contributes the BIM model, Microsoft Excel facilitates the construction schedule preparation for 4D BIM scheduling. Synchro 4D Pro integrates both the BIM model and construction schedule, enabling the visualization of 4D BIM planning and scheduling and fostering communication among construction professionals. Subsequently, the 4D BIM scheduling can undergo analysis in the construction management context to discern its benefits, limitations, and potential future directions, ultimately leading to the formulation of 4D planning and scheduling in construction management.

Figure 1. A workflow of BIM 4D scheduling adoption in construction management

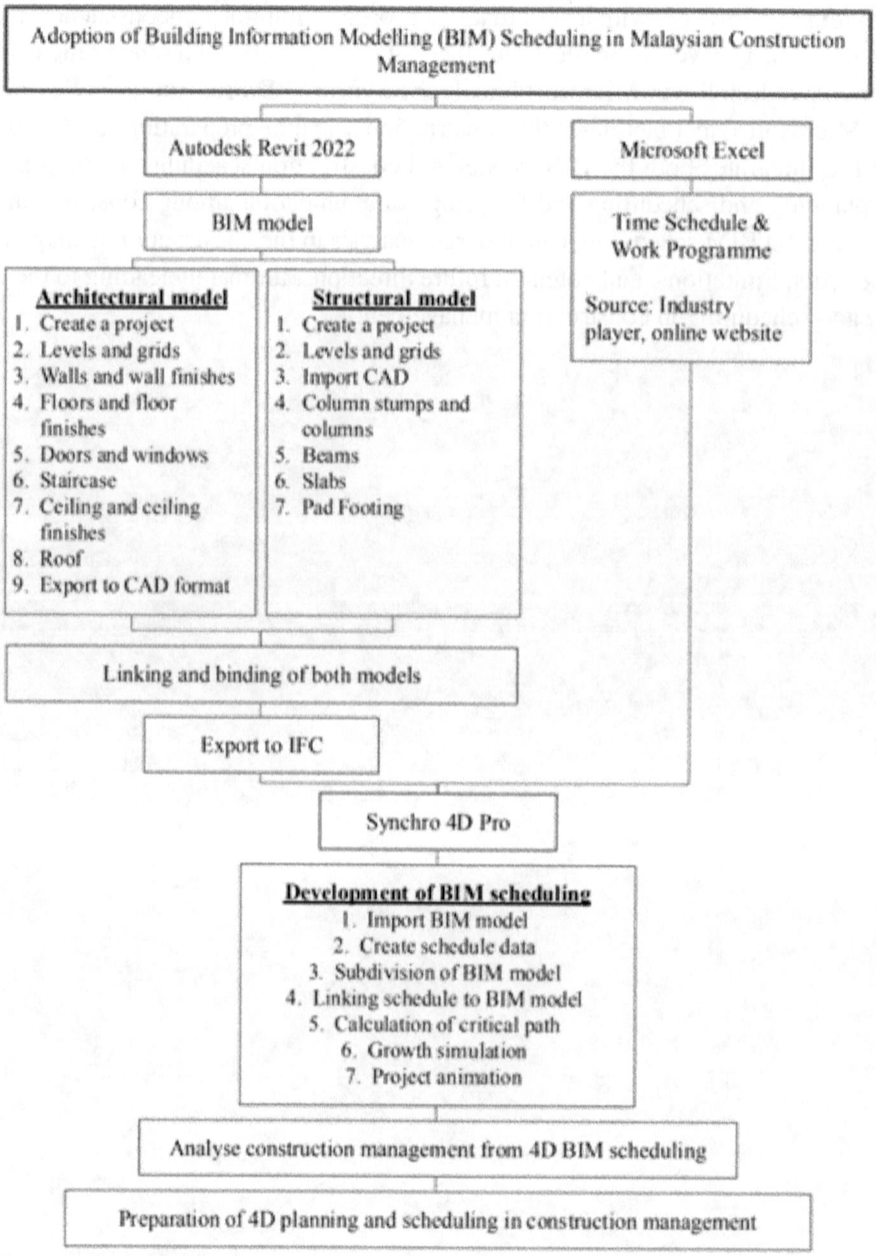

CONCLUSION AND RECOMMENDATIONS

In summary, this research successfully showcases the implementation of BIM scheduling in Malaysian construction management. Undoubtedly, 4D BIM scheduling holds the potential to enhance the efficiency of construction management by offering visualisations that intuitively capture the construction process and fostering improved collaboration among construction professionals and stakeholders through effective communication. The development of BIM 4D scheduling for residential buildings requires both a BIM model from Autodesk Revit and a construction schedule from Microsoft Excel, with Synchro 4D Pro serving as the platform for executing 4D BIM scheduling. Achieving LOD 200-300 in the BIM model is essential to precisely describe dimensions and relative positions. The linking and binding of the BIM model and construction schedule, followed by export to an IFC file for BIM 4D scheduling, form crucial steps in the process. Despite the similarities with previous research, limitations related to individuals, software, time, and budget constraints have been identified, hindering the creation of 4D BIM models during the data gathering phase. Information constraints predominantly stem from the lack of correlation between resources and hours in 4D models, while software constraints arise from the data level limitations allowing only plans to be updated or synchronised. These constraints, highlighted in the analysis of 4D BIM scheduling in construction management, indicate common challenges faced by the Malaysian construction management. The development of a BIM scheduling framework establishes a theoretical and conceptual foundation for planning and scheduling in Malaysian construction management.

The BIM software landscape in Malaysia is continually advancing with the introduction of new features. Anticipated future developments suggest that BIM scheduling will play an even more pivotal role, particularly as construction processes transition toward enhanced communication practices before, during, and after construction. The utilisation of virtual models in BIM scheduling is poised to minimize waste across various stages of a building's life cycle and facilitate the creation of more precise timelines. In short, this research has paved the way for the integration of BIM scheduling practices within Malaysian construction management. The insights gleaned from this study can provide a valuable framework for future researchers aiming to explore BIM scheduling adoption in diverse building types such as commercial structures, schools, hotels, mixed-use developments, and more. Additionally, it is suggested that forthcoming studies delve into the analysis of various BIM scheduling software options to comprehensively assess their adoption in Malaysian construction management. These findings open up new avenues of exploration, suggesting that employing a 3D model supported by network infrastructure could enable collaborative 4D building planning.

REFERENCES

Abanda, F. H., Vidalakis, C., Oti, A. H., & Tah, J. H. M. (2015). A critical analysis of Building Information Modelling systems used in construction projects. *Advances in Engineering Software*, 90, 183–201. 10.1016/j.advengsoft.2015.08.009

Abbasi, S., Taghizade, K., & Noorzai, E. (2020). BIM-based combination of takt time and discrete event simulation for implementing just in time in construction scheduling under constraints. *Journal of Construction Engineering and Management*, 146(12), 04020143. 10.1061/(ASCE)CO.1943-7862.0001940

Al-Ashmori, Y. Y., Othman, I., Rahmawati, Y., Amran, Y. M., Sabah, S. A., Rafindadi, A. D. U., & Mikić, M. (2020). BIM benefits and its influence on the BIM implementation in Malaysia. *Ain Shams Engineering Journal*, 11(4), 1013–1019. 10.1016/j.asej.2020.02.002

Alzarrad, M. A., Moynihan, G. P., Parajuli, A., & Mehra, M. (2021). 4D BIM Simulation Guideline for Construction Visualization and Analysis of Renovation Projects: A Case Study. *Frontiers in Built Environment*, 7, 617031. 10.3389/fbuil.2021.617031

Büchmann-Slorup, R., & Andersson, N. (2010). BIM-based scheduling of Construction–A comparative analysis of prevailing and BIM-based scheduling processes. In *Proc., 27 th Int. Conf. of the CIB W78* (pp. 113-123). Research Gate.

Chand, S. (2014). Meetings: Types, Purpose, Advantages and Disadvantages. Your Article Library. https://www.yourarticlelibrary.com/businesscommunication/meetings-types-purpose-advantages-and-disadvantages/27671

Charlesraj, V. P. C., & Dinesh, T. (2020). Status of 4D BIM implementation in Indian construction. In *ISARC. Proceedings of the International Symposium on Automation and Robotics in Construction* (Vol. 37, pp. 199-206). IAARC Publications.

Daga, L. (2022). *4D BIM - Collaboration of Schedule with 3D Model.*

Ghaffarianhoseini, A., Tookey, J., Ghaffarianhoseini, A., Naismith, N., Azhar, S., Efimova, O., & Raahemifar, K. (2017). Building Information Modelling (BIM) uptake: Clear benefits, understanding its implementation, risks and challenges. *Renewable & Sustainable Energy Reviews*, 75, 1046–1053. 10.1016/j.rser.2016.11.083

Giretti, A., Carbonari, A., Novembri, G., & Robuffo, F. (2012). Estimation of job-site work progress through on-site monitoring. In *Proceedings of the 29th International Symposium of Automation and Robotics in Construction, ISARC 2012*, Kochi,India 10.22260/ISARC2012/0089

Gledson, B. J., & Greenwood, D. J. (2016). Surveying the extent and use of 4D BIM in the UK. *Journal of Information Technology in Construction*, 21, 57–71.

Hadavi, P., & Tavakolan, M. (2018) 4D modeling and BIM-based Project planning using constraint-based simulation for implementation in Iran considering an actual case study. *Construction Research Congress, CRC 2018*, (pp. 542–551). ASCE. 10.1061/9780784481264.053

Harris, F., McCaffer, R., Baldwin, A., & Edum-Fotwe, F. (2021). *Modern construction management.* John Wiley & Sons.

Huhnt, W., Richter, S., Wallner, S., Habashi, T., & Krämer, T. (2010). Data management for animation of construction processes. *Advanced Engineering Informatics*, 24(4), 404–416. 10.1016/j.aei.2010.07.009

Kassem, M., Brogden, T., & Dawood, N. (2012). BIM and 4D planning: A holistic study of the barriers and drivers to widespread adoption. *Journal of Construction Engineering and Project Management*, 2(4), 1–10. 10.6106/JCEPM.2012.2.4.001

Kim, C., Son, H., & Kim, C. (2013). Automated construction progress measurement using a 4D building information model and 3D data. *Automation in Construction*, 31, 75–82. 10.1016/j.autcon.2012.11.041

Lymath, A. (2014) *The Top Five Barriers to BIM Implementation*. NBS. https://www.thenbs.com/knowledge/the-top-five-barriers-to-bimimplementation

Migilinskasa, D., Popovb, V., Juoceviciusc, V., & Ustinovichiu, L. (2013). The benefits, obstacles and problems of practical BIM implementation. *Proceedings of the 11th International Conference on Modern Building Materials, Structures and Techniques*. Elsevier. 10.1016/j.proeng.2013.04.097

Sam, S. (2022). *3D BIM Explained: The Foundation of a BIM Journey*. 3D Design Bureau. https://3ddesignbureau.com/blog/3d-bim-explained-the-foundationof-a-bim-journey/

Singh, H. (2021). *5 Tips for Implementing 4D BIM for Construction Scheduling*. https://www.hitechcaddservices.com/news/tips-to-implement-4d-bim-forconstruction-scheduling/

Soman, R. K. (2019). Modelling construction scheduling constraints using Shapes Constraint Language (SHACL). In *2019 European Conference on Computing in Construction* (pp. 351-358). IEEE. 10.35490/EC3.2019.170

Steel, J., Drogemuller, R., & Toth, B. (2012). Model interoperability in building information modelling. *Software & Systems Modeling*, 11(1), 99–109. 10.1007/s10270-010-0178-4

Stonemark, S. (2022). *How Construction Schedules Keep Projects on Time and Budget*. Stonemark Construction Management. https://stonemarkcm.com/blog/how-construction-schedules-keep-projects-on-time-and-budget/

Succar, B., & Kassem, M. (2015). Macro-BIM adoption: Conceptual structures. *Automation in Construction*, 57, 64–79. 10.1016/j.autcon.2015.04.018

VIATechnik. (2022). *4D BIM and Scheduling*. VIATechnik. https://www.viatechnik.com/services/construction-administration/4d-bim-scheduling/

Wang, Sh. H., Wei-Chih, W., Kun-Chi, W., & Shih-Yu, Sh. (2015). Applying building information modeling to support fire safety management. *Automation in Construction*, 59, 2015. 10.1016/j.autcon.2015.02.001

Westland, J. (2006). *The Project Management Life Cycle: A Complete Step-by-step Methodology for Initiating Planning Executing and Closing the Project*. Kogan Page Limited.

Zhang, J., & Li, D. (2010). Research on 4D virtual construction and dynamic management system based on BIM. In *Proceedings of the International Conference on Computing in Civil and Building Engineering, ICCBE* (pp. 78- 83). IEEE.

Zhou, W., Georgakis, P., Heesom, D., & Feng, X. (2012). Model-based groupware solution for distributed real-time collaborative 4D planning via teamwork. *Journal of Computing in Civil Engineering*, 26(5), 597–611. 10.1061/(ASCE)CP.1943-5487.0000153

Chapter 11
Educational Dynamics in Urban Study:
Unraveling Student Preferences and Expectations in BQS Study

Kenn Jhun Kam
https://orcid.org/0000-0003-4927-4611
Taylor's University, Malaysia

Wei Vern Tan
Taylor's University, Malaysia

Tze Shwan Lim
https://orcid.org/0000-0002-6291-5287
Taylor's University, Malaysia

Lam Tatt Soon
Taylor's University, Malaysia

Chin Ai Ling Shirley
Taylor's University, Malaysia

ABSTRACT

Understanding the factors influencing students' decisions in choosing a university for tertiary education is important. The factors under consideration in this study include employability, academic quality, campus environment and facilities, reputation of university, geographic location of university, family influences and peers' influences. These insights can help universities to enhance programme quality and remain competitive in the education field. The chapter's finding provides university with insight into identifying students' preferences to enroll in the bachelor of quantity surveying (BQS) degree programme. The proposed research method for this study utilized the questionnaire survey that will be distributed to the targeted respondents of this study to gather data and information. This study is expected to analyse the determinants influencing students' decision to enroll in Taylor's University BQS degree programme, an analysis of students' experiences versus expectations of Taylor's University BQS degree programme,

DOI: 10.4018/979-8-3693-1650-4.ch011

and practical ways to improve students' enrolment rate.

BACKGROUND

Since 1824, Malaysia has undergone various reforms, and during British Colonialism, Malaysia's schooling system was structured. Education was generally given through a non-formal pedagogy in 1824 around religious instruction, value and morality instruction, spiritual instruction, martial arts instruction, and handicraft instruction. The Carr-Saunders Commission consolidated King Edward VII College of Medicine and Raffles College in 1947, which led to the establishment of the University of Malaya in Singapore in 1949. A second campus was later established in Kuala Lumpur. Following the separation of these two campuses in 1962, the University of Singapore in Singapore and the University of Malaya in Kuala Lumpur were established separately. (Chang, 2017).

According to Cheema (2020), the Malaysia Education Blueprint (MEB) 2013-2025 was developed to evolve the country's education system sustainably and effectively. At present, there are two types of higher educational services in Malaysia. There are public higher education institutions (PUHEs) and private higher education institutions (PHEIs). The tertiary education system in Malaysia is one of the most significant factors that contribute to the country's development. Malaysia's government invests around 5% of its GDP in education, an amount that is above the regional and world averages (Education Malaysia, n.d.). Additionally, higher education has a significant role to play in generating knowledge and raising qualified manpower for the country.

In Malaysia, tertiary education is provided by three types of institutions: public universities, private universities, and university colleges. Malaysia currently has 20 public universities, 36 polytechnics, and 105 community colleges. Aside from public Higher Education Institutions (HEIs), Malaysia has 434 private HEIs (PHEIs), which are divided into four categories: 54 universities, 39 university colleges, 331 colleges, and 10 international branch campuses (IBC) (Ministry of Higher Education Malaysia, 2022).

Nonetheless, the rise in the public demand for tertiary education has resulted in a monopolistically competitive higher education market. Consequently, competition for student enrolment exists between public and private higher education institutions (PUHEs and PHEIs). PHEIs primarily enrol students who have been denied admission to public universities, and most of these institutions compete fiercely for admission of students. Due to the highly competitive environment, student enrolments have declined by an estimated 20%, primarily at smaller private colleges that enrol 400 to 500 students (Zalina, 2003). These institutions have had to compete to draw students from both domestic and international regions because of the continued growth of private institutions of higher education, including international institutions.

Private Higher Education Institutions (PHEIs) in Malaysia face numerous challenges that impact their ability to provide quality education and maintain financial stability. One significant challenge revolves around meeting accreditation standards set by regulatory bodies like the Malaysian Qualification Agency (MQA), with some PHEIs failing to meet these standards, raising concerns about the quality of education they offer (Abdullah et al., 2021). Financial sustainability is another major issue, as PHEIs heavily rely on private funding and face intense competition, leading to struggles with high debt burdens and inadequate financial management. The reduction of government student loans has further strained their financial viability. Additionally, challenges related to inadequate infrastructure, mismanagement, and issues with course accreditation exacerbate the difficulties faced by PHEIs, sometimes resulting in closures and financial crises (Tajudeen, 2023). Overall, these challenges highlight the complex landscape

PHEIs navigate in Malaysia, requiring concerted efforts to address issues related to quality assurance, financial management, and regulatory compliance.

Taylor's University is Malaysia's top private university, ranked 332 in the QS World University Rankings 2022 and 53 in the QS Asia University Rankings 2022. In 1969, George Archibald Taylor and George Leighton Taylor founded it as a college. It was granted 'University College' status in 2006 (TheStar, 2011), resulting in two distinct identities under the brand's tertiary arm - Taylor's College and Taylor's University College. Taylor's University Lakeside Campus in Subang Jaya, Malaysia, began construction in early 2007 and was completed in January 2010. The Ministry of Higher Education Malaysia officially recognized it as a full-fledged university in September 2010.

Without a doubt, people see higher education as a stepping stone to better job opportunities in the future (Lowden et al., 2011). Even though students have the option of choosing a profession of interest and a university for their tertiary education, it is critical to understand that it is complicated and not at all straightforward. This is because enrolment in desired career courses necessitates one's effort and time. As a result, making appropriate career choices implies less waste of educational and training resources (Audit Commission, 1993). Students have preferences when it comes to selecting a university or college for their postsecondary education. As mentioned by Lichtenstein and Slovic (2006), preferences in psychology describe a person's attitude toward a group of objects, and those preferences are often reflected in a process of explicit decision-making that is usually based upon specific criteria. The term "preferences" in this research paper is defined as the students' choice of university based on its attributes.

The students' preferences in selecting a university are very important. They would have a significant influence on student selection and success. The students' preferences in selecting a university can be divided into three types. The students' preferences on selecting a university depend on personal factors such as individual characteristics, background and personality traits, abilities, interests, and so on; The students' preferences in selecting a university depend on social factors such as family background, educational background and other personal characteristics; The students' preferences on selecting a university depend on external factors such as culture, lifestyle, environment, etc. Hagel and Shaw (2008) stated that factors influencing students' university choice include academic reputation, campus facilities, location of the campus, tuition fees, and track availability. Ciriaci and Muscio (2011) concur that one of the key determinants of a student's choice of future study was the institution's reputation and its favourable impact on employability after graduation. The referenced literature delves into students' preferences for higher education institutions, specifically highlighting factors such as academic quality, campus environment and facilities, employability, family influences, geographic location of universities, peers' influences, and university reputation. Numerous studies underscore the importance of these factors in shaping students' decisions when selecting higher education institutions in Malaysia. However, a notable research gap exists regarding the preferences of students enrolled in Taylor's University BQS Programme.

PROBLEM STATEMENT

Despite the pivotal role of quantity surveyors in the construction industry, applications to the program have notably declined. While applications for the Bachelor of Quantity Surveying (BQS) program at Taylor's soared to 134 in 2019, a noticeable decline was witnessed between 2020 and 2022. As the industry rebounds from the pandemic, there is an urgent need to bolster the enrolment of quantity surveying students to meet the escalating demand for professionals in this field. The rapid growth of Malaysia's construction industry necessitates a significant presence of technical and professional personnel, including quantity surveyors, according to Works Minister Datuk Seri Alexander Nanta Linggi. He emphasized the crucial role of quantity surveyors in managing finances and building contracts throughout construction projects. As of 2022, there are only 5874 registered quantity surveyors in Malaysia, a significant shortfall compared to the projected need of 14000 by 2020. The labour market now encompasses three distinct generations: Generation X, born between the early 1960s and late 1970s; Generation Y, born in the 1980s and 1990s; and Generation Z, born between 1977 and 2012. Yeo (2018) highlights that as older Generation X members prepare for retirement, it is imperative that younger quantity surveyors from Generation Y and Z step in to fill the impending employment gap. Despite this critical need for quantity surveyors, there has been a noticeable dearth of research addressing the dwindling interest in this profession. Additionally, researchers have underscored the significance of understanding the motivators influencing career choices to cultivate enthusiasm and interest among young individuals contemplating a profession (Ekung & Okonkwo, 2018). The decision-making process, especially regarding career choices, is a significant aspect of individuals' lives regardless of age. While some decisions might appear straightforward, others, like choosing a career path, can be complex, particularly for students during their undergraduate education. Given this context, the primary objective of this research is to discern the factors that impact students' decisions to pursue quantity surveying courses at Taylor's University. The study seeks to shed light on the evolving preferences and considerations of students in selecting this field of study. Notably, Taylor's University is recognized for offering one of the best quantity surveying programs in Malaysia, ranking 53rd in the QS Asia University Rankings 2022. Furthermore, the research emphasizes the broader implications of the declining interest in quantity surveying. For any professional body to thrive and evolve, it is crucial to attract and nurture new members, particularly from younger generations. A reduction in the number of quantity surveying students poses a significant challenge to the future of the profession.

LITERATURE REVIEW

This literature review will focuson analyzing previous scholars' findings regarding students' preferences for higher education institutions, along with related aspects like Taylor's University's accreditation and facilities. The goal is to extract gaps in existing research for further exploration. Ultimately, it will conclude with a summary synthesizing the key insights gained from the reviewed literature.

Academic Quality

The referenced literature suggests that academic quality plays a crucial role in students' decisions when choosing higher education institutions. Various studies, including those by Ramalu et al. (2013), Ruslan et al. (2014), and Husain et al. (2018), consistently highlight the significance of academic quality in influencing students' preferences. Factors such as teaching effectiveness, support, assessment, and learning opportunities are encompassed within the notion of academic quality. These studies indicate that the perceived academic quality of an institution significantly impacts students' confidence and decisions in selecting a university or college. While these studies focus on different universities and colleges in Malaysia, they highlighted the importance of academic quality as a key determinant in students' choices. However, it's noted that the studies were conducted at different times and targeted different institutions, suggesting a need for updated research as student preferences might have evolved over time. Nonetheless, the collective findings emphasize the strong influence of academic quality on students' preferences for higher education institutions.

Campus Environment and Facilities

The availability and quality of teaching facilities represent a significant factor influencing students' choices when selecting higher education institutions, according to various studies. Researchers such as Migin et al. (2014), Ivy (2008), Joseph & Joseph (1998), and others have highlighted the importance of well-equipped libraries, computer labs, lecture halls, and overall campus environment as critical considerations for students. Studies by Sidin et al. (2003), Fernandez (2010), Khairani & Razak (2013), Ramalu et al. (2013), Husain et al. (2018), and Awang (2015) examined different aspects of campus facilities and their influence on students' preferences at various universities in Malaysia, emphasizing factors like teaching quality, sports facilities, library resources, and overall campus environment. These studies collectively emphasized the crucial role of campus facilities in shaping students' decisions regarding higher education institutions. However, despite extensive research on this topic across different universities in Malaysia, there is a gap in understanding the preferences of students enrolled in Taylor's University BQS program. Consequently, this gap in research led to the hypothesis that campus facilities significantly impact the decision-making process of students when choosing higher education institutions, indicating a need for further investigation specifically targeting Taylor's University BQS students.

Employability

Employability, encompassing skills, and qualities vital for graduates' career success, holds immense significance for individuals, and economies globally. Higher education institutions play a crucial role in enhancing employability (Kenayathulla et al., 2019). Students often pursue tertiary education aspiring for improved job prospects, emphasizing the crucial link between university experience and future employability (Sevier et al., 1998; Chris et al., 2005). The reputation of a university is intricately tied to employability, influencing students' choices due to the perceived market perception and the expectations employers hold for graduates (Awang, 2015). Research by Ramalu et al. (2013), Awang (2015) and Husain et al. (2018) highlights the substantial impact of employability on students' preferences in selecting higher education institutions in Malaysia. These studies underscored the significance of job placement success and university reputation in students' decision-making. However, despite comprehensive studies on this topic across various Malaysian universities, there remains a gap in understanding the preferences of students enrolled in Taylor's University BQS program. This research gap prompts the hypothesis that employability significantly shapes students' choices in higher education institutions, warranting further investigation, particularly focusing on Taylor's University BQS students, to deepen insights in this area.

Family Influences

Family influence plays a substantial role in shaping the values and educational choices of children, especially in Asian cultures where parental guidance strongly impacts higher education decisions (Nebor, 1986; Leong, 1985; Cabrera et al., 2000). Cultural values emphasizing respect for elders and aspirations for higher socioeconomic status drive Asian parents' expectations for their children to attend prestigious universities and pursue specialized fields (Tang, 1997; Barringer et al., 1993). Despite the autonomy of post-secondary students, studies reveal their reliance on family advice for selecting higher education institutions. Awang highlights family influence as the second most significant factor for private university choices in Malaysia. However, Fernandez (2010) and Ramalu et al. (2013) found varying perceptions among students regarding the importance of parental advice in university selection, ranking it lower among the influencing factors. Additionally, Husain et al. (2018) observed differing levels of influence from significant people, including family members, among students in private university colleges. While previous research delved into diverse universities in Malaysia, none specifically targeted Taylor's University BQS students. This study aims to explore the relationship between family influences and students' preferences in university selection, particularly within Taylor's University, recognizing the potential impact of family influences on these students' choices.

Geographic Location of University

The geographic location of a university holds varying influence on students' choices in higher education institutions. Proximity to home often appeals to some students for its convenience and familiarity, reducing the uncertainty of the unknown (Persson, 2014). Conversely, others prioritize reputable institutions despite the distance, opting to relocate for the duration of their courses, a critical decision considering the residency commitment of at least three years (Persson, 2014). Studies, including those by Jackson (1982) and Shanka et al. (2005) highlight the impact of geographic location on students' choices, emphasizing the preference for institutions closer to their residences. However, conflicting findings emerged from research by Fernandez (2010), Ramalu et al. (2013), and Awang (2015), indicating that the geographic location of higher education institutions might not significantly influence student preferences. Ramalu et al. (2013) ranked institution location as the least influential factor among the ten factors considered in their study. These studies varied in their targeted populations, focusing on different student demographics and geographic regions in Malaysia. Notably, no research has investigated the link between student preferences and university location specifically at Taylor's University. This study aims to explore this relationship, recognizing the potential impact of geographic location on the preferences of Taylor's University students in their selection of higher education institutions.

Peers' Influences

The influence of peers on students' decisions regarding higher education institutions has been widely acknowledged. Friends and classmates are noted to play significant roles akin to parental influence in college attendance decisions (Hossler, 1995). Classmates not only impact educational levels but also offer valuable insights into future study options (Ancheh et al., 2007). Paulsen (1990) highlighted friends as a major information source for students considering college, while Rowe (2002) found friends and peers to hold sway in the college selection process. Contrary to these observations, studies by Fernandez (2010), Ramalu et al. (2013), and Husain et al. (2018) suggest that peer influence may not significantly sway students' higher education choices. Fernandez (2010) found peer recommendations to be of minimal importance, with only 8% considering them crucial. Ramalu et al. (2013) ranked friend advice relatively low among 23 factors, and Husain et al. placed peer influence third among four factors. Despite research across various Malaysian universities, none focused on Taylor's University BQS students. Thus, this study aims to examine the relationship between peer influence and students' preferences in university selection, specifically at Taylor's University, to gauge the impact of peers on students' decisions within this academic context.

Reputation of University

Multiple studies underscore the pivotal role of a university's reputation in shaping students' choices among higher education institutions in Malaysia. Keling's research revealed the reputation's significance in drawing international students to Malaysian universities, impacting their preferences among institutions (Keling, 2006). Similarly, Nagaraj et al. (2008) highlighted reputation as a top influencer for students opting for the esteemed University of Malaya, acknowledged as Malaysia's best university (Daniel, 2022). Studies by Ancheh et al. (2007) and Khairani & Razak (2013) further emphasized the paramount importance of a university's recognizability and reputation, particularly in public colleges and universities in Malaysia. Ramalu et al.'s extensive survey reinforced this, highlighting that a university's image significantly shapes individuals' opinions and choices. Awang's study highlighted the critical nature of reputation, ranking it highest among eleven factors influencing enrolments in private universities in Sarawak, particularly among Sarawakian and foreign students. In summary, these studies show how important a university's reputation is in influencing students' choices for higher education in Malaysia. It affects preferences and enrolment decisions in various regions and universities.

AIM AND OBJECTIVES

This research paper aims to offer universities, specifically Taylor's University, more understanding of what influences students' choices when selecting where to pursue a Bachelor of Quantity Surveying degree. It also seeks to explore the expectations of these students on the Bachelor of Quantity Surveying (Hons) Degree Programme. Upon identifying the issues in the literature review, the subsequent research objectives have been formulated.

1. To identify and prioritize the determinants influencing students' decision to enrol in Taylor's University Bachelor of Quantity Surveying (Hons) Degree Programme.
2. To analyse the students' expectations versus experiences of Taylor's University Bachelor of Quantity Surveying (Hons) Degree Programme.
3. To suggest practical ways to improve the students' enrolment rate in Taylor's University Bachelor of Quantity Surveying (Hons) Degree Programme.

RESEARCH METHODOLOGY

Data Collection Method

The research method chosen for this study is quantitative, focusing on numerical data to explore the correlation between variables. This approach involves using numerical data to assess how various factors impact individuals physically, mentally, or intellectually (Patten, 1997; Shenmare, n.d.). To gather data from the target population, a standardized questionnaire survey serves as the primary method. This questionnaire employs Likert scales, multiple-choice questions, and numerical ratings, ensuring a measurable assessment of responses. Moreover, the study employs both nominal and interval scales to measure and analyse the collected data from the respondents. The study aimed to collect data from 137 Taylor's University BQS students, representing a portion of the total 210 students. Ultimately, 140 responses were gathered and will be utilized for data analysis.

Design of Questionnaire

This research study uses two questionnaires to address different objectives. The first questionnaire survey focuses on objective 1 which is the determinants influencing students' decision to enrol in Taylors University BQS Programme, and objective 2 which analyses the students' expectations versus experiences of BQS Programme at Taylor's University. The second questionnaire survey which is linked to objective 3, provides practical ways for the top two factors with the highest mean value differences identified in the first questionnaire. This ensures a clear and organized approach to collecting and analysing data for this research study.

This research study's survey questionnaire 1 comprises four sections. The first section of the survey questionnaire gathers information about the demographic details of the target respondents, while the subsequent three sections are designed to address the research study's objectives. The questionnaire's layout is outlined as follows: -

A. Section A: Demographic Information
B. Section B: Determinants influencing students' decision to enrol in Taylor's University BQS Programme
C. Section C: Students' expectations versus experiences of BQS Programme at Taylor's University
D. Section D: Practical ways to improve the students' enrolment rate in BQS Programme at Taylor's University

Section A serves the purpose of collecting the respondeht's profile details. It consists of four questions related to the respondent's information, including their current semester, gender, age, and nationality. The options provided to the respondents for selection utilize the usage of both nominal scale and interval scale of measurement. For instance, a nominal scale is utilized to determine the respondent's gender, while an interval scale is utilized to determine the respondent's current semester, age, and nationality.

Section B serves the purpose of collecting the determinants influencing the respondents' decisions to enrol in Taylor's University BQS Programme. It consists of several questions related to the determinants influencing their decisions. The options provided to the respondents for selection utilize the usage of an ordinal scale of measurement from 1 to 5 (Not important at all to extremely important) to indicate the level of importance of each determinant.

Section D serves the purpose of collecting the respondents' experiences regarding to Taylor's University BQS Programme. It consists of several questions related to their experiences. The options provided to the respondents for selection utilize the usage of an ordinal scale of measurement from 1 to 5 (Not important at all to extremely important) to indicate the level of importance of each experience.

This research study's survey questionnaire 2 comprises two sections. The first section of the survey questionnaire gathers information about the demographic details of the target respondents, while the subsequent three sections are designed to address the research study's objective 3. The questionnaire's layout is outlined as follows: -

A. Section A: Demographic Information
B. Section B: Practical Suggestions to Improve Enrolment Rate

Section A serves the purpose of collecting the respondent's profile details. It consists of four questions related to the respondent's information, including their current semester, gender, age, and nationality. The options provided to the respondents for selection utilize the usage of both nominal scale and interval scale of measurement. For instance, a nominal scale is utilized to determine the respondent's gender, while an interval scale is utilized to determine the respondent's current semester, age, and nationality.

Section B serves the purpose of collecting the respondents' opinions on the practical suggestions to improve the enrolment rate It consists of several questions related to the practical suggestions. The options presented to respondents for selection utilize an ordinal scale of measurement ranging from 1 to 5 (Highest preference to least preferred) to assess the level of preference for each practical suggestion.

Data Analysis Method

The research study employed a descriptive analysis technique to examine the gathered data. Descriptive analyses were conducted using the Statistical Package for Social Science (SPSS), a statistical software. The analysis methods used for evaluating the collected data include the analysis of mean and ranking analysis.

FINDINGS

Key Findings in Relation to Objective 1

Table 1. Descriptive analysis on the determinants influencing students' decision to enroll in Taylor's University BQS programme

Variables	Mean	Rank
Academic Quality	4.21	2
Campus Environment and Facilities	4.04	3
Employability	4.28	1
Family Influences	3.26	6
Geographic Location of University	3.56	5
Peers' Influence	2.84	7
Reputation of University	3.91	4

This section focuses on calculating the mean values of determinants impacting students' enrolment decisions in Taylor's University BQS Programme. These determinants will be ranked based on their mean values derived from descriptive analysis. The analysis based on Table 1 illustrates the prioritization of determinants influencing students' enrolment choices at Taylor's University BQS Programme. Employability emerged as the most significant factor, securing the top spot with a mean value of 4.28. This signifies a strong inclination towards universities that promise better job prospects post-graduation, supported by Awang's (2015) research emphasizing employability's pivotal role in students' selection of private university. Following closely, Academic Quality ranked second at 4.21, indicating the respondents' high regard for educational standards. Ruslan et al.'s (2014) study similarly highlighted academic quality as a critical determinant in university selection. Campus Environment and Facilities secured the third spot at 4.04, indicating the value placed on the physical campus. Migin et al.'s (2014) study highlighted this significance of facilities in students' university choices. The Reputation of University was ranked fourth at 3.91, reflecting the preference for esteemed institutions. Khairani and Razak (2013) also emphasized the role of university reputation in influencing students' decisions. The Geographic Location of University ranked fifth with a mean rating of 3.56, aligning with Awang's (2015) study, highlighting its relatively lower significance in students' selection criteria. Family Influences and Peers' Influences were rated lower in importance, securing sixth and seventh positions, respectively, with mean ratings of 3.26 and 2.84. Ruslan et al.'s (2014) research suggested that family influence has less impact on students' university choices, while Ramalu et al.'s (2013) findings supported the lower priority of peers' advice in decision-making.

Key Findings in Relation to Objective 2

This part investigates the difference between what respondents initially expected and what they experienced in Taylor's University BQS program to uncover any gaps. This comparison is done by subtracting the average value of experiences from the initial expectations. If the result is negative, it suggests that the experiences were below their initial expectations.

Academic Quality

Table 2. Comparison of mean value between the expectations and experiences of the respondents on the academic quality of Taylor's University BQS programme

Variables	Difference of Mean	Rank
Quality of lecturers	(0.67)	1
Range of academic resources available	(0.38)	4
Academic accreditation	0.92	9
Internship	(0.30)	6
Field trips	(0.62)	2
Group projects or team-based assignments	(0.34)	5
Latest developments and trends	(0.45)	3
Quality of course content and materials	(0.11)	8
Academic support	(0.20)	7

The quality of lecturers has the most substantial difference in mean at 0.67, ranking first. This suggests the lecturers' quality didn't meet the initial expectations. Concerns raised included slide-based teaching, a desire for more engaging methods, improved slide quality, and a need for more interactive learning. Field trips ranked second with a mean difference of 0.62, indicating experiences fell below expectations. Issues included unsatisfactory accommodation, concerns about cleanliness and food quality, and lack of transportation, posing challenges for non-drivers. Conversely, academic accreditation had a notably positive mean difference of 0.92, exceeding respondents' expectations. They highlighted that accreditation addressed students' needs and concerns effectively.

Campus Environment and Facilities

Table 3. Comparison of mean value between the expectations and experiences of the respondents on the campus environment and facilities of Taylor's University BQS programme

Variables	Difference of Mean	Rank
Quality of facilities	(0.31)	8
Study environment	(0.35)	7
Gym	0.44	10
Safety of the environment	(0.42)	5
Disability-friendly facilities	(0.08)	9
Cleanliness and maintenance of the environment	(0.37)	6
Quality of internet connectivity	(1.14)	1
Food and beverage options	(1.02)	2
Parking spaces	(0.95)	3
Shuttle bus	(0.48)	4

Internet quality showed the highest difference in mean at 1.14, ranking first. This suggests respondents found the internet quality lower than expected, facing issues like disconnections, slow speeds, and Wi-Fi coverage problems. Some also noted inconvenience in re-entering login credentials. Food and beverage options ranked second with a difference of mean at 1.02, indicating lower satisfaction than expectations. Concerns were about food quality, pricing, and portion sizes not justifying costs. Some found it more cost-effective to eat off-campus but were constrained by short break times between classes. Conversely, the gym showed a positive mean difference at 0.44, surpassing respondents' expectations, indicating they had better experiences than initial expectations.

Employability

Table 4. Comparison of mean value between the expectations and experiences of the respondents on the employability of Taylor's university BQS programme

Variables	Difference of Mean	Rank
Alumni network	(0.76)	1
Industry connections and partnership	(0.70)	2
Career counselling and job placement assistance	(0.62)	3

The strongest difference was found in the alumni network, ranking first with a mean difference of 0.76. This indicates respondents' experiences were lower than their initial expectations, expressing concerns about the absence of a well-established alumni network within Taylor's University BQS Programme. Instead, alumni connections primarily relied on student-organized society activities. Next, industry connections and partnerships had the second-highest mean difference at 0.70. Respondents acknowledged lecturer-industry sharing sessions but expressed a need for more programs involving industry players and students to enhance practical experiences and connections. Lastly, career counselling and job placement assistance showed the lowest mean difference at 0.62 compared to other factors.

Geographic Location of University

Table 5. Comparison of mean value between the expectations and experiences of the respondents on the geographic location of Taylor's University

Variables	Difference of Mean	Rank
Transportation accessibility	(0.84)	1
Topography and geographic features	(0.18)	2

The analysis shows that transportation accessibility had the highest difference in mean at 0.84, indicating that respondents found transportation accessibility to be lower than their initial expectations. Respondents noted that the closest BRT station required a 15-minute walk, which was seen as inconvenient, especially when carrying heavy bags. Additionally, the topography and geographic features had a lower mean difference at 0.18 among all the variables analysed.

Reputation of University

Table 6. Comparison of mean value between the expectations and experiences of the respondents on the reputation of Taylor's University

Variables	Difference of Mean	Rank
Reputation of the BQS Programme at Taylor's University globally	(0.38)	1
Reputation of the BQS Programme at Taylor's University locally	(0.36)	2
Overall reputation of Taylor's University	(0.22)	3

The analysis showed that the global reputation of the BQS Programme at Taylor's University had the highest difference in mean at 0.38, indicating that respondents found their experiences lower than their initial expectations. Similarly, the reputation of the BQS Programme among local employers and professionals had the second highest difference in mean at 0.36. However, the overall reputation of the university had the lowest difference in mean at 0.22. While no specific issues were highlighted by respondents, addressing these differences is essential to enhance Taylor's University's reputation.

Key Findings in Relation to Objective 3

This section concentrates on analysing practical methods to enhance the enrolment rate of Taylor's University BQS Programme. The proposed strategies are derived from the two highest-ranking variables in each area with the highest difference in mean values. The frequency analysis was conducted to determine the most effective way to improve the student's enrolment rate.

Academic Quality

Quality of Lecturers

Table 7. Frequency analysis on the practical ways to improve the quality of lecturers

Practical ways	Frequency	Rank
Incorporate technology for enhancement	32	1
Professional development workshop	14	2
Guest speaker series	12	3
Professional teaching communities	14	2

Firstly, integrating technology such as multimedia resources and virtual reality simulations enhances learning experiences and comprehension. Secondly, conducting professional development workshops focusing on skills like presentation design and communication benefits lecturers. Thirdly, introducing guest speaker series enriches the curriculum by providing real-world insights. Finally, establishing teaching communities promotes collaboration among lecturers, fostering the development of more engaging teaching methods.

Field Trip

Table 8. Frequency analysis on the practical ways to improve the field trip

Practical ways	Frequency	Rank
Partnership with accredited facilities	30	1
Pre-Trip Assessment	22	2
Feedback Mechanism	13	3
Student Involvement	7	4

Respondents identified partnership with accredited facilities as the most effective measure, ranking it 1st with a frequency of 30. Following closely, the pre-trip assessment was ranked 2nd with a frequency of 22, indicating its perceived effectiveness. Additionally, feedback mechanisms were ranked 3rd with a frequency of 13. On the other hand, student involvement received the lowest frequency, ranking 4th with a frequency of 7, suggesting it might be considered less effective compared to other measures.

Campus Environment and Facilities

Quality of Internet Connectivity

Table 9. Frequency analysis on the practical ways to improve the quality of internet connectivity

Practical ways	Frequency	Rank
Network infrastructure upgrade	29	1
Wi-Fi signal optimization	28	2
Bandwidth management and prioritization	8	3
Network monitoring and troubleshooting	7	4

The survey ranked network infrastructure upgrades as the most effective measure, with a frequency of 29, addressing the need for improved coverage. Additionally, the pre-trip assessment was ranked 2nd with a frequency of 28, indicating its perceived effectiveness in improving internet quality. Furthermore, bandwidth management and prioritization were ranked 3rd with a frequency of 8. Lastly, network monitoring and troubleshooting received the lowest frequency at 4th place with a frequency of 7, suggesting it might be considered less effective compared to other measures.

Food and Beverage Options

Table 10. Frequency analysis on the practical ways to improve the food and beverage options

Practical ways	Frequency	Rank
Collaboration with culinary program	17	2
Collaboration with food trucks	19	1
Collaboration with food vending machine	19	1
Launching new restaurants	17	2

Both collaboration with food trucks and food vending machines were ranked as the most effective measures, holding the 1st position with a frequency of 19. In contrast, collaboration with culinary programs and launching new restaurants were ranked 2nd with a frequency of 17, suggesting they might be considered less effective compared to other measures.

Employability

Alumni Network

Table 11. Frequency analysis on the practical ways to improve the alumni network

Practical ways	Frequency	Rank
Alumni workshops and seminars	26	1
Alumni networking dinner and social gathering	17	3
Alumni job fairs and recruitment events	19	2
Online alumni networking platforms and communities	10	4

Respondents identified alumni workshops and seminars as the most effective measure to enhance the alumni network, ranking it 1st with a frequency of 26. Additionally, alumni job fairs and recruitment events were ranked 2nd with a frequency of 19, indicating perceived effectiveness. Furthermore, alumni networking dinners and social gatherings were ranked 3rd with a frequency of 17. Conversely, online alumni networking platforms and communities received the lowest frequency, ranking 4th with a frequency of 10, suggesting they might be considered less effective compared to other measures.

Industry Connections and Partnerships

Table 12. Frequency analysis on the practical ways to improve the industry connection and partnership

Practical ways	Frequency	Rank
Industry speaker series	14	2
Hackathons or innovation challenges	8	3
Industry research symposia	7	4
Site visits to construction companies	43	1

Respondents ranked site visits to construction companies as the most effective measure, holding the 1st position with a frequency of 43. Additionally, the industry speaker series was ranked 2nd with a frequency of 14, indicating perceived effectiveness. Furthermore, Hackathons or innovation challenges were ranked 3rd with a frequency of 8. Conversely, industry research symposia received the lowest frequency, ranking 4th with a frequency of 7, suggesting they might be considered less effective compared to other measures in improving industry connections and partnerships.

Geographic Location of University

Transportation Accessibility

Given the current scenario where the nearest station involves a 15-minute walk, one proposed solution involves adding an extra bus route. This addition aims to significantly assist students who depend on BRT (Bus Rapid Transit) services, potentially encouraging more students to utilize public transportation, which aligns with environmental concerns.

Topography and Geographic Features

Table 13. Frequency analysis on the practical ways to improve topography and geographic features

Practical ways	Frequency	Rank
Sustainable Landscaping	15	3
Preserve Natural Areas	20	1
Outdoor Education and Recreation	20	1
Recreational and Leisure Spaces	17	2

Respondents ranked preserving natural areas and promoting outdoor education and recreation as the most effective measures, both securing the 1st rank with a frequency of 20. Additionally, recreational and leisure spaces were ranked 2nd with a frequency of 17, indicating perceived effectiveness. Lastly, sustainable landscaping received the lowest frequency, ranking 3rd with a frequency of 15, suggesting it might be considered less effective compared to other measures in enhancing and maintaining topography and geographic features.

Reputation of University (Reputation of Taylor's BQS Programme)

Table 14. Frequency analysis on the practical ways to improve enhance and uphold the reputation of Taylor's BQS programme

Practical ways	Frequency	Rank
Alumni engagement and success stories	17	3
Global experiences and study abroad opportunities	26	1
Cutting-edge facilities and resources	10	4
High quality of faculty	19	2

Respondents ranked global experiences and study abroad opportunities as the most effective measures, both securing the 1st rank with a frequency of 26. Additionally, the high quality of faculty was ranked 2nd with a frequency of 19, indicating perceived effectiveness. Furthermore, respondents ranked alumni engagement and success stories 3rd with a frequency of 17. Lastly, cutting-edge facilities and resources received the lowest frequency, ranking 4th with a frequency of 10, suggesting it might be considered less effective compared to other measures in enhancing and upholding the reputation of Taylor's BQS Programme.

CONCLUSION

The analysis indicates that employability stands as the most influential determinant affecting students' decision to enrol in Taylor's University BQS program, while peer influence holds the least significance among the determinants considered. The analysis reveals varying perceptions between expected and experienced aspects across different domains. In academics, the emphasis on lecturer quality surpasses academic accreditation, notably with a substantial gap between expected and actual experiences for lecturers. Campus facilities highlight a considerable contrast in anticipated significance between internet quality and gym facilities, showing the greatest disparity in internet quality. Employability expectations prioritize industry connections over alumni networks, resulting in a larger gap in experienced satisfaction. Transportation accessibility, despite its anticipated importance, exhibits the highest variance in experience, contrasting with minimal differences in topography. Lastly, while the global reputation of the BQS program is highly anticipated, the overall university reputation displays the least variance between expectations and experiences. Priority enhancements included incorporating technology for lecturers' quality, partnering with accredited facilities for field trips, upgrading network infrastructure for better internet, and collaborating with food trucks for improved dining. Strategies also focused on alumni workshops, industry site visits, and preserving natural areas for enhanced engagement. Improving transportation via additional bus routes and emphasizing global experiences for the university reputation were recommended.

ACKNOWLEDGMENT

The researchers of this article would like to thank the journal's editors and the anonymous reviewers for their positive feedback and suggestions for the improvement of this journal.

This research received no specific grant from any funding agency in the public, commercial, or not-for-profit sectors.

REFERENCES

Abdullah, N. R. W., Rosli, N. A., & Ramasamy, R. (2021). Private Higher Education Institutions (PHEIs) in Malaysia: An Assessment of Service Quality and Students' Satisfaction. *International Journal of Management, Accounting, Governance and Education, 1*(2).

Ancheh, K. S. B., Krishnan, A., & Nurtjahja, O. (2007). Evaluative criteria for selection of private universities and colleges in Malaysia. *Journal of International Management Studies*, 2(1), 1–11.

Audit Commission. (1993). *Unfinished Business. Fulltime education courses for 16-19 year olds*. HMSO.

Awang, Y. A. (2015, January). *Students' Factors Preference Of Choosing Private University In Sarawak, Malaysia. Swinburne University Technology, Sarawak, Malaysia. Proceedings of the Asia Pacific Conference on Business and Social Sciences 2015*, Kuala Lumpur. ISBN 978-0-9925622-2-9.

Barringer, H. R., Gardner, R. W., & Levin, M. J. (1993). *Asians and Pacific Islanders in the United States*. Russell Sage Foundation.

Cabrera, A. F., & Nasa, S. M. L. (2000). Understanding the college-choice process. In A. F. Cabrera & S. M. L. Nasa (Eds.), *Understanding the college choice of disadvantaged students* (pp. 5-22). San Francisco: Jossey-Bass Publishers.

Cheema, S. (2020, November 11). *Everything you need to know about Malaysia's Education Blueprint 2013-2025*. Mashable SE Asia. https://sea.mashable.com/culture/13175/everything-you-need-to-know-about-Malaysias-education-blueprint-2013-2025

Chris Corker and Sarah Holland (2005). Introducing students to employability, skills and reflection: A case study from history. *Student Engagement and Experience Journal, 4*(1). 10.7190/seej.v4i1.113

Ciriaci, D., & Muscio, A. (2011). University choice, research quality and graduates' employability: Evidence from Italian national survey data. *AlmaLaurea Working Papers, 49*, 1-14.

Daniel, T. (2022). Rankings released! QS World University Rankings: Asia 2023. https://www.qs.com/rankings-released-qs-world-university-rankings-asia-2023/\

Ekung, S., & Okonkwo, E. (2018). Trends Influences on Career Choice In Quantity Surveying And Its Implications. *International Journal of Education (IJE), 3*(1), March 2015.

Fernandez, J. L. (2010). An Exploratory Study Of Factors Influencing The Decision Of Students To Study At Universiti Sains Malaysia. Universiti Sains Malaysia, Penang. *Kajian Malaysia*, 28(2), 2010.

Hagel, P., & Shaw, R. (2008). The influence of delivery mode on consumer choice of university. *European Advances in Consumer Research*, 8, 531–536.

Hossler, D. E. F. (1995). *D Wallers, and M.K.Moss (EDS), New Directions Institutional Research, N.88: Evaluating and responding to college guidebooks and rankings*. Jossey-Bass, Inc.

Husain, S., Safiei, H., Ghazali, A., Nazir, N. N. M., Redzuan, N. I. N., & Ghani, F. A. (2018). Factors Affecting Students' Preferences Towards Private University College in Malaysia. *Universiti Tenaga Nasional, Malaysia. Global Business and Management Research: An International Journal, 10*(3).

Ivy, J. (2008). A New Higher Education Marketing Mix: The 7Ps for MBA Marketing. *International Journal of Educational Management*, 22(4), 288–299. 10.1108/09513540810875635

Jackson, G. A. (1982). Public efficiency and private choice in higher education. *Educational Evaluation and Policy Analysis*, 4(2), 237–247. 10.3102/01623737004002237

Joseph, M., & Joseph, B. (1998). Identifying need of potential students in tertiary education for strategy development. *Quality Assurance in Education*, 6(2), 90–96. 10.1108/09684889810205741

Kenayathulla, H., B., Ahmad, N., A., & Idris, A., R. (2019, August 14). Gaps between competence and importance of employability skills: evidence from Malaysia. Higher Education Evaluation and Development, 13(2), 97-112.

Khairani, A. Z., & Razak, N. N. (2013). Assessing Factors Influencing Students' Choice of Malaysian Public University: A Rasch Model Analysis. Universiti Sains Malaysia. *International Journal of Applied Psychology*, 3(1), 19–24.

Leong, F. T. L. (1985). Career development of Asian Americans. *Journal of College Student Personnel*, 26, 539–546.

Lichtenstein, S., & Slovic, P. (2006). *The construction of preference*. Cambridge University Press. 10.1017/CBO9780511618031

Lowden, K., Hall, S., Elliot, D., & Lewin, J. (2011). *Employers' perceptions of the employability skills of new graduates*. University of Glasgow. https://www.educationandemployers.org/wp-content/uploads/2014/06/employability_skills_as_pdf_-_final_online_version.pdf

Migin, M. W., Falahat, M., Yajid, M. S., & Khatibi, A. (2014). Impacts of Institutional Characteristics on International Students' Choice of Private Higher Education Institutions in Malaysia. *Higher Education Studies*, 5(1), 31–42.

Ministry of Higher Education Malaysia. (2022, April 22). *Higher Education Report: [MALAYSIA]. Paper presented at the UNESCO World Higher Education Conference*. UNESCO. https://whec2022.net/resources/Country%20report%20-%20Malaysia.pdf

Nagaraj, S. S. M., Noor, I. M. J., Diana, A. W., & Mirzaei, T. (2008). *How do undergraduates choose their university? A study of first year University of Malaysia students*. (FEA Working Paper No. 2008-8) University of Malaya.

Nebor, J. N. (1986). Parental Influence and Involvement on Reading Achievement. *Information Analyses (070)*. https://files.eric.ed.gov/fulltext/ED286150.pdf

Patten, S. B. (1997). Performance of the Composite International Diagnostic Interview Short Form for Major Depression in Community and Clinical Samples. *Health Canada*, 18(3), 131.9375257

Paulsen, M. B. (1990). ASHE-ERIC Higher Education Report: Vol. 6. *College choice: Understanding student enrollment behavior*. George Washington University.

Persson, M. (2014). *Socio-geographic Effect On Higher Education Choice*. XVIII ISA World Congress of Sociology 13-19/7 - 2014, Yokohama, Japan.

Ramalu, J. C. D., Bakar, N. A., & Nijar, N. N. (2013). Factors That Determine Students' Preferences In Selecting Higher Learning Institution. *Infrastructure University Kuala Lumpur Research Journal, 1*(1).

Ruslan, R., Ariffin, K. H. K., Islam, M. A., & Zaidi, N. I. M. (2014). Determinants Students' Selection of Higher Education Institutions in Malaysia. Universiti Malaysia Perlis. *Advances in Environmental Biology, 8*(9), 406-416.

Shenmare, K. (n.d.). Quantitative Research. *Wall Street Mojo*. https://www.wallstreetmojo.com/quantitative-research/

Sidin, S., Hussin, S., & Soon, T. (2003). An exploratory study of factors influencing the college choice decision of undergraduate students in Malaysia. *Asia-Pacific Management Review.*, 8, 259–280.

Tajudeen, A. B. A. (2023). The challenges faced by private higher education in Malaysia from the 1990s to COVID-19. Sejarah: Journal of History Department. *University of Malaya*, 32(1), 166–187. 10.22452/sejarah.vol32no1.10166

Tang, M., & Fouad, N. (1997). Understanding Asian Americans' career choices: A social cognitive approach to examine the factors affecting their choices. Paper presented at the Annual Conference of American Psychological Association, Chicago.

The Star. (2011, February 20). University's First Chancellor. *TheStar.*

Yeo, H. C. (2018, June). Generational Differences Between Employees In Quantity Surveying Firms. *Universiti Teknologi Malaysia.*

Zalina, Y. (2003, March 10). IPTS, IPTA: Perang Berebut Pelajar Kini Bermula. *Utusan Malaysia.*

KEY TERMS AND DEFINITIONS

Enrolment Rate: The percentage of eligible people who sign up for a program or school.
Quantity Surveying: Managing costs and resources for construction projects.
Students' Expectations: What students hope or anticipate from their education.
Students' Experiences: The things students go through and learn during their education.
Students' Preferences: The choices and likes of students about their education.
Tertiary Education: Higher education beyond high school, like college or university.

Chapter 12
Research Methodology for Cost Modelling Development on Smart Living Features Implementation in Malaysian Housing

Koh Fung Chieng
https://orcid.org/0009-0007-6477-4928
Universiti Teknologi Malaysia, Malaysia

Fara Diva Mustapa
Universiti Teknologi Malaysia, Malaysia

ABSTRACT

The development of smart living housing in Malaysia, despite its 'people-centric' concept that focuses on resident well-being, faces implementation challenges. Notably, uncertainties related to the monetary implications of integrating smart living features into conventional houses. Consequently, such a situation highlights the need for a decision-making tool – a cost model, to address these uncertainties during feasibility studies. Thus, this chapter proposes the research methodology through a PRISMA-guided literature review to develop the cost model per se. By employing the interpretivism constructivism paradigm, the study utilizes a sequential explanatory mix-method. Quantitatively, an open-ended questionnaire survey identifies the cost drivers, while qualitatively a case studies document survey gathers the cost data. The content analysis legitimizes the claims of both findings for cost modeling development. The research establishes relationships between the cost drivers and cost data, facilitating a monetary implication study to manage uncertainties in Smart Living development.

DOI: 10.4018/979-8-3693-1650-4.ch012

INTRODUCTION TO SMART LIVING HOUSING DEVELOPMENT IN SMART CITY

Smart City is an innovative concept of sustainable town planning initiated through Agenda 2023 Decade of Action (Sharifi et al., 2024). It consists of 6 elements – Smart Governance, Smart Economy, Smart Mobility, Smart Infrastructure, Smart Social, and Smart Living (Giffinger & Gudrun, 2010). Each of the elements carries its roles and complements one another in promoting Smart City. Among the elements, the Smart Living deem to be more important as it focuses in building a shelter that prioritizes in accommodating residents' well-being (Ministry of Housing & Local Government, 2019; Visutsak & Daoudi, 2017) and comply to the fundamental Maslow's Hierarchy of Human Needs (Trivedi & Mehta, 2014). Interestingly, Smart Living concept houses are proposed to cater national issues like aging population and empty nest syndrome (B & B, 2023; Creaney et al., 2021; Khazanah Research Institute, 2021; Leeraphong et al., 2015; Visutsak & Daoudi, 2017a). The initiative deems relevant as the example of Smart Living houses that adhere to the well-being are equipped with assisted washrooms in universal design, water-saving sanitary fittings and motion sensor detectors that allows elderly convenience independence. Unlike the Green buildings, Smart Living concept covers all social, economic, and environmental as the 3 pillars of sustainability. Thus, it is desirable to develop Smart houses as more than just shelters capable of meeting the safety, tranquility, contentment, and well-being of the occupants compared to what can be provided by conventional houses (Wu et al., 2023).

Problem Statement: Action Research Development and the Gap Pertains

Many recent studies from developed countries have proven the workability of Smart Living through their Smart City initiative (Creaney et al., 2021; Visutsak & Daoudi, 2017b; Wahab et al., 2018). Thus, more developing countries have been incorporating the Smart Living concept aggressively with their fiscal planning in the spirit of balanced national growth (Organisation for Economic Co-operation and Development (OECD), 2022), which includes Malaysia. Considering the advantages, the Malaysian policy makers have been proactive in promoting Smart Living development to participate with the revolution of the global housing market (Creaney et al., 2021; Sabrina Abdul Latif et al., 2020). Therefore, while various execution plans and frameworks have been published since 2018, these efforts primarily stem from policy making and development standards. However, achieving successful implementation will necessitate incorporating insights from action research findings. Integrating practical observations and outcomes from action-oriented research can bridge the gap between theoretical frameworks and real-world application, leading to more effective and sustainable outcomes in practice.

From the research standpoint, the research pertaining to Smart Living is still surface (Creaney et al., 2021; Foo, 2018; Ha et al., 2020; Michele Koch, 2021; Sabrina Abdul Latif et al., 2020). Mainly regarding the market's perspective, obstacles, and expectations, theoretical suggestions and implementation, technical aspects of smart features (such as software and hardware for the Internet of Things), a comparison of the green and smart concepts, and conceptual touches on design solutions were all heavily discussed. Moreover, since 2012, local setbacks have persisted due to several reasons, including role overlaps between government agencies leading to gaps in enforcement and policing, limited industry participation in government initiatives, absence of clear guidance modes, and a focus on comparing relative costs and advantages of different technologies. These factors have collectively hindered the smooth progress and effective implementation of initiatives in the relevant sectors. (Che Maznah et al., 2021; Ma et al., 2016; Morozova & Yatsechko, 2022; Wahab et al., 2018). In summary, there is a clear pattern of issues stemming from uncertainties in development activities and concerns about the monetary implications of implementing Smart Living features. These challenges highlight the need for strategic solutions and informed decision-making processes to address these ongoing issues effectively.

Besides, literature reveals the ambiguous nature of Malaysian Smart Living execution plans and standards. There is a missing link between the policing and enforcing where it lacks decision-making tools regarding monetary implication study on features implementation and development activities. A Smart Living housing is a practically conventional house with additional Smart Living features that comply with all pillars of sustainability to serve the occupants. However, Qian et al., (2015) reinforced the addition of features that bring 'extra works' (read: additional development cost components) that are unprecedented among conventional stakeholders. Additionally, conventional houses have been expensive due to the compliance to governance requirements (National Housing Department, 2016). Therefore, there is a critical need for a decision-making tool focused on studying monetary implications, commonly known as a cost model. Existing frameworks, blueprints, and action plans lack a specific costing-related model that facilitates practical implementation. Without such a tool, stakeholders are left to navigate monetary uncertainties and face unpredictable financial risks during their investment decision-making processes. Developing and implementing a robust cost model can significantly mitigate these challenges and enhance the feasibility and success of Smart Living initiatives (Ashworth & Perera, 2015; Gehner, 2009; Qian et al., 2015).

Proposed Solution: Exploring the Cost Model as a Decision-Making Tool and the Research Methodology

The integration of a cost model as a decision-making tool in Smart Living housing development holds immense potential for managing uncertainties and enhancing efficient decision-making during investment phases. There are plenty of research projects that focused on managing uncertainties and successfully utilized models as their decision support systems (Burhanuddin et al., 2011; Dawadi et al., 2021; Munce et al., 2021; Škerlič et al., 2016; Vila-Henninger et al., 2022; Zhang et al., 2019). In the context of Smart Living housing development, a cost model consists of cost drivers and cost data. For Smart Living features implementation on housing development, the cost drivers act as the Smart features that are also known as the anticipated development cost components. While for cost data, it is the cost per yardstick (RM/yardstick) on each Smart features/additional development cost components from current industry practitioners' data. The two variables are placed together to create the relationship between them, thus exhibiting the monetary implications of study. The cost model acts as a pivotal guide in navigating the complexities of financial uncertainties associated with the implementation of Smart Living features. By analyzing cost data and identifying key cost drivers, stakeholders can make informed decisions aligned with project objectives and financial feasibility. However, developing a robust cost model requires a systematic and structured research methodology (Ashworth & Perera, 2015).

This methodology serves as a bridge to bridge the practical-research gap, as highlighted by Miles (2017). The methodology involves thorough literature review, data collection from relevant sources, and data analysis techniques to develop a practical and effective cost model tailored to the specific needs of Smart Living projects. By integrating practical insights with research rigor, the methodology aims to provide actionable strategies for stakeholders to make informed decisions and navigate financial challenges effectively in Smart Living initiatives. According to Alok & Mishra (2017), research methodology serves as a comprehensive approach to achieving research objectives effectively. It is a structured strategy crucial for aligning problem statements with research questions and objectives, while also ensuring consistency with research philosophies, reasoning, and data evaluation processes. The hypothesis posited suggests that a cost model, as a decision-making tool, can effectively manage uncertainties and facilitate efficient decision-making. Therefore, a well-rounded research methodology becomes indispensable. This paper aims to elucidate the adopted research methodology through supporting literature, specifically focusing on the development of a cost model. Importantly, this research seeks to highlight methodological avenues for advancing new knowledge claims in the field of Smart Living housing and its cost model development.

In summary, the Smart Living concept extends beyond traditional housing by integrating features that promote occupants' well-being, especially amidst demographic shifts. While global trends and policies advocate for Smart Living adoption, challenges persist within the Malaysian context. This paper underscores gaps in execution plans, emphasizes the necessity of a decision-making tool, and assesses the current state of Smart Living development in the country. Recognizing the cost model's relevance as a pivotal decision-making tool for managing uncertainties in Smart Living housing development, there is a need for a robust research methodology to bridge the practical-research gap (Miles, 2017). To delineate the research methodology for developing the cost model, the following research questions are proposed in codes Q1 to Q6:

Q1: What are the research philosophies paradigm?
Q2: What are the research approaches?
Q3: What is the research method?
Q4: How is the research strategy?
Q5: When is the research time horizon?
Q6: What are the research techniques and analysis?

LITERATURE REVIEW ON RESEARCH METHODOLOGY

According to M. N. K. Saunders et al. (2016), There are 6 important components – philosophy, approach, method, strategy, time horizon and technique, within the scope of social sciences research methodology. Sutrisna (2009) and Dawadi et al., (2021) reinforced the need to involve all the components with rationale for efficient evaluation of data before presenting any conclusion. For this research, the research methodology is depicted thoroughly in the form of research onion. This structuring is adopted from the work of M. Saunders, Lewis, & Thornhill (2007) as shown in Figure 1. The sorts of instruments applied, and underpinning reasons will be described as well. Intentionally, referencing these sequential layers of approach allows this research to be covered systematically & comprehensively.

Figure 1. Research methodology onion

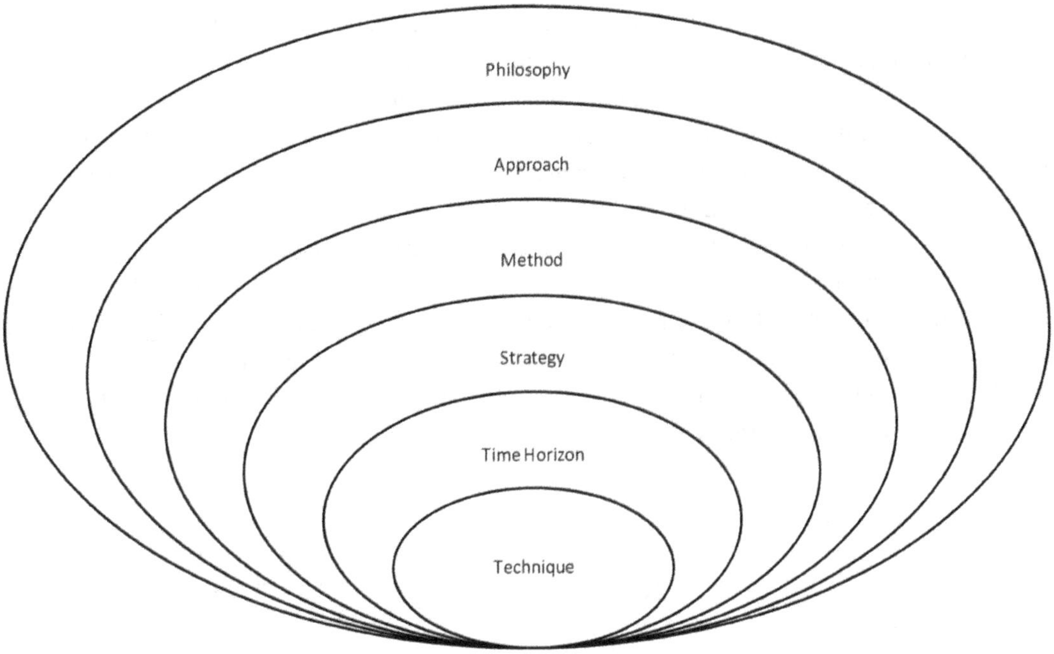

According to Kulatunga et al., (2007), the philosophy helps identify the appropriate procedures for research at the beginning stage by specifying the needed evidence, how to collect it, and how to interpret it. After scoping on research philosophy, M. N. K. Saunders et al., (2016) explained the need for a research approach. According to M. N. K. Saunders et al., (2016), the research approach explicit presentation of findings and the extent of research concerns is raised. It is typically portrayed as two contrasting prospects – induction and deduction, yet it may alternatively be abductive. Janiszewski & van Osselaer (2022) explained its potential in exceptional discoveries that change the arc of consumer research as it simply leverages the original perceptions with real-world identification to augment the theory via abduction approach (Halpin & Richard, 2021). Following is research method, as Creswell (2014), M. N. K. Saunders et al. (2016) and Taherdoost (2022) thoroughly explained and agreed on the three (3) research methods, namely quantitative, qualitative, and mixed-method.

When it comes to time-horizon, there are two main methods: longitudinal and cross-sectional. However, the cross-sectional approach is best for explaining initial findings and developing further insights. This is because it allows for analyzing and comparing subjects in the sample based on how often they occur and how they relate to each other in that particular context (Zangirolami-Raimundo, Echeimberg, & Leone, 2018; Setia, 2016). For research strategy, Dawadi, Shrestha, & Giri (2021), Munce, Guetterman, & Jaglal (2021), Acquah, Naude, & Soni (2021), Othman, Steen, & Fleet (2020) and Subedi (2016) commonly agreed on research strategy is the action took to attain the objectives of the research. For example, surveys and interviews. Subsequently, it led to the need for research techniques. Research technique is known as the instrument to collect the data required (M. N. K. Saunders et al., 2016). In conclusion, there are 6 layers in Saunders' research unions that are sequential by nature that will assist researchers to develop a research methodology comprehensively.

METHOD OF THIS STUDY

The study utilizes a research methodology focused on conducting a literature review within the supporting literature category to thoroughly address the research problems. This approach involves using a structured strategy aligned with research philosophies, reasoning, and data evaluation principles. This research paper is guided by PRISMA Statement through 'Preferred Reporting Items for Systematic Literature Review and Meta-Analysis' (Moher et al., 2009) and 'The PRISMA 2020 Statement: An Updated Guideline for Reporting Systematic Reviews'(Liberati et al., 2009). The PRISMA method has been widely used in the field of built environment and it greatly helps researchers identify articles more accurately. This method also covers three explicit approaches, namely (a) identify large databases of scientific academic literature through keywords and search strategies; (b) filter inclusion and exclusion criteria; and (c) conduct eligibility processes in evaluating relevant literature to analyze data from the studies. However, this paper simply presents the supporting literature by content analysis instead of comprehensive systematic literature review.

Figure 2. Prisma procedures

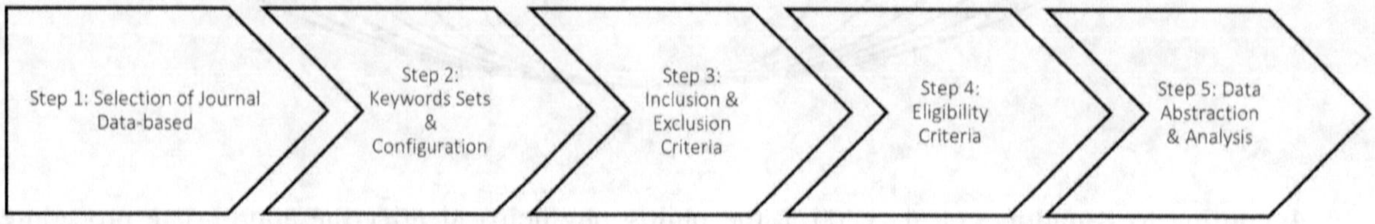

Adopted from the Work of Liberati et al. (2009)

Step 1: Selection of Journal Database

Scopus became the main source of data for the literature review. A large number of journals were indexed by Scopus compared to PubMed, WOS, and Google Scholar as it allows the operation of four main search techniques covering Boolean operators, phrases, truncation, and wildcards. Table 1 and 2 shows the search terms and search string:

Table 1. Search terms

Process	Body of Knowledge	Field Area	Context
• Development • Implementation	• Cost Modeling • Decision-making Tool • Decision Support System • Smart Living Features • Feasibility Studies • Research Methodology • Research Philosophy • Research Technique • Research Approach	• Smart Living Development • Sustainable Construction • Green Building Development	• Developers/ Stakeholders • Construction Industry • Housing Project

The search terms are categorized into 4 – process, body of knowledge, field area and context. The key process involves the development of Smart Living housing and the implementation of Smart Living features onto conventional housing. Both processes encompass the body of knowledge on cost modeling, decision-making tool, decision support system, the Smart Living features, feasibility studies concerns, research methodology, research philosophy, research technique, and research approach. Due to the infancy of Malaysian Smart Living and Smart City, the field areas have to include the preceding Green development and sustainable concept construction. The circumstances that form the setting for the research is through the perspective of developers/stakeholders, housing project and construction industry. The following Table 2 sets the search strings lead by the setting of search terms from Table 1.

Table 2. Search strings

Source	Journal Database	Search String
Primary	Scopus	TITLE-ABS-KEY • Development/Implementation; • Cost Modeling/Decision-making Tools/ Decision-Support System/ Smart Living Features/Feasibility Studies/Research Methodology/ Research Philosophy/Research Technique/Research Approach; • Sustainable Construction/Smart Living Development/Green Building Development; • Developers/Stakeholders/Construction
Secondary	(a) Sustainability Journal; (b) Journal of Engineering, Design and Technology; (c) Journal of Cleaner Production; (d) GeoJournal; (e) Smart and Sustainable Built Environment	The combination of keywords outlined in row above were formulated interchangeably as some journal did not support all the key search techniques

According to Table 2, there are primary and secondary source journal databases. The primary database is Scopus with the search strings of development/implementation, cost modeling/decision-making tools/decision-support system/smart living features/feasibility studies/research methodology/research philosophy/research technique/research approach, sustainable construction/smart living development/green building development, and developers/stakeholders/construction. The secondary journal database includes Sustainability Journal, Journal of Engineering, Design and Technology, Journal of Cleaner Production, GeoJournal with Smart, and Sustainable Built Environment. Keyword combinations are used interchangeably as some journals do not support all major search terms. The following Step 2 is to discuss the keywords set and configuration.

Step 2: Keywords Set and Configurations

The list of keywords can be found in Tables 1 and 2, and the acquisition of synonyms is consistent with the research question and research objective. As secondary sources, the keywords were arranged interchangeably. A total number of n=33 (n=23 (primary source) and n=10 (secondary source)) articles were retrieved, arranged, and managed within 'Mendeley' for screening.

Step 3: Inclusion and Exclusion Criteria

After the screening process through Step 2: Keywords set and Configuration, the n = 33 articles were filtered according to the predetermined inclusion and exclusion criteria as in Table 3.

Table 3. Inclusion and exclusion criteria

Aspect/Criteria	Inclusive	Exclusive
Source	• Journal article and conference paper	• Book chapter and Book series
Type of study	• Project, Sustainable, Green & Smart concept development, including framework protocol, and modeling relevant to this development	• Typical conventional development to be excluded
Time Horizon	• Project development/studies profile from 5 to 10 years, 2013- 2023	• Project development/studies profile longer than 10 years, dated < 2013
Intervention	• Summarize Research Methodology, Research Philosophy, Research Technique & Research Approach	• N/A
Outcome Measure	• Establish the findings as per research questions: Research philosophies paradigm, approaches, method, strategy, time horizon, techniques, and analysis	• N/A
Language	• English	• Non-English

For sources, only article journals and conference papers with empirical evidence were included, excluding book chapters and book series. In terms of study type, it focuses on sustainability project,, the Green and Smart concept development, which include framework, protocol, and modeling, without considering conventional development. In terms of time horizon, articles published in the 5 to 10 years (between 2013 to 2023) were selected due to early availability of most of the literature from abroad practices and the publication of the gray literature from Malaysia Ministry of Housing regarding Smart Living, which is only available after 2018.

Next, the intervention involves for this research summarizing issues and additional cost components/ transaction cost components for the Smart Living development, most of which originate from abroad practices and several from local practices. Then, the outcome measure is used to establish the findings as per research questions: research philosophies paradigm, approaches, method, strategy, time horizon, techniques, and analysis. The search activities focused merely on articles published in the English language and this is important to avoid any complexity and confusion in interpreting non-English publications. Finally, only n = 12 articles remained in this study.

Step 4: Eligibility Criteria

Eligibility screening of the remaining 12 articles was carried out by including only relevant articles to be used for decision-making tool development. The eligibility criteria centered on research methodology, decision-making system, and cost modeling development. Therefore, the final stage of the review resulted in the exclusion of 6 articles because there was a lack of focus on the eligibility criteria per se. In conclusion, only 6 articles were left eligible for review and analysis.

Step 5: Extraction and Analysis

Following the 6 articles eligible for review (refer appendix), content analysis was conducted to categorize the findings as per 6 research questions for a clearer understanding of research methodology in developing the cost model for Smart Living housing development.

FINDINGS AND DISCUSSION ON RESEARCH METHODOLOGY FOR COST MODEL DEVELOPMENT

The study summarizes six articles obtained through PRISMA literature review procedures (Liberati et al., 2009), as outlined in the appendix. These articles discuss decision-making models for managing uncertainties in development, relevant to developing a cost model for Smart Living in Malaysian conventional housing. To clarify the research methodology, the findings are structured using a research onion framework (adapted from M. Saunders, Lewis, & Thornhill, 2007), along with coding of - Q1: Research Philosophies Paradigm, Q2: Research Approach, Q3: Research Method, Q4: Research Strategy, Q5: Research Time Horizon, and Q6: Research Technique and Analysis.

According to the appendix, findings from the 6 articles deduce as follows: For code Q1: Research Philosophies Paradigm, 4 research are interpretivism by motivation and 2 are constructivism-interpretivism; For code Q2: Research Approaches, all 6 articles are abductive approaches; For code Q3: Research Method, there are there are 1 qualitative research, 3 explanatory sequential mixed-method research, and 2 sequential mixed-method; For code Q4: Research Strategies, 4 adopted survey, and 2 case studies; For code Q5: Research Time Horizon, 4 opted longitudinal and 2 opted cross-sectional; and for code Q6: Technique and Analysis, 4 conducted questionnaire and semi-structured interviews with 2 conducted semi-structured interview only. The following subsections further explain the findings by codes.

Q1: Research Philosophies Paradigm

The research objective is to develop a cost model for simulating monetary aspects of implementing Smart Living features in conventional housing. It serves as an initial guide for benchmarking costs, aiding in cost control, and planning within the feasibility study of Smart Living projects. The proposed decision-making tool aims to handle uncertainties inherent in investment decisions, providing support at crucial decision-making junctures. Consequently, the research aligns with an epistemological assumption, emphasizing the underlying principles and beliefs that guide the acquisition of knowledge in this study (Boon et al., 2022).

In this research, assumptions regarding Smart Living features are grounded in the theory of the Smart Living concept, encompassing considerations of features, methods, validity, scope, and the differentiation between justified belief and opinion. The validation of this data involves experts and multiple layers of data validation. The primary goal remains the development of a foundational cost model for implementing Smart Living features in Malaysian housing. However, uncertainties arise in determining the specific features and the subsequent financial implications post-implementation. Consequently, the epistemological framework for this study is rooted in interpretivism and constructivism paradigms.

Guba & Lincoln (1994) stated through "Competing Paradigms in Qualitative Research" handbook, interpretivism and constructivism are presented as residing on different levels and facing uncertainties in practicality. Interpretivism is described as pertaining to data analysis, while constructivism is often associated with knowledge production, raising ontological concerns. However, Van Der Walt (2020) argues that both interpretivism and constructivism are viable epistemological stances, particularly in the realm of social science. Embracing either or both of these paradigms prompts researchers to reflect on their underlying assumptions. This reflection, aligned with the issue of inconsistent implementation of Smart features and their monetary implications, allows for rational considerations in interpreting research results. Van Der Walt's perspective encourages the presentation of results as negotiable constructs rather than models representing definite social realities (Mark, 2003). Consequently, the research aims to amalgamate diverse approaches within the Smart Living concept, acknowledging differences in opinions. The focus shifts toward reasoning and employing multiple methods to reflect various aspects of the choices in Smart Living features implementation, tailored to fit the feasibility study. This approach acknowledges the complex nature of Smart Living features and aims to capture nuanced perspectives through a multifaceted research methodology.

In conclusion, the interpretivist-constructivist approach in this research leads to findings that are socially constructed products influenced by stakeholders. The practicality of Smart features for Malaysian housing and the financial implications of their implementation will be continually (re)constructed through ongoing and invisible interactions once uncertainties are effectively managed. The interpretivist-constructivist nature of the research findings positions them to anticipate future trends in Smart features and the development of Smart Living housing to be more practical and responsive to evolving needs. Notably, 2 out of 6 articles in the study share a similar paradigm, and 4 out of 4 revolve heavily on interpretivism as detailed in appendix. It is considered acceptable due to the background of the 4 research are industrial based which is different in nature with the construction industry. With the research philosophy established, the subsequent subsection delves into exploring the research approach.

Q2: Research Approach

The research approach, according to M. N. K. Saunders et al. (2016), is the explicit presentation of findings and the extent of research concerns, often framed as either inductive or deductive, but alternatively abductive. This diverse range of approaches holds the potential for transformative discoveries that can reshape the landscape of consumer research by combining original perceptions with real-world identification, particularly through an abductive lens. To explore current trends in Smart Living features implementation, identify additional development cost components, and create a cost model for a monetary implication study, an abductive approach is employed. Abduction enables the simultaneous consideration of both deductive and inductive methods, seeking the most fitting explanations for the current market situation without exclusively concluding or developing theory. The objective is not to challenge the novelty of Smart Living but to gather a broader spectrum of possibilities related to features implementation and development cost components. Crucially, all the articles in appendix align with and adhere to a similar abductive approach. This approach, by its nature, enhances the research's capacity to capture the complexity of Smart Living features and their associated costs, offering a more comprehensive understanding of the dynamic market conditions.

Q3: Research Method

The research method, as highlighted by Mishra and Alok (2017), is integral for thoroughly addressing research problems. Drawing insights from the six articles in the appendix, this study recommends an explanatory sequential mixed-methods approach to construct the cost model for implementing Smart Living features in Malaysian housing, as supported by Acquah et al., (2021), Dawadi et al., (2021), and Munce et al., (2021). The efficacy of this method has been demonstrated in previous research, particularly in defining programs, supply chains, and understanding the relationships between variable drivers and collected data in various fields of study. This success suggests the potential applicability of the same method for the current research.

The explanatory sequential mixed-methods approach involves a sequential process of qualitative and quantitative data collection and analysis. This combination aims to complement the respective shortcomings of each approach, directing the results toward practicality in line with the demands of the research. Given the objective of developing a cost model for a monetary implication study, this method proved to be crucial. It allows for the exploration of potential cost drivers through quantitative data, specifically related to Smart Living and sustainable features, and later filters these findings through qualitative aspects to establish relationships in the cost implications of different feature choices. In this context, the cost drivers are the anticipated development costs associated with Smart Living and sustainable features, while the cost data represent numerical reflections measured by specific yardsticks.

For this research, it begins with the quantitative method and subsequently the qualitative method. For mixed-method research, the connective point between the quantitative and qualitative is known as the 'point of interface'. The point of interface is the relationship built between the cost drivers and cost data. The rationale lies in identifying the point of interface by exploring every possibility of features implementation as the anticipated development cost components. The diagram in Figure 3 below displays the explanatory design that interprets the results (Subedi, 2016).

Figure 3. Explanatory sequential mixed-method

Figure 3 illustrates the sequential phase of the entire research methodology. The quantitative aspect involves identifying cost drivers, while the qualitative component focuses on gathering cost data. These two sets of data are integrated at a defined interface point before progressing to the data interpretation stage, specifically in the context of developing cost implication studies for this research. In essence, the approach taken is data-driven, closely aligning with qualitative perspectives without absolute rigidity, yet maintaining a parallel structure with predetermined outcomes from the quantitative data. Importantly, the justification for this methodology is strengthened by the practices observed in the six articles in the appendix. Three out of six articles share similar sequential mixed-method practices, with only one adopting a qualitative-oriented approach. This alignment with existing practices provides further support for the viability and workability of the chosen research methodology.

Q4: Research Strategy

Research strategy the research strategy refers to the actions undertaken to achieve the research objectives. The appendix highlights those 4 articles employed a survey approach, while 2 articles utilized case studies. Considering this pattern, the current research adopts a survey and case study strategy throughout the explanatory sequential mixed-method methodology. Aligning with the expected output, philosophical stance, and abductive approach, the survey strategy will be employed to gather information on cost drivers, specifically Smart Living features, as the anticipated development cost components. Subsequently, the case study strategy will be utilized in the form of a document survey to conduct interviews with project experts, aiming to obtain and validate cost data.

In the survey phase, an open-ended questionnaire survey is chosen for the preliminary data collection. Quantitatively, this approach seeks to collect a comprehensive range of cost drivers covering a broad population and focusing on common favorable types of Smart Living features. The obtained data will then undergo a saturation level analysis before being subjected to another series of Relative Important Index (RII) (Rajgor et al., 2016). This process aims to include only practical features in the list of cost drivers for the subsequent Cost Modeling simulation. Following the open-ended questionnaire survey, a 5-point Likert Scale is employed to rate the practicality of each feature. This rating system serves to select and include prominent cost drivers suitable for the Cost Modeling simulation. Overall, the chosen survey and case study strategy is tailored to comprehensively address the research objectives while ensuring the practicality and relevance of the identified cost drivers.

The subsequent procedure justifies the integration of case studies by utilizing them as a reference in the form of a document survey for conducting expert interviews. Given the subjective and competitive nature of Smart Living features implementation, this approach is deemed necessary to gather judgmental yet unbiased inputs from a broader population. The collected data will undergo evaluation and filtering to reach a consensus and enhance data validity. This pragmatic method aims to identify common Smart Living features, minimize uncertainty in terms of cost data, and establish an efficient relationship for cost implication simulations.

In summary, the inputs obtained will be presented during the interview engagement to prevent data dispersion. Subsequently, data gathered from the questionnaire survey will be rigorously evaluated and filtered for practicality, guided by insights from the document surveys conducted on case studies. This reciprocal process ensures that the developed cost model aligns with market needs and remains relevant to real-world industry practices. The integration of case studies serves as a valuable mechanism for grounding the research in practical and industry-relevant insights, enhancing the robustness of the overall research methodology.

Q5: Research Time Horizon

In considering the research time horizon, the appendix indicates that out of the 6 articles, 4 adopted a longitudinal time horizon, while 2 articles chose a cross-sectional approach. Given that many models developed require rich data, especially in complex fields such as engineering and supply-demand with numerous variables acting as cost drivers, the selection of a time horizon becomes crucial. Contrary to the prevailing trend, this research opts for a cross-sectional time horizon. A cross-sectional study involves collecting data from a population at a specific point in time (M. N. K. Saunders et al., 2016), and it is analytical in nature as it establishes comparisons between subjects based on absolute and relative frequency in that particular period. In this context, the focus is on developing only two variables which are cost drivers and cost data. By adopting a cross-sectional time horizon, the research aims for a simpler yet efficient simulation that captures the current moment without the complexities associated with tracking changes over an extended duration. This approach is deemed practical for the objectives of the study, emphasizing the immediate and comprehensive understanding of cost implications related to Smart Living features implementation.

Q6: Research Technique and Analysis

The research employs various techniques to collect the necessary data, including a systematic literature review, open-ended questionnaire, and a document survey on case studies through semi-structured interviews. These techniques are strategically utilized in different phases of the research process, aligning with the practices observed in previous works presented in the appendix. The systematic literature review serves as a tool to investigate previous and current milestones in Smart Living development, providing a comprehensive understanding of the existing body of knowledge in the field. Following this, the open-ended questionnaire, a quantitative method, is employed to collect data on cost drivers. This survey allows for the systematic gathering of information related to the anticipated development cost components of Smart Living features. The third technique involves a document survey on case studies conducted through semi-structured interviews. This method is instrumental in collecting and validating cost data. By engaging with experts and referring to documented case studies, the research ensures a robust and reliable foundation for the cost data used in the subsequent phases of the study. In conclusion, these chosen techniques not only align with the methodologies observed in the appendix but also contribute to a comprehensive and multi-faceted approach in gathering data essential for the development of the cost model for Smart Living features implementation in Malaysian housing.

COST MODEL DEVELOPMENT

The discussion from Q1 to Q6 has led to proposing a research methodology guiding data collection and determining the cost data and drivers for Smart Living features. Next, this study leverages Ashworth's Cost Modeling theory to explain cost model development within this context, offering deeper insights into cost dynamics in Smart Living housing. Ashworth's theory provides a structured framework for understanding and managing costs in construction projects, particularly relevant for integrating cost models into feasibility studies for Smart Living housing projects. Ashworth's theory also aligns cost planning and control with the builder's Business Strategy Model, enhancing decision-making regarding the cost implications of Smart Living feature implementation in Malaysian housing projects.

Ferry and Brandon (2007) and Seeley (1996) elucidate the cost model as a symbolic representation of a system, capturing the intricacies of factors influencing its costs. Seeley (1996) emphasizes that a cost model goes beyond representation; it is a procedural framework developed to yield reliable outputs based on a predetermined series of input data. According to Beroe Inc (2021), a cost model serves as a systematic tool for businesses to determine the best production costs for their products. Its advantages are manifold. Firstly, it enhances supplier relationship management by promoting transparency and a deep understanding of cost structures. Secondly, it facilitates informed negotiations and precise estimations through comprehensive consideration of all variables. Thirdly, it aids in profit optimization by identifying and addressing spending irregularities that could impede growth. Ultimately, it empowers businesses to make strategic decisions, offering a competitive advantage in cost management and resource allocation.

According to Ashworth & Perera (2015), the nature of a cost model is multifaceted, defined by two primary classifications: deterministic and probabilistic. In the deterministic realm, the model assumes that values can be precisely ascribed to all variables, either known or accurately predictable at the initiation of the project. This type of model necessitates a robust foundation of solid historical data to develop accurate estimations, providing a clear and deterministic view of the cost landscape. Conversely, the probabilistic type acknowledges the inherent uncertainty in some variable values, recognizing that these can only be estimated. This approach is particularly pertinent for projects in their exploratory phases or those embracing innovative approaches, allowing for flexibility and adaptability as uncertainties unfold during the project life cycle.

Examining the three primary types of cost models, we first encounter the designers' cost model, which is a traditional and commonly used approach. This model estimates future costs based on elements or units from past building projects, providing historical context for cost projections. In contrast, the Constructor's Production Model is created during the tendering phase before construction begins. It focuses on simulating the construction process itself, detailing operations and activities to determine costs during construction. Lastly, the Mathematical Model aims to identify key variables describing costs, simplifying the model by reducing priceable units for a more efficient representation of cost drivers.

Cost models serve various functions depending on the specific needs at different stages of a construction project. Design optimization is crucial to ensure cost-effective building designs, commonly used in design economics and cost planning for comparisons between completed projects and proposed alternatives. Tender prediction forecasts the likely tender sum accepted by a client, aiding in selecting the most economical proposal during tendering. Whole life costing considers a project's entire lifespan, including operational phases, for informed long-term financial decisions. Resource-based models help contractors estimate costs related to resource utilization, improving project efficiency through accurate estimations.

In essence, the multifaceted nature of cost models positions them as indispensable tools across various facets of the construction industry. They contribute significantly to decision-making, planning, and overall project success, offering valuable insights into the complex dynamics of construction cost management. Whether applied for design optimization, tender prediction, cash flow management, whole life costing, or resource-based analysis, cost models are instrumental in navigating the intricate landscape of construction projects, ensuring their financial viability and success. Following Figure 4 illustrates the steps for cost model development, Table 4 shows the cost drivers and respective yardsticks, and accompanied with Figure 5 that shows the sample cost model for Johor mid-landed conventional housing development cost components.

The following Figure 4 illustrates 6 steps procedures (A to F) of cost model development. In this context, the cost drivers are the Smart features or known as the anticipated development cost components. While for cost data, it is the cost per yardstick (RM/yardstick) on each Smart features/additional development cost components from current industry practitioners' data. The procedures start with document surveys by collecting similar basis case studies and categorizing them by nature of landed/high-rise and affordable/mid-range housing projects. Particulars on regions and states need to be consistent because it will influence the cost data information. Next, identify the cost drivers and cost data. Finally, it is to find the mean value for the cost data. The two variables are placed together to create the relationship between them thus exhibiting the monetary implications.

Figure 4. Ashworth's cost model theory

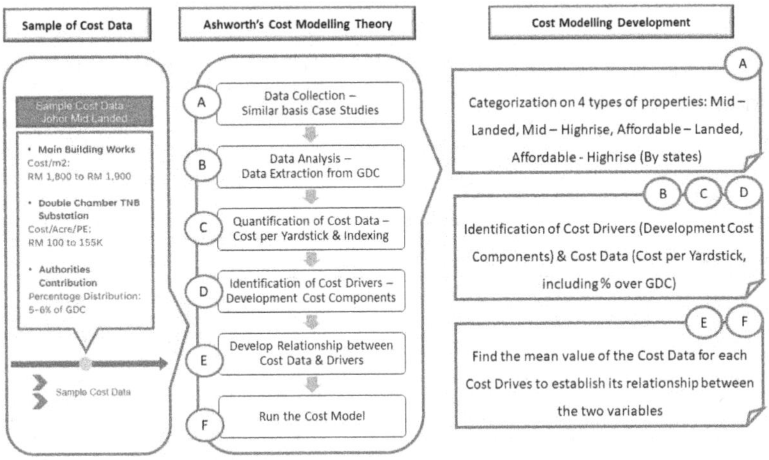

The significance of yardsticks in the context of cost models, particularly concerning Smart Living features as cost drivers, cannot be overstated. Each cost driver, representing distinct Smart Living features, is associated with a specific yardstick, essentially serving as a unit of measurement for its cost. This practice is important to ensure uniformity and clarity in cost estimations. For instance, the Smart Living yardstick of recycling bins for landed housing (e.g., terrace houses), is calculated through the cost per item or set (RM/unit). In contrast, for a Smart Living feature like a recycling chute in high-rise buildings (e.g., apartments), the yardstick takes the form of the cost per length (RM/m). These variations in yardsticks, whether per item or per length, have direct implications on the costing of the gross development cost. Therefore, establishing a comprehensive list of cost drivers accompanied by their respective yardsticks is not only advisable, but it is critical in ensuring precision and accuracy in cost estimations for Smart Living features, which in turn contribute to the overall success and financial viability of a construction project. This meticulous approach facilitates transparent and standardized cost assessments, essential for effective decision-making throughout the project life cycle. The following Table 4 shows the cost drivers and respective yardsticks.

Table 4. Cost drivers: Smart living features with respective yardsticks

Cost Drivers: Smart Living Features	Yardstick	Cost Drivers: Smart Living Features	Yardstick
Video Monitoring System	RM/Item	Solid waste Management System	RM/Nr
Smart Security Lock System	RM/Nr	Solar Panel on Building Roof	RM/kW
Security-to-Household Intercom System	RM/Nr	Recreational Facilities & Ancillaries	RM/M^2
Universal Design: Ramp & Assisted Washroom (ground floor)	RM/Item	Entrance Landscape	RM/M^2
Wheelchair Lift at Stairs	RM/Item	Gated Guarded Community	RM/Item
Occupancy, Motion & Sensory based Appliances System	RM/Nr	Smart Pole with Panic Button	RM/Nr
Smart Sensor Cloth Dryer	RM/M^2	Pedestrian Friendly Walkway with Roof	RM/M

continued on following page

Table 4. Continued

Cost Drivers: Smart Living Features	Yardstick	Cost Drivers: Smart Living Features	Yardstick
Fire Detector	RM/M^2	Cycling Friendly Street	RM/M
Safety Panic Button with Siren	RM/M^2	Universal Design: truncated dome tactile strip & alike	RM/M^2
Portable Parcel Box/Compartment at Front Gate	RM/Item	Public Transport Facilities - Sheltered Bus Stop	RM/M^2
Garbage & Recycling Bin/Chute	RM/Nr	School & Education Centre	RM/M^2
Daylight-oriented Structure	RM/Item	Health & Medical Centre	RM/M^2
Acoustical Environmental Structure	RM/Item	Sport Facilities	RM/M^2
Real-Time Water Quality Checker	RM/Nr	Retail Facilities	RM/M^2
Real-Time Air Quality Checker	RM/Nr	Religious Building - Surau/Masjid	RM/M^2
Remote Control Blinding/Shades	RM/M^2	Digital Board/LED Display (Dimension: 5m x 8m)	RM/Nr
Telemetering Waste Collection System	RM/Item	Farming/Tree Cover/Tree Inventory	RM/M^2
Garden Works & Landscaping	RM/Item	Disaster Risk Management (Consultation)	RM/Item
Private Farming Proximity	RM/Item	Hill Slope Structure	RM/Item
Water Closet Items - Faucet Aerator	RM/Nr	Flood Mitigation Structure/Facilities/Convertible Community Hall	RM/M^2
Water Closet Items - Dual Flush Toilet	RM/Item	Street light with Solar PV	RM/Nr
Light-emitting Diode (LED) type Lighting System	RM/M^2	Street light Pollution Filter/Reflector/Shielded Fixture	RM/Nr
Smart Thermostat	RM/M^2	Solar Panel on Pedestrian Walkway Roof	RM/Nr
Inverter Energy Saving Air-conditioner	RM/Nr	Smart Tourism Area	RM/M^2
Polarized Glass Panel	RM/Nr	EV Charging Port (22kW)	RM/Nr
Rainwater Harvesting System	RM/Nr	EV Charging Point at Home (10kW)	RM/Nr
Grey Water System	RM/Item	Sheltered Bus Stop	RM/Nr

As depicted in Figure 4, the subsequent step in cost model development is illustrated in Figure 5, showcasing a sample of cost model development specifically for conventional housing development cost items, including land cost, hard cost, soft cost, and transaction cost. In this particular analysis, a cumulative total of 16 case studies pertaining to mid-landed conventional housing in Johor has been meticulously examined. To ensure data consistency, the costs from these 16 case studies have been uniformly indexed to the year 2019 using the Tender Price Index (TPI). Additionally, the property type has been maintained, focusing exclusively on mid-landed structures. Once these constant variables are established, the extraction of both cost data and cost drivers from the 16 case studies is undertaken.

Concerning the cost data, a mean value is calculated for each cost driver, thereby generating a comprehensive range for each cost item. For instance, the yardstick for land cost is measured in RM/acre, amounting to RM 2,163,363.72 per acre. The hard cost for the main building work is quantified in RM/m2, resulting in an estimate of RM 3,029.00 per meter square. In the case of a double-chamber TNB substation, the yardstick is in RM/acre, amounting to RM 125,946.00 per acre. Furthermore, the soft cost, coupled with transaction cost, involves authorities' contributions distributed as a percentage across the entire gross development cost. This percentage distribution falls within the range of 5.00% to 6.00%. This meticulous process of cost model development ensures not only the precision of estimates but also provides a robust framework for understanding and projecting the financial implications associated with the various components of conventional housing development.

Figure 5. Cost model development sample: Johor Mid-Landed Housing Project

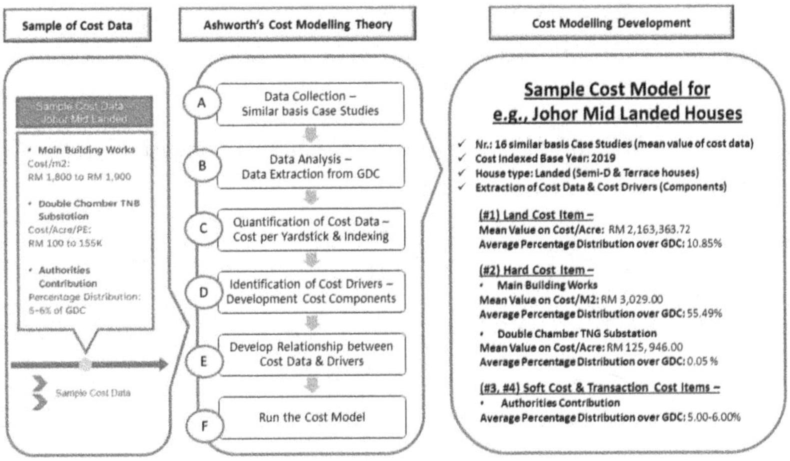

In Figure 5, a tangible illustration is presented, offering a sample of the cost model development for conventional housing development cost items. However, the transition to Smart Living housing development introduces a distinctive paradigm, where the cost drivers are intricately linked to itemized Smart Living features, as exemplified in Table 4. The methodology employed for conventional housing cost model development is mirrored for Smart Living features. Each Smart Living feature, meticulously itemized in Table 4, undergoes a similar process wherein mean value calculations are applied. This results in obtaining a numerical price for each Smart Living feature, providing insights into their individual cost implications.

Crucially, this process does not just stop at numerical values; it extends to encompass the percentage distribution of these Smart Living features over the entire gross development cost. The goal is to ascertain the proportional contribution of each Smart Living feature to the overall financial landscape of the project. By embracing a consistent and systematic approach, akin to the one employed for conventional housing, this methodology ensures that the cost model development for Smart Living housing is not only accurate but also aligns with industry standards, while fostering a comprehensive understanding of the financial dynamics associated with the integration of Smart Living features into the development landscape.

CONCLUSION

The research aimed to establish an effective research methodology for developing cost modeling as a decision-making tool in Smart Living housing development. Through content analysis of six articles using the PRISMA methodology, a research methodology was proposed in the form of a research onion (Figure 6). The chosen methodology, explanatory sequential mixed-methods, was found to be comprehensive. To delve into the economic realities of implementing Smart Living features and their perceived value, the research philosophy aligns with constructivism and interpretivism, using qualitative and quantitative approaches. An abductive approach is employed, utilizing sequential explanatory mixed-methods to address inductive and deductive concerns. The study adopts a cross-sectional time horizon to analyze evolving Smart Living frameworks. Initially, a quantitative approach identifies commonly practiced features as cost drivers, followed by qualitative screening to ascertain cost data. Both findings undergo content analysis to validate feature claims for cost model development. This approach guides data collection and selection effectively.

Figure 6. Research methodology for cost modeling development on smart living features implementation in Malaysian housing

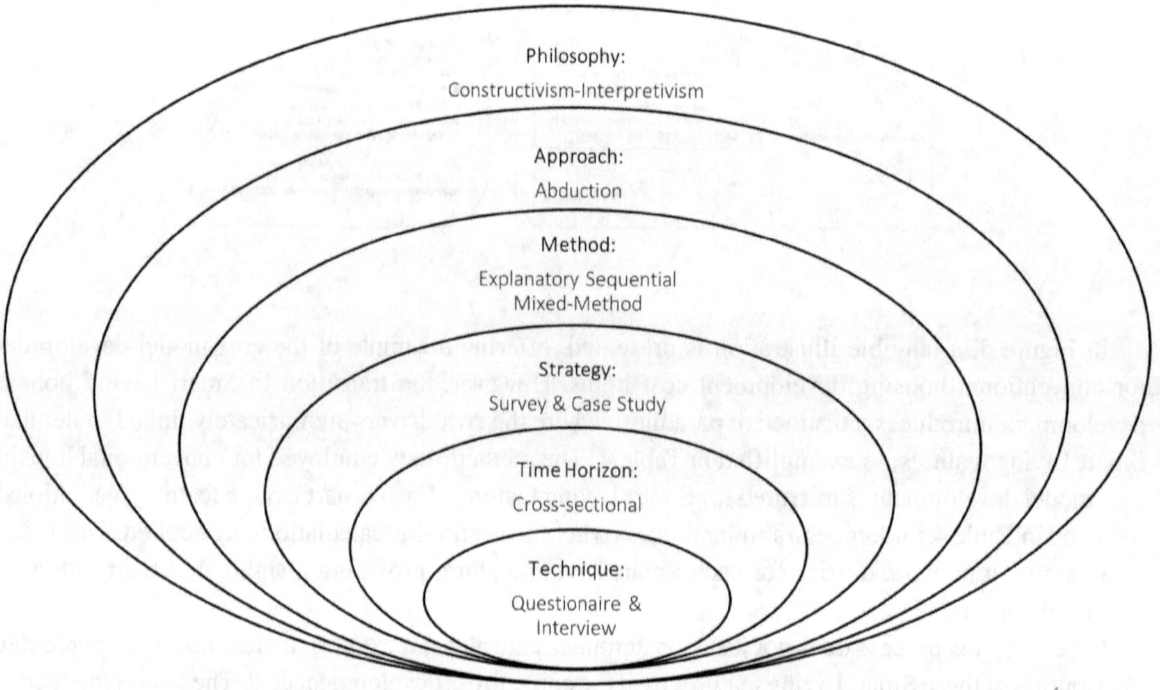

Revisiting Ashworth's Cost Modeling Theory, the developed cost model for integrating Smart Living features into conventional housing is classified as a probabilistic model. This classification acknowledges the fluctuating values of cost drivers over time, within an acceptable range unless specified otherwise. This adaptability allows the model to estimate changing costs, aligning with uncertainties in cost data for each feature. The cost model serves as both a designers' tool, offering estimations pre-development, and a constructors' tool, aiding tendering by detailing construction costs.

Transparent listing of Smart Living features as cost drivers facilitates identifying itemized cost data, aiding stakeholders in monetary studies for different scenarios. This is crucial for managing uncertainties and risks in cost implications. The iterative procedures align with the paper's goal of understanding monetary implications in integrating Smart Living features. Ashworth's Theory guides this process effectively under uncertainties, bridging practical knowledge gaps.

Smart Living improves occupants' well-being beyond traditional housing, though challenges persist in Malaysia, noted in the paper. It identifies execution gaps, decision-making tool needs, and current Smart Living development. Recognizing the cost model's relevance in managing monetary implications, a robust research methodology is crucial. The proposed methodology, combining quantitative and qualitative methods, establishes a strong cost driver-cost data relationship, forming the cost model basis. This structured strategy aligns with research philosophies, contributing to Smart Living challenges, understanding and methodological advancements.

REFERENCES

Acquah, I. S. K., Naude, M. J., & Soni, S. (2021). How the dimensions of culture influence supply chain collaboration: An explanatory sequential mixed-methods investigation. *Revista de Gestao*, 28(3), 241–262. 10.1108/REGE-11-2020-0105

Alok, S., & Mishra, S. B. (2017). *Handbook of Research Methodology*. Educreation Publishing. 10.1097/00003465-199001000-00018

Ashworth, A., & Perera, S. (2015). Cost Modelling. In *Cost Studies of Buildings* (6th ed.). Routledge. 10.4324/9781315708867

B, J. C., & B, M. Z. (2023). Application of Aging-Appropriate Smart Home. In *Proceedings of the 2023 2nd International Conference on Social Sciences and Humanities and Arts (SSHA 2023)*. Atlantis Press SARL. 10.2991/978-2-38476-062-6

Boon, M., Orozco, M., & Sivakumar, K. (2022). Epistemological and educational issues in teaching practice-oriented scientific research: Roles for philosophers of science. *European Journal for Philosophy of Science*, 12(1), 16. 10.1007/s13194-022-00447-z

Burhanuddin, M. A., Halawani, S. M., & Ahmad, A. R. (2011). A costing analysis for Decision Making Grid model in failure-based maintenance. *Advances in Decision Sciences*, 2011. 10.1155/2011/205039

Che Maznah, M. I., Fatin Najwa, M. N., Nur Kamaliah, M., Jeffery, L., Sahithi, A. S., & Preece, C. N. (2021). Sustainable Township and Sustainable Home: Public Perceptions. *Journal of the Society of Automotive Engineers Malaysia*, 5(3), 331–347. 10.56381/jsaem.v5i3.176

Creaney, R., Reid, L., & Currie, M. (2021). The Contribution of Healthcare Smart Homes to Older Peoples' Wellbeing: A New Conceptual Framework. *Wellbeing, Space and Society, 2*(June 2020), 100031. 10.1016/j.wss.2021.100031

Creswell, J. W. (2014). The Selection of a Research Approach. *Research Design*, 3–23. https://doi.org/45593:01

Dawadi, S., Shrestha, S., & Giri, R. A. (2021). Mixed-Methods Research: A Discussion on its Types, Challenges, and Criticisms. *Journal of Practical Studies in Education*, 2(2), 25–36. 10.46809/jpse.v2i2.20

Ferry and Brandon. (2007). *Ferry and Brandon's Cost Planning of Buildings, 8th Edition*. Wiley. https://www.wiley.com/en-us/Ferry+and+Brandon%27s+Cost+Planning+of+Buildings%2C+8th+Edition-p-9781405130707

Foo, C. H. (2018). The Impact of Capital Contributions and Compliance Costs on Housing Affordability. *Building and Investment*, (August), 34–38.

Gehner, E. (2009). Real Estate Development Strategies and Their Impact on the Risk Profile of a Project. *CIB Joint International Symposium*, (pp. 853–863). IEEE.

Giffinger, R., & Gudrun, H. (2010). Smart Cities Ranking: An Effective Instrument for the Positioning of the Cities? *Journal of the Centre of Land Policy and Valuation*, 4(12), 7–25. 10.5821/ace.v4i12.2483

Guba, E. G., & Lincoln, Y. S. (1994). Competing Paradigms in Qualitative Research. In *Handbook of qualitative research* (pp. 105–117). Sage Publications, Inc.

Ha, C. Y., Radzi, I., & Khoo, T. J. (2020). The Barriers of Implementing Green Building in Penang Construction Industry. *Progress in Energy and Environment, 12*(March), 1–10. https://www.akademiabaru.com/doc/progeeV12_P1_10.pdf

Halpin, M., & Richard, N. (2021). An invitation to analytic abduction. *Methods in Psychology (Online)*, 5, 100052. 10.1016/j.metip.2021.100052

Janiszewski, C., & van Osselaer, S. M. J. (2022). Abductive Theory Construction. *Journal of Consumer Psychology*, 32(1), 175–193. 10.1002/jcpy.1280

Khazanah Research Institute. (2021). *Building Resilience: Towards inclusive social protection in Malaysia*. Khazanah Research Institute.

Kulatunga, K., Amaratunga, D., & Haigh, R. (2007). Researching construction client and innovation: methodological perspective. *7th International Postgraduate Research Conference in the Built and Human Environment*. Research Gate.

Leeraphong, A., Papasratorn, B., & Chongsuphajaisiddhi, V. (2015). A Study on Factors Influencing Elderly Intention to Use Smart Home in Thailand : A Pilot Study. *The 10th International Conference on E-Business*, 1–10.

Liberati, A., Altman, D. G., Tetzlaff, J., Mulrow, C., Gotzsche, P. C., Ioannidis, J. P. A., Clarke, M., Devereaux, P. J., Kleijnen, J., & Moher, D. (2009). The PRISMA Statement for Reporting Systematic Reviews and Meta-Analyses of Studies that Evaluate Healthcare Interventions: Explanation and Elaboration. *BMJ, 339*(jul21 1), b2700–b2700. 10.1136/bmj.b2700

Ma, Z., Badi, A., & Jørgensen, B. N. (2016). Market opportunities and barriers for smart buildings. *2016 IEEE Green Energy and Systems Conference, IGSEC 2016*. IEEE. 10.1109/IGESC.2016.7790078

Mark, N. P. (2003). Book Review: Accountability in Social Research: Issues and Debates. *Sociological Methods & Research*, 31(3), 430–434. 10.1177/0049124102239084

Michele Koch. (2021). GreenPrint Survey Finds Consumers Want to Buy Eco-Friendly Products, but Don't Know How to Identify Them. *Businesswire*, 50–51.

Miles, D. A. (2017). A Taxonomy of Research Gaps: Identifying and Defining the Seven Research Gaps. *Journal of Research Methods and Strategies*, 2017, 1–15.

Ministry of Housing & Local Government. (2019). Malaysia Smart City Framework. *Malaysia Smart City Framework, September*, 251. https://drive.google.com/file/d/1lgnk6rGDx7ci08-MjXNTiAyaW-PfVwWs/view%0Ahttps://www.thesundaily.my/local/ministry-launches-malaysia-smart-city-framework-BN1395377

Moher, D., Liberati, A., Tetzlaff, J., & Altman, D. G. (2009). Preferred reporting items for systematic reviews and meta-analyses: The PRISMA statement. *BMJ (Clinical Research Ed.)*, 339(7716), 332–336. 10.1136/bmj.b253519622551

Morozova, I. A., & Yatsechko, S. S. (2022). The Risks of Smart Cities and the Perspectives of Their Management Based on Corporate Social Responsibility in the Interests of Sustainable Development. *Risks*, 10(2), 34. 10.3390/risks10020034

Munce, S. E. P., Guetterman, T. C., & Jaglal, S. B. (2021). Using the Exploratory Sequential Design for Complex Intervention Development: Example of the Development of a Self-Management Program for Spinal Cord Injury. *Journal of Mixed Methods Research*, 15(1), 37–60. 10.1177/1558689820901936

National Housing Department. (2016). *National Housing Policy*. NHD.

Organisation for Economic Co-operation and Development (OECD). (2022). *Achieving the SDGs in cities and regions. Promoting local and regional development*. OECD. https://www.oecd.org/about/impact/achieving-sdgs -in-cities-and-regions.htm

Othman, S., Steen, M., & Fleet, J.-A. (2020). A sequential explanatory mixed methods study design: An example of how to integrate data in a midwifery research project. *Journal of Nursing Education and Practice*, 11(2), 75. 10.5430/jnep.v11n2p75

Qian, Q. K., Chan, E. H. W., & Khalid, A. G. (2015). Challenges in Delivering Green Building Projects: Unearthing the Transaction Costs (TCs). *Sustainability (Basel)*, 7(4), 3615–3636. 10.3390/su7043615

Rajgor, M., Paresh, C., Dhruv, P., Chirag, P., & Dhrmesh, B. (2016). RII & IMPI : Effective Techniques for Finding Delay in Construction Project. *International Research Journal of Engineering and Technology*, 3(1), 1173–1177.

Sabrina Abdul Latif, N., Muhammad Rizwan Majeed, K., Rozzani, N., & Khadijah Saleh, S. (2020). Factors Affecting Housing Prices in Malaysia: A Literature Review. *International Journal of Asian Social Science*, 10(1), 63–67. 10.18488/journal.1.2020.101.63.67

Saunders, M., Lewis, P., & Thornhill, A. (2007). Research Methods for Bussiness Students. In *Pearson* (4th ed.).

Saunders, M. N. K., Lewis, P., & Thornhill, A. (2016). Understanding Research Philosophy and Approaches to Theory Development. Research Methods for Business Students. In *Harlow: Pearson Education* (Issue January).

Seeley, I. H. (1996). *Building Economics*. Macmillan Education UK., 10.1007/978-1-349-13757-2

Setia, M. S. (2016). Methodology Series Module 3: Cross-Sectional Studies. *Indian Journal of Dermatology*, 61(3), 261–264. 10.4103/0019-5154.18241027293245

Sharifi, A., Allam, Z., Bibri, S. E., & Khavarian-Garmsir, A. R. (2024). Smart cities and sustainable development goals (SDGs): A systematic literature review of co-benefits and trade-offs. *Cities, 146*, 104659. 10.1016/j.cities.2023.104659

Škerlič, S., Muha, R., & Logožar, K. (2016). A Decision-Making Model for Controlling Logistic Costs. *Tehnicki Vjesnik (Strojarski Fakultet)*, 23(1), 145–156. 10.17559/TV-20141015121023

Subedi, D. (2016). Explanatory Sequential Mixed Method Design as the Third Research Community of Knowledge Claim. *American Journal of Educational Research, 4*(7), 570–577. https://doi.org/10.12691/education-4-7-10

Sutrisna, M. (2009). Research Methodology in Doctoral Research: Understanding the Meaning of Conducting Qualitative Research. *Association of Researchers in Construction Management (ARCOM). Doctoral Workshop*, 2(May), 48–57.

Taherdoost, H. (2022). What are Different Research Approaches? Comprehensive Review of Qualitative, Quantitative, and Mixed Method Research, Their Applications, Types, and Limitations. *Journal of Management Science & Engineering Research*, 5(1), 53–63. 10.30564/jmser.v5i1.4538

Trivedi, A. J., & Mehta, A. (2014). Maslow's Hierarchy of Needs. A Theory of Human Motivation. Psychological. *Psychological Review*, 7(38), 2014. https://kalyan-city.blogspot.com/2010/06/maslow-hierarchy-of-needs-theory-of.html

Van Der Walt, J. L. (2020). Interpretivism-Constructivism as a Research Method in the Humanities and Social Sciences-More to It Than Meets the Eye. *International Journal of Philosophy*, 8(1), 2333–5769. 10.15640/ijpt.v8n1a5

Vila-Henninger, L., Dupuy, C., Van Ingelgom, V., Caprioli, M., Teuber, F., Pennetreau, D., Bussi, M., & Le Gall, C. (2022). Abductive Coding: Theory Building and Qualitative (Re)Analysis. In *Sociological Methods and Research*. 10.1177/00491241211067508

Visutsak, P., & Daoudi, M. (2017a). The smart home for the elderly: Perceptions, technologies and psychological accessibilities: The requirements analysis for the elderly in Thailand. *ICAT 2017 - 26th International Conference on Information, Communication and Automation Technologies, Proceedings, 2017-Decem*(October 2017), (pp. 1–6). IEEE. 10.1109/ICAT.2017.8171625

Visutsak, P., & Daoudi, M. (2017b). The smart home for the elderly: Perceptions, technologies and psychological accessibilities: The requirements analysis for the elderly in Thailand. *ICAT 2017 - 26th International Conference on Information, Communication and Automation Technologies, Proceedings, 2017-Decem*(October), (pp. 1–6). IEEE. 10.1109/ICAT.2017.8171625

Wahab, E., Shamsuddin, A., Abdullah, N., & Yi, K. S. (2018). *A Study on the Smart Home Adoption in Malaysia : A Foresight Perspective. 112th The IRES International Conference*, Jeju Island, South Korea.

Wu, S. Z., Zhen, F., Wu, X., Guo, L., Zhou, X., Zhao, Y., Mao, W., Hu, J., & Liu, Z. (2023). Roundtable discussion: Rethinking and exploring smart city development. *Frontiers of Urban and Rural Planning*, 1(1), 1–10. 10.1007/s44243-022-00003-6

Zangirolami-Raimundo, J., Echeimberg, J. de O., & Leone, C. (2018). Research methodology topics: Cross-sectional studies. *Journal of Human Growth and Development*, 28(3), 356–360. 10.7322/jhgd.152198

Zhang, H., Kou, G., & Peng, Y. (2019). Soft Consensus Cost Models for Group Decision Making and Economic Interpretations. *European Journal of Operational Research*, 277(3), 964–980. 10.1016/j.ejor.2019.03.009

APPENDIX

These articles were selected as supporting literature in determining the research methodology for cost model development.

Table 5. Six articles selected form PRISMA as supporting literature for this research

Research Title	Q1: Research Philosophies Paradigm	Q2: Research Approach	Q3: Research Method	Q4: Research Strategy	Q5: Research Time Horizon	Q6: Research Technique & Analysis
Soft Consensus Cost Models for Group Decision Making and Economic Interpretations	Epistemology – Interpretivism	Abductive	Sequential mixed-method	Survey	Longitudinal	Semi-structured Interview
Analysis of Overhead Cost Behavior: Case Study on decision-making approach	Epistemology – Constructivism Interpretivism	Abductive	Sequential mixed-method	Survey	Longitudinal	Questionnaire & Semi-structured Interview
A Decision-Making Model for controlling Logistics Costs	Epistemology – Constructivism Interpretivism	Abductive	Qualitative	Case Studies	Longitudinal	Semi-structured Interview
Costing for Decision Making: Activity-based Costing vs. Theory of Constraints	Epistemology – Interpretivism	Abductive	Explanatory sequential mixed-method	Survey	Cross-Sectional	Questionnaire & Semi-structured Interview
A Costing Analysis for Decision Making Grid Model in Failure-Based Maintenance	Epistemology – Interpretivism	Abductive	Explanatory sequential mixed-method	Case Studies	Longitudinal	Questionnaire & Semi-structured Interview
The Importance of the Cost Information in Making Decision	Epistemology – Interpretivism	Abductive	Explanatory sequential mixed-method	Survey	Cross-Sectional	Questionnaire & Semi-structured Interview

Chapter 13
Review of Quantity Surveying Training in China Between 2002 to 2022 Based on CiteSpace

Dongmei Huangfu
https://orcid.org/0009-0006-3866-1076
Taylor's University, Malaysia

Yun Fah Chang
Taylor's University, Malaysia

Sok Li Lim
Taylor's University, Malaysia

ABSTRACT

China's construction sector embarked on a journey of transformation with a "new engineering" strategy, beginning in 2016. Notably, two areas that have captured the attention of both industry and academics are the enhanced quality of quantity surveyors (QS) and the sustainable growth of the QS industry. This study focuses on the literature related to personnel training for QS over the last two decades. Utilizing the CiteSpace software for analysis, the authors examined the research status, focus, and trends in QS personnel training. The findings indicate growing scholarly interest in the training of QS personnel, with strong research focus on areas encompassing the definition of quantity surveying, higher vocational education, personnel training, innovation, and development. Furthermore, emerging areas such as new engineering, vocational education, curriculum ideology, political aspects, and BIM are anticipated to attract a lot of attention in the future.

INTRODUCTION

Recently, new engineering education reforms have been developed with the goal of cultivating fresh talent for the development of new technologies and industries. And also, with the development and national policy guidance of the construction industry, the industry has been reformed. This is especially so with the increased industry's usage of prefabricated construction technology and building information

DOI: 10.4018/979-8-3693-1650-4.ch013

modelling (BIM) technology, which not only has a great impact on traditional construction technology but also on quantity surveying management and talent training (She, 2021).

Reform and Innovation of Talent Training Modes

In order to adapt to the new era and cultivate useful talents, the Ministry of Education is aggressively promoting the "new engineering" building plan. Meanwhile, Fudan Consensus, Tianda Action, and Beijing Guide were developed sequentially. Besides, "Notice on the Development of New Engineering Research and Practice" (MOHURD, 2017) as well as "Notice on the Promotion of New Engineering Research and Practice Projects" (MOHURD, 2017) are issued in order to explore a Chinese model or Chinese experience in engineering education. The documents released a list of the first batch of "new engineering" research and practice projects, including 202 "new engineering" comprehensive reform projects and 410 "new engineering" professional reform projects (MOHURD, 2017). Among them, the professional reform category of "new engineering" covers 19 project groups with hot "new engineering," such as artificial intelligence, big data, and intelligent manufacturing (Wang & Zhu, 2021). That is to say, with the concept of "new engineering" proposed by China for the first time in 2016, the Ministry of Education actively promotes the construction of "new engineering," which is a major strategic choice for engineering education reform against the backdrop of a new scientific and technological revolution (Wang & Zhu, 2021), a new industrial revolution, and a new economy (MOHURD, 2017). It is also a new idea and a new way for China's future engineering education (Zhang & Wang, 2020). The aim of "new engineering" construction is to cultivate new talents for the development of new technology and new industries. Therefore, the most critical task in the construction of "new engineering" is the reform and innovation of talent training modes (Wang & Zhu, 2021).

Changes of industry demand for quantity surveyors

Quantity surveying (QS) is a prominent building specialty that emerged as a result of China's market and economic development. The increasing expansion of the construction sector propels the QS industry forward. Moreover, the comprehensive quality improvement of QS personnel is the core element of the QS industry's sustainable and rapid development (Hou, 2021). With the industry's change, employment requirements for QS fresh graduates' abilities continue to improve, and the shortcomings of the traditional talent training model are also highlighted (Zhang & Wang, 2020).

As Wu & Li (2020) stated, with the development of the market economy, the value of the QS consulting industry and quantity surveyors has become more obvious. Enterprises will require whole-process QS consulting service capacity and enterprise informationization construction levels in order to respond to future development (Wu & Li., 2020). The rise of the whole process business and overseas markets opened up new growth opportunities for the QS consulting industry.

As Zhu et al. (2020) pointed out, China's QS industry has a good development trend and strong market demand, but there are also problems such as unbalanced development and talent shortages in the industry development. The training and management of QS talents are related to the quality of QS service and the level of professional level, closely related to the changes in the QS industry and the level of transformation and upgrading of construction industry.

PROBLEM STATEMENT

A rising corpus of research concentrating on talent training in quantity surveying has emerged in the context of the construction industry's transition and the strategic development of "new engineering." Some researchers have focused on the discussion of "Development of QS industry" (Liu & Wu, 2020), "Current situation of quantity surveyors" (Zhu, Zhang & Li, 2020), "Ability of quantity surveyors" (Ye & Yan, 2018), "Talent market demand and professional ability" (Wu & Li, 2020), "Reformation of the Training Mode" (Yu, Chen & Xiong, 2020), "Curriculum System Design" (Zhao, Li, Jiang & Yi, 2017), "Cultivation of quantity surveyors based on the OBE concept" (Ying & Zhao, 2021), "Cultivation mechanism of students' practical ability" (Yu, Xu & Qiu, 2020), "Vocational Education and Training (VET) of QS" (Ken, 2016), "Incorporation of BIM into QS undergraduate curriculum" (S & Do, 2018), "Private university QS specialty" (Zhao et al., 2020; Liu, 2019), "College professional personnel training" (She, 2021; Wang & Zhu, 2021; Sun, 2019; Bao & Wang, 2020).

While these studies overlooked present challenges and potential progress in this study area, fewer reviews of QS training have been published in the last decade.

RESEARCH QUESTIONS

The objective of this paper is to learn about China's current research situation on QS personnel training and explore the future research focus and trend. It is hoped that this study provides a reference for related research on QS personnel training. Following this objective, some specific research questions are proposed:

1. What is the status and distribution of these studies?
2. What will the discussion focus on?
3. What are the current research hotspots and development trends?

It is critical to identify and visualise the present state, hot themes, and trends in this study field of China's QS staff training.

RESEARCH OBJECTIVES

The study's particular aims are as follows: (i) Conduct bibliometric analysis with Cite Space to examine important research, identify significant academics or institutions involved in QS training research, and clarify their research environment. (ii) Identify and display research hotspots in QS training-related disciplines. (iii) Forecast research trends in QS training-related disciplines.

RESEARCH SIGNIFICANCE

The significance of this work are is as follows: (i) It employs bibliometric and scientometric analysis to elicit detailed information from each study. (ii) A systematic summary and analysis of QS training research published between 2002 and 2023 is conducted, which not only identifies and visualises key scholars or institutions and research hot spots through quantitative analysis but also discusses the development trend of QS training research through qualitative analysis.

Based on this assumption, the CNKI database was used for literature retrieval in this study, with the keywords "quantity surveying" and "personnel training" as the keywords. The literature review approach with Cite Space was employed. It is envisaged that this study will serve as a reference for future research on QS training.

METHOD

In order to identify and visualise China's current research status, research hotspots, and future research focus and trend of QS personnel training, this study adopts the holistic evaluation method, combining quantitative evaluation with Cite Space and qualitative evaluation to eliminate biassed conclusions and subjective interpretation. Cite Space software was developed by Professor Chen Chaomei of Drexel University in the US (Hou & Hu, 2013), which can analyse the research progress of a certain discipline or field and identify research hotspots and future development trends (Hou & Hu, 2013). A three-stage methodology was developed to assess published research in the subject of QS training. Figure 1 shows the flow of the research.

Figure 1. Flow of the research

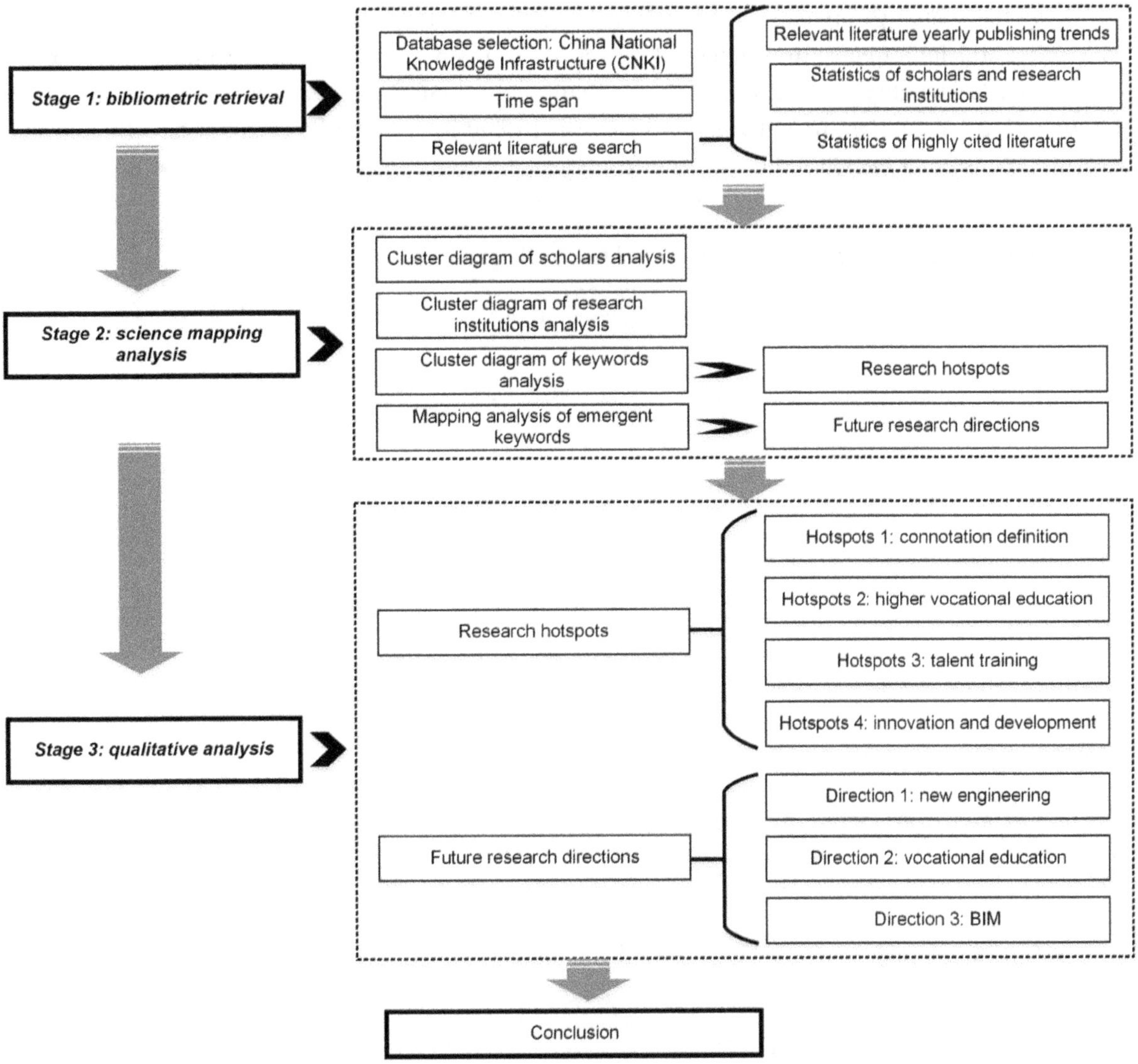

Stage 1: Bibliometric Retrieval

According to the reference retrieval database China National Knowledge Infrastructure (CNKI), selecting "quantity surveying" and "talent training" as retrieval subjects in July 2023 produces a total of 1832 relevant references, which include 1562 academic journal articles and 89 theses. The papers were published between 1997 and 2023, indicating that considerable literature on this topic first appeared in China in 1997.

Using the search parameters, 1752 relevant articles were found between 2002 and 2022. In terms of yearly publishing trends, the number of relevant research papers increased overall, from 10 in 2002 to 204 in 2019 (Figure 2). The academic community increasingly values meaningful research on the training of QS skills. On the other hand, the shifting trend in research literature is directly tied to the growth processes of the QS industry and institutions.

Figure 2. CNKI general trend chart of related research literature

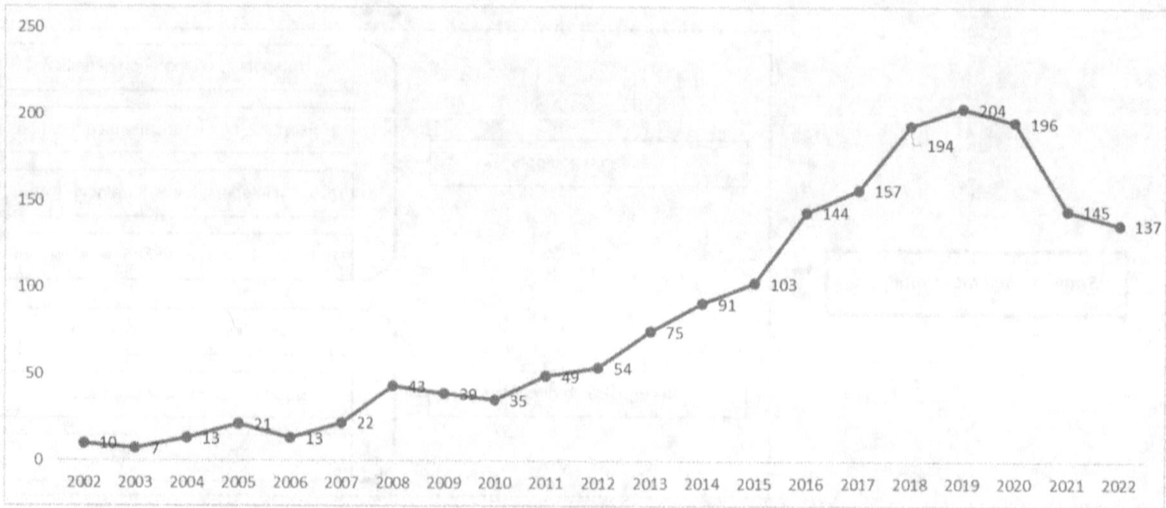

Source: Data retrieved from CNKI. https://kns.cnki.net/kns8/Visual/Center

QS specialty is an interdisciplinary major that combines civil engineering technology, project management, economic management, and other discipline expertise. It emerged very late in China (Wang et al., 2024). Following the adoption of the engineer system in 1996, the QS was formally included in the catalogue of China's higher education majors in 1998. In 2000, China's real estate business grew fast, increasing the demand for expert construction workers (Wang et al., 2024). The QS major might also grow significantly. As a result, the QS major has been introduced to the project management majors of many undergraduate schools, and the QS major at vocational institutions is expanding (She, 2021).

Prior to 2012, several schools and universities established "QS direction," focusing on the project management major or teaching quantity surveyors as one of the training objectives for project management professionals. In order to address the demands of social growth, China's Ministry of Education permitted select schools and universities to establish QS majors outside of the undergraduate major catalogue (Shen & Sun, 2022). In 2012, in response to national economic and social development demands, China's Ministry of Education introduced the QS major to the undergraduate major catalogue management (Shen & Sun, 2022). As a result, it explains that research on staff training connected to QS has been conducted since 2002.

Stage 2: Science Mapping Analysis

The Cluster View-static and Show Merged Network were used to view and study scientific literature in order to create a logically and cohesively ordered knowledge map (Shen & Sun, 2022). The reference literature from 2002 to 2022 was used as a database, and information such as titles, scholars, abstracts, and keywords was extracted to facilitate the examination of hotspots and trends for this issue. This information was then transformed into a format that the Cite Space programme could understand. In terms of parameter selection, we picked a time frame of 2002-2022, with a one-year interval. Figure 3 shows the science mapping analysis process with Cite Space.

Figure 3. Science mapping analysis process with Cite Space

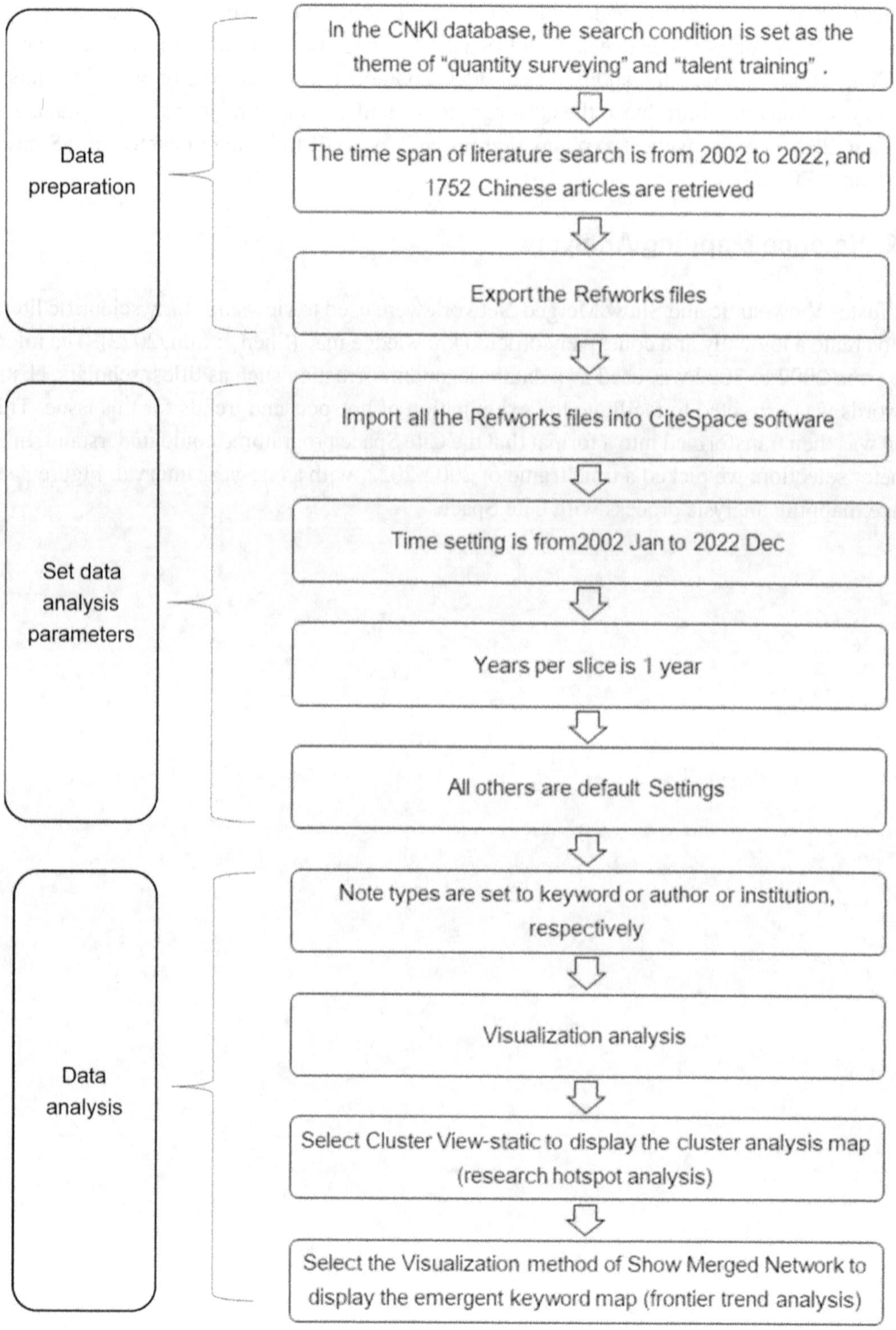

Stage 3: Qualitative Analysis

Qualitative analysis is a systematic procedure that involves numerous people making joint decisions to classify 1752 relevant articles based on key words, topics, and content.

BIBLIOMETRIC ANALYSIS OUTCOMES

Statistics of Scholars and Research Institutions

To better understand the role of research scholars and research institutions with a high number of publications in the development of QS talents, this paper conducted statistical analyses on the number of publications by scholars with a high number of publications (Figure 2) and institutions with a high number of publications (Figure 4). Figure 2 shows that Yu Jingbo and Li Junhong are among the top ten scholars in terms of published articles. In addition, Yin Yilin and Liu Yuanfang contributed nine publications, while Yan Ling and Huang Chen published eight.

Figure 4. CNKI chart of high published scholars

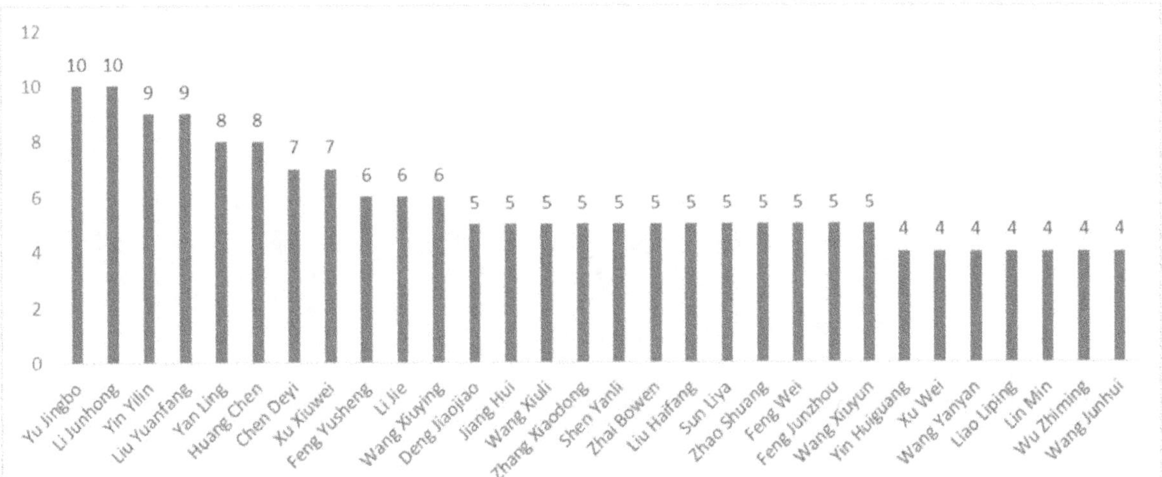

Source: Data retrieved from CNKI. https://kns.cnki.net/kns8/Visual/Center

Figure 5 shows that the top 25 institutions produce a minimum of 9 papers and a maximum of 26, with all 25 being universities. Tianjin University of Technology, Fujian Institute of Engineering, Hunan Vocational and Technical College of Urban Construction, and Chongqing University were placed first through fourth in terms of article publication. Furthermore, according to the top 25 colleges and universities, both regular undergraduate colleges and higher vocational colleges conduct ongoing and in-depth studies in this sector, although other research institutions show minimal interest in this topic. Furthermore, an examination of the cooperative network of research institutions reveals that cooperative research across research institutions is limited, and a research force has yet to materialise.

Figure 5. Top twenty-five publications of CNKI

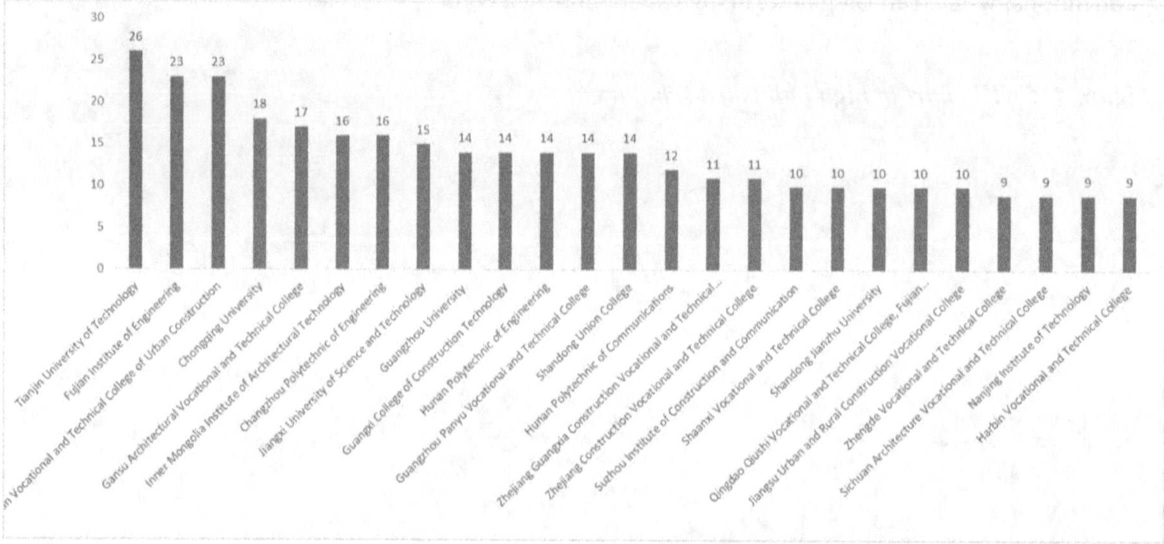

Source: Data retrieved from CNKI. https://kns.cnki.net/kns8/Visual/Center

In addition, cluster analysis of research institutes and academics was performed using Cite Space software. Figure 6 shows a clustering pattern of research institutions, and hot universities with high publications include Tianjin University of Technology, Fujian Institute of Engineering, and so forth.

Figure 6. Cite space cluster diagram of research institutions

Furthermore, as shown in Figure 7, the researchers with the most publications include Li Junhong, Yin Yilin, Liu Yuanfang, Yan Ling, Liu Yuanfang, Li Jie, and others.

Figure 7. Cite space cluster diagram of scholars

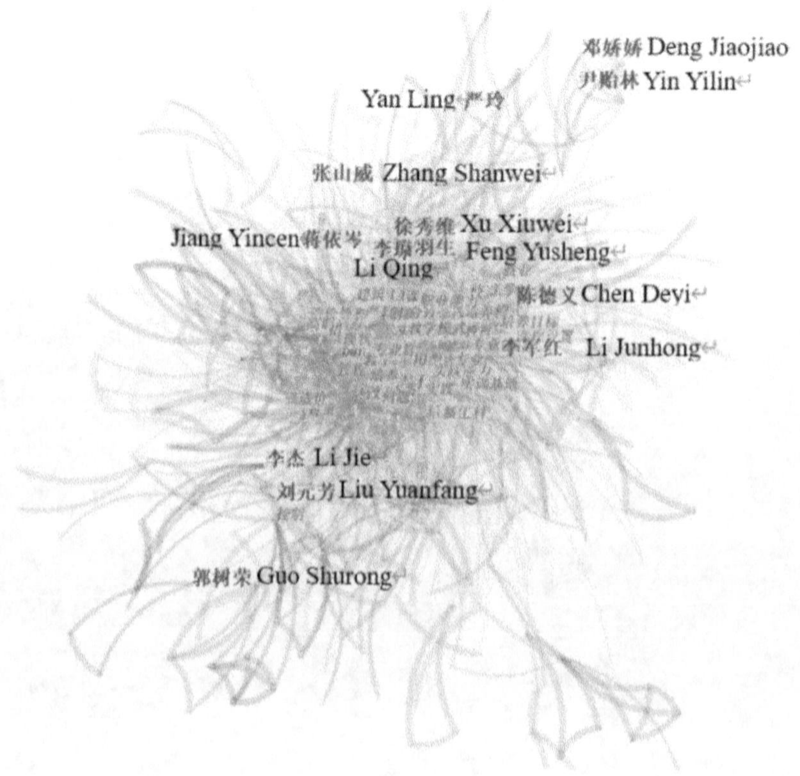

To summarise, whether the CNKI measurement visualisation analysis chart or the Cite Space software cluster analysis chart are used, it is clear that relevant research on QS staff training is receiving increasing interest in the academic community. Furthermore, universities, as well as higher vocational institutes, do ongoing and in-depth study in this area. Further review of the literature revealed that the majority of research in this sector is conducted independently, with little crossover or collaborative research between universities and higher vocational institutions. As a result, this study path is cross-comparison research, which may be viewed as a research complement to this academic research topic.

Statistics of Highly Cited Literature

Table 1 displays information on the top ten referenced literature over the last two decades. As shown in Table 1, the research began late, and the academics who conducted the research on QS staff training were mostly restricted to the scope of universities, thus the citation frequency was low. Second, about three-fifths of the top ten referenced literatures were produced between 2015 and 2016, indicating that after more than 10 years of research (2002–2016), researchers have developed relatively mature literature materials that serve as a reference for ongoing academic research. Third, some of the top ten most referenced pieces of literature were published in CSSCI publications, while the remainder were also published in Peking University and CSCD core journals, demonstrating the impact of related research on QS talent training.

Table 1. Top ten cited journal references in the past two decades

IDX	Author	Title	Journal	Time	Citation
1	Hao Li	Application research of BIM technology integrated into college engineering specialty teaching	Civil Construction Engineering Information Technology		104
2	Zheng Xiaoxia, Xu Zhichao, Yin Yilin	Research on the impact of BIM on the training of engineering cost professionals in colleges and universities and countermeasures	Construction Economy		93
3	Li Maoying, Zeng Qingjun	Reform and practice of professional course teaching with the integration of course, certificate, post and competition	Vocational Technical Education		67
4	Yin Yilin, Bai Juan	Exploration and Practice of the training mode of applied engineering Cost Professionals -- Taking Tianjin University of Technology as an example	Engineering Science		57
5	Yan Ling, Huo Shuangshuang, Deng Jiaojiao	Competency-based Workshop Practice Teaching Research -- A case study of Engineering Cost major in Tianjin University of Technology	Modern Educational Technology		51
6	Lin Xibao, Yu Jingbo, Liu Meiyun	A study on the integration of professional teaching standards and vocational skill grade standards in higher vocational colleges under the 1+X certificate system	Research on Vocational Education		44
7	Liu Yuanfang, Yan Ling, Yin Yilin	Innovation and practice of characteristic talents training system of engineering cost specialty	Higher Education Research in China		43
8	Yan Heping, Luo Guoji, Chen Juan et al.	Research on the problems and countermeasures of BIM application in engineering cost management -- Taking the application of BIM technology in engineering cost industry of Hunan Province as an example	Accounting and Finance		42
9	Liu Haifang	Research on the application of BIM technology in engineering cost specialty in colleges and universities	Engineering Economy		40
10	Liao Liping	Development status and countermeasures of engineering cost consulting enterprises	Enterprise Economy		39

Source: Data retrieved from CNKI. https://kns.cnki.net/kns8/Visual/Center

Hotspots of the QS Studies

Key words are crucial parts of academic papers that provide a condensed portrayal of the publication's main essence (Xie & Chen, 2021). Keyword analysis can reveal some information about a field's overall research direction and research hotspots (Xie & Chen, 2021; Wang, 2018). Clustering and analysing keywords from QS staff training literature samples using Cite Space software. Figure 8 shows a network map.

Figure 8. Cite space keyword cluster diagram

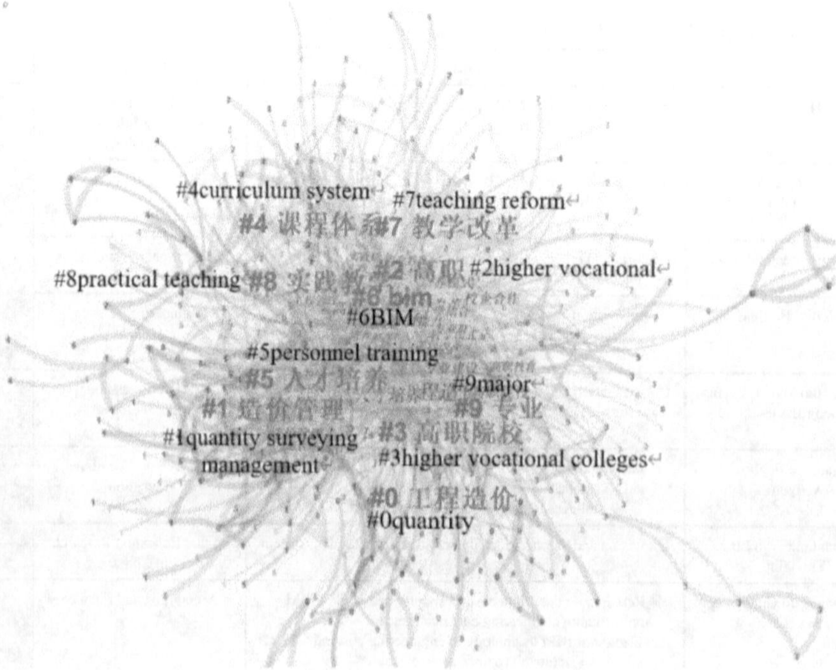

By comparing the keyword cluster diagram, useful information regarding high-frequency terms with co-occurrence frequencies greater than 15 is filtered out. The data comprises the frequency and centrality of keywords, which indicate the relevance of nodes in the network. The larger the centrality, the stronger the effect in the co-occurrence network (Xie & Chen, 2021). To some extent, a keyword serves as a research centre, linking several subjects. The ten keywords are prioritised based on frequency and centrality: quantity surveying, quantity surveying management, higher vocational, higher vocational colleges, curriculum system, personnel training, BIM, teaching reform, practical teaching, and major. Based on the foregoing analysis, an in-depth review of relevant research direction literature was carried out. Finally, QS talent training has four research hotspots: connotation definition, higher vocational education, talent training, innovation, and development.

FINDINGS OF THE QUALITATIVE ANALYSIS

The scientometric review enables the speedy identification of prominent scholars, institutions, and hot subjects in the field of QS personnel training.

Research Hotspots 1: Connotation Definition

Through the analysis of the top high-frequency keywords "quantity surveying," "quantity surveying management," and "major," it can be seen that the connotation definition of QS is a hot spot in academic research (She, 2021). Following the adoption of the engineer system in 1996, QS was legally included in China's higher education majors catalogue in 1998, a rather late development in China. With the rapid expansion of China's real estate business in 2000, the construction industry faced an increased demand for professional expertise, and the QS major thrived (She, 2021). As a result, the QS major has been added to the project management major of many undergraduate colleges, and the QS major of vocational colleges is mushrooming (She, 2021). Approved by the Ministry of Education in 2002, Tianjin University of Technology (Tianjin Institute of Technology before 2004) was the first university to independently set up a QS major in China (Yin & Bai, 2015). The Ministry of Education designated the QS major at Fujian Institute of Technology as the pilot major for national higher vocational education reform in October 2001. Through further exploration, Fujian Institute of Technology became the first university in Fujian Province, approved by the provincial and national ministries of education in 2004. And only five universities in China formally recruit QS undergraduate students (Liu, Yan, & Yin, 2005). QS major was introduced to undergraduate major catalogue management in 2012 to meet the demands of national economic and social growth (Yin & Bai, 2015). Yin Yilin and Bai Juan explained in their article that QS professionals are committed to applied talents training, and the school will pay attention to the students' ability to analyse and solve problems to meet the project's needs with good professional quality, systemic QS management knowledge and skills, a solid professional foundation, strong adaptability, and applicable talents with innovative and entrepreneurial ability (Yin & Bai, 2015). Liu Yuanfang, Yan Ling, and Yin Yilin explored the problems related to the QS professionals' training through investigation and research, which proposed that in order to meet the employers' new requirements for QS professionals' knowledge, ability, and quality in the new era, it needs to develop QS professionals with high-quality characteristics (Liu, Yan, &Yin, 2005).

To sum up, QS is a major product proposed and developed in order to better adjust to the market's talent demands (She, 2021). To respond to the continuous division of labour and differentiation of occupations, education in modern society has gradually formed two types of education: academic and vocational education (Zhuang, 2014). It can be seen that the QS major is more inclined towards vocational education.

Research Hotspots 2: Higher Vocational Education

In 2019, China's State Council released the "National Vocational Education Reform Implementation Plan," often known as "Vocational Education Article 20" (Hou, 2021). The plan proposed two types of education: vocational education and general education, both of which are equally significant positions. The plan aims to begin a new adventure in vocational education reform by presenting a road map, timeline, and specifications for deepening the reform (Hou, 2021). Furthermore, to realise the "2035 goals" and "2050 important vision." It is also projected that by 2022, a considerable proportion of traditional undergraduate colleges and universities will transition to application-oriented programmes. "The newly upgraded universities in 2018 will focus on vocational education at the undergraduate level," Lou Xuqing, deputy inspector of the Ministry's Development Planning Department, said during the news conference. Simultaneously, the transfer of independent institutions should evolve into application-focused undergraduate universities (Su, Xia, & Tao, 2024). According to Lou, the Ministry of Education will continue to support the development of application-oriented undergraduate universities, push the college enrollment plan to favour talents urgently required for industrial development, and increase the proportion of application-oriented, technical, and compound talents. The new higher education enrollment strategy is primarily focused on the development of applied and technical skills. Furthermore, one of the plan's specific indicators is that, beginning in 2019, vocational colleges and application-oriented undergraduate universities will implement the pilot system of "academic certificate + certificate of several vocational skills" (hereinafter referred to as the pilot system of 1+X certificate). The relevant units in the Ministry of Housing and Urban-Rural Development's "13th Five-Year Plan of QS Undertakings Development" emphasised the need to enhance the vocational education system. Actively adapt to the revision of the national vocational qualification system and continue to develop the QS vocational qualification system (Su, Xia, & Tao, 2024). To build a continuing education and training system that is consistent with the QS specialist features, to innovate continuing education modes and methods, and to increase continuing education quality (MOHURD, 2017).

In the keyword cluster analysis, there are two important clusters: "higher vocational" and "higher vocational education," which include the research topics of vocational education, curriculum setting, school-enterprise cooperation, vocational ability, employment type, work-study combination, and so on. This also reflects the hot topic of research on higher vocational education. Lin Xibao et al. (2020) explored the current situation and influencing factors of the integration of professional teaching standards and vocational skill grade standards in higher vocational colleges under the 1+X certificate system. It is necessary to actively explore the implementation path of the 1+X certificate system, optimise the training programme for professional talents, and innovate the personnel training and vocational skills training models (Lin, Yu, & Liu, 2020; Zhao & Luo, 2019). Zhang Huiqing (2016) studied the structural adjustment of higher vocational majors based on the evolution of industrial structure and pointed out that higher vocational education serves regional economic development and is endogenous to the needs of regional industries. The industrial structure determines the talent employment structure, which in turn dictates the professional structure of higher vocational education (Su, Xia, & Tao, 2024). Therefore, the industrial structure change will inevitably promote professional structure adjustment (Zhang, 2016; Zhang, 2017).

Research Hotspots 3: Talent Training

"The 13th Five-Year Plan for the Development of QS Business" outlined the main objectives: it needs to improve the talent training mechanism led by the QS qualification system, strengthen continuing education and professional training, and promote the overall quality of QS personnel through creating 100 professional leading talents and 5,000 gold quantity surveyors (Su, Xia, & Tao, 2024). In order to cultivate and train a group with high quality and compound talents who are proficient in business, good at management, and with ability, moral integrity, international vision, and strategic thinking (Hou, 2021). In terms of market and industry, it is vital to develop professional talent training and reserve mechanisms, build a professional talent echelon, and form the talent training normalisation in QS management institutions (Hou, 2021). The QS consulting enterprises need to play a core role in personnel training and strengthen the establishment of knowledge management systems. From the perspective of colleges and universities, they should guide the professional talent cultivation, formulate the strategic plan for the QS professionals' training and development, support the QS professionals' training, strengthen the QS major teaching guidance, actively guide colleges and universities to participate in the key topics' study of QS management, innovate the talent training and selection models, and vigorously promote cooperation between universities and enterprises (Su, Xia, & Tao, 2024). The QS consulting enterprises need to play an active role in the talent training of colleges and universities and explore the talent training mechanisms of the integration of industry, education, and research (MOHURD, 2017).

In the keyword cluster analysis, there are two important clusters: "curriculum system" and "talent training." Liu Yuanfang et al. (2005) systematically discussed the goal, scheme, mode, and mechanism of training the characteristic talents of the higher quantity surveying specialty and pointed out that in order to train the high-quality characteristic talents of the quantity surveying specialty, we must take the scientific outlook on development as guidance, deepen the teaching reform, and innovate the objective and mode of personnel training. At the same time, it is emphasised that the curriculum should also reflect the latest development trend in the quantity surveying discipline (Liu, Yan, & Yin, 2005). Yin Yilin and Bai Juan took Tianjin University of Technology as an example to analyse the quality characteristics of applied engineering professionals. From two aspects of teaching system construction: team training and construction, Yin Yilin and Bai Juan systematically expounded on the construction of an applied quantity surveying professional personnel training mode. In the tide of the fine development of the quantity surveying consulting industry, higher professional education must make timely changes to the mode of education and talent training scheme so that there are more opportunities for development (Su, Xia, & Tao, 2024). The major is committed to applied talents training. The school needs to pay attention to the students' ability to analyse and solve problems and train them to adapt to engineering construction's needs with good professional quality, systemic quantity surveying management knowledge and skills, a solid professional foundation, strong adaptability, and applicable talents with innovative and entrepreneurial ability. The quantity surveying professionals' training should not only timely meet the industry's requirements but also lead the industry's development with a certain degree of foresight (Yin & Bai, 2015).

Research Hotspots 4: Innovation and Development

The "13th Five-Year Plan for the Development of QS Business" states that during the "13th Five-Year Plan" period (2016-2020), the new type of urbanisation and the "One Belt and Road Initiative" will provide new impetus and vitality to the development of fixed asset investment and the construction industry. The demand for institutional reform, transformation, and upgrading of the construction industry is increasing (MOHURD, 2017; Su, Xia, & Tao, 2024). In addition, the Central Urban Work Conference clearly proposed to implement major public facilities and infrastructure projects, strengthen the construction of urban rail transit, sponge cities, and urban underground comprehensive pipe corridors, speed up the renovation of rundown towns and dilapidated buildings, and orderly promote the comprehensive renovation of old residential areas and project maintenance. New innovation points, growth poles, and growth belts are constantly forming in the QS consulting industry (MOHURD, 2017).

What's more, it is pointed out that we should promote the innovation and development of the QS consulting industry. Expand the scope of QS consulting business, optimise the business structure, carry out all-round business expansion in the service stage, service level, and service field, and explore and study the market development of new businesses such as building carbon measurement and information engineering valuation (Su, Xia, & Tao, 2024). It needs to develop the technical standard and contract model for the whole-process QS consultation service. Moreover, it also needs to promote comprehensive project management services with QS management as the core and provide consulting services such as investment and financing management, investment control, and design optimisation for PPP development and the general contracting mode of project management (Su, Xia, & Tao, 2024). In addition, it needs to promote transformation and upgrading with information technology innovation and extend to the high-end of the engineering consulting value chain with BIM, big data, cloud technology, and other advanced information technology to enhance the value of QS consulting services (MOHURD, 2017).

In the keyword cluster analysis, "BIM," "teaching reform," and "practical teaching" all reflect the research hotspots of innovation and development. Li Shu et al. (2020) proposed that the QS consulting industry developed rapidly, sprouting many new directions and businesses, and the development momentum is full. The implementation of the "One Belt and Road Initiative," the development of new urbanisation, and the promotion of information construction have all brought new opportunities for the industry's development. Construction industry transformation and upgrading, production mode, and the change of management model also put forward "compound" new requirements on QS talents (Li, Cao, & Shao, 2020). Wu Lin et al. (2020) pointed out that there is a huge crisis behind the favourable conditions for the development of China's construction industry and QS consulting industry. On the one hand, the slowdown in fixed asset investment growth will affect the development space of traditional QS consulting businesses. On the other hand, the emergence of whole-process businesses and overseas markets opens up new growth opportunities for the QS consulting industry. The development of big data, artificial intelligence, and other information technologies not only provides powerful tools for the innovation and development of the industry but also poses a great threat to the traditional business form (Hou, 2021). In the face of the above opportunities and challenges, the demand for QS personnel may appear to be polarised; the lack of core technical ability and innovative thinking of QS personnel will be eliminated in the increasingly fierce competition, and the QS personnel with strong comprehensive professional ability and innovative thinking ability will win the competition (Wu & Li, 2020).

DISCUSSION OF FUTURE TREND OF QS RESEARCH

The dynamic change of keywords can more accurately analyse and judge a new research frontier in the field (Wang & Fu, 2020). Using Cite Space to extract mutation terms from literature information and trace the research frontiers and development trends in this field. Figure 9 shows the emergence keywords of the research on QS personnel training (the emergence intensity ranks in the in the top 25). As can be seen from Figure 7, the earliest emergent words occurred in 2002, which were "WTO," "cultivation goal," and "countermeasure." Among the 25 keywords, the one with the one with the longest duration was "cultivation goal," which lasted for 12 years. Followed by "employment," which lasted for 11 years. In addition, the strongest strength is "combination of work and study" (Strength 8.36). "integration of production and education," "curriculum reform," "new engineering," "vocational education," "curriculum ideological and political," and "BIM" have emerged until 2022. Through the analysis of the above situation, the focus of the early research focuses on the QS major itself, with more focus on professional construction. With the development of the quantity surveying industry and the university itself, the research is more focused on new engineering, vocational education, reform, and other aspects. Especially since 2019, the research has focused on new engineering, vocational education, ideological and political curriculum, BIM, and other topics, and the research on these topics will continue.

Figure 9. Map of emergent keywords (Top 25)

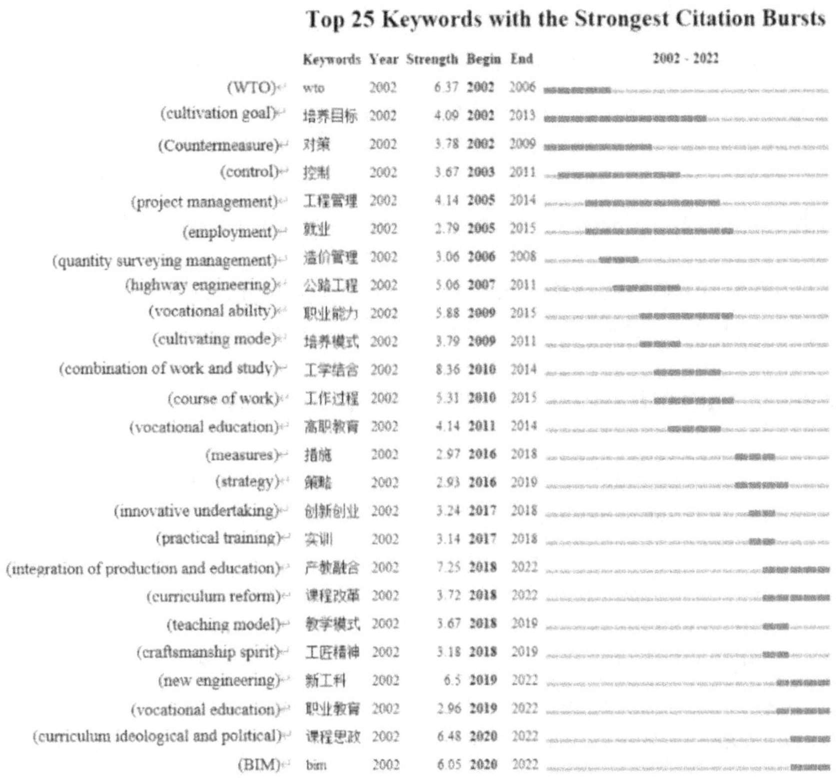

Future Research Directions 1: New Engineering

In order to adapt to the new era and cultivate useful talents, the Ministry of Education actively promotes the plan of "new engineering" construction. Meanwhile, Fudan Consensus, Tianda Action, and Beijing Guide were formed in succession. Besides, "Notice on the Development of New Engineering Research and Practice" as well as "Notice on the Promotion of New Engineering Research and Practice Projects" are issued in order to explore a Chinese model or Chinese experience in engineering education (Hou, 2021). The documents released a list of the first batch of "new engineering" research and practice projects, including 202 "new engineering" comprehensive reform projects and 410 "new engineering" professional reform projects. Among them, the professional reform category of "new engineering" covers 19 project groups with hot "new engineering," such as artificial intelligence, big data, and intelligent manufacturing (Su, Xia, & Tao, 2024). That is to say, with the concept of "new engineering" proposed by China for the first time in 2016, the Ministry of Education actively promotes the construction of "new engineering," which is a major strategic choice for engineering education reform under the background of a new scientific and technological revolution, a new industrial revolution, and a new economy. It is also a new idea and a new way for China's future engineering education (Zhang & Wang, 2020). The aim of "new engineering" construction is to cultivate new talents for the development of new technology and new industries.

In the context of the transformation and upgrading of the construction industry and the strategic construction of "new engineering," there are more and more studies on the talent training of quantity surveying majors (Su, Xia, & Tao, 2024). In the exploration of the "new engineering" path driven by "new business form," scholars have recently discussed the derivation background, practical significance, and talent training connotation of the concept from multiple perspectives (Ye, Kong, & Zhang, 2018). However, the innovation path of engineering colleges or specialties promoted by "new engineering," such as specialty settings, knowledge systems, curriculum modules, and teaching methods, still needs further discussion.

Therefore, the most critical task in the construction of "new engineering" is the reform and innovation of talent training modes, as well as QS talent training, which is the future research trend.

Future Research Directions 2: Vocational Education

From 2006 to 2020, the state and local governments carried out a series of policies on the construction of higher vocational education. During this period, higher vocational education not only completed the transformation from grade to category but also achieved a great improvement in scale and quality. It is also the basis for future adjustments to education policy (Su, Xia, & Tao, 2024).

In October 2021, the General Offices of the CPC Central Committee and the State Council issued "Opinions on Promoting the High-quality Development of Modern Vocational Education," guiding the direction for the high-quality development of vocational education (Shen & Sun, 2022). On the whole, there are still some problems in the research results of the high-quality development of vocational education in China, such as a small number, scattered results, and not prominent enough high-quality results (Su, Xia, & Tao, 2024). In the future, we should further improve the high-quality research system of vocational education in China from the perspectives of broadening the research horizon, strengthening comparative research, expanding the research team, and promoting the high-quality development of vocational education (Shen & Sun, 2022).

Under the influence of multiple factors such as technology, the economy, and society, the demand for skills is in dynamic change, and promoting skill transformation has become a highly concerned measure of the international community (Su, Xia, & Tao, 2024). Based on this, the practical way of using higher vocational education to promote skill transformation is put forward (Wei & He, 2024).

From the literature, scholars have begun to pay attention to the impact of vertical education mismatch, and overeducation has once become the focus of people's attention (Guo et al., 2019).

Moreover, under the emergence of artificial intelligence, in the current era, traditional teaching methods are out of date (Lyu & Liu, 2023). The lack of AI infrastructure is also an important factor hindering the integration of AI technology with education. Based on this, it is the need to explore the integration of vocational education and artificial intelligence (Lyu & Liu, 2023).

As QS major belongs to the professional skill specialty, therefore, for the future research of QS talent training, there is need to be with vocational education.

Future Research Directions 3: BIM

Information technology has become a very important tool for the industrialization of construction. The rapid development of information technology as represented by cloud computing, BIM, the internet, the Internet of Things, and big data provides the technical foundation for the development of lean, modern, and smart construction (Akin et al., 2022). Research on BIM has continued to grow in recent years, thereby attracting the attention of a large number of researchers and practitioners. The BIM research trend mainly includes "issues in BIM software and application standards," "islanding of BIM applications across all stages of the building life cycle," "BIM talent problems," and "BIM application in different stages of construction throughout the throughout the whole life cycle" (Clyde et al., 2023).

From the perspective of research trends, the deep integration of BIM technology with artificial intelligence, the Internet of Things, cloud computing, big data, and other new generation information technologies will become the future development direction to realise the deepening application of railway engineering collaborative design, construction management, visual management of operation and maintenance stages, railway life cycle management, BIM application standardisation, and other engineering fields (Wang, 2023).

With the rapid development of urban underground space, the requirements for underground space information are getting higher and higher. Building information modelling (BIM) is introduced into the field of underground space as a means of informationization (He et al., 2022). At present, 3D geological modelling, simulation, performance evaluation, digital twin, and the integration of 3D laser scanning, the Internet of Things, virtual simulation, and other emerging technologies are the key research directions of BIM in the field of underground space (He et al., 2022).

In recent years, it has been found that research on the building energy saving aspect is mainly about "BIM technology," "sensitivity analysis," "multi-objective optimization," "building life cycle," "renewable energy technology," "energy saving transformation," and other research hotspots in the field (Wang et al., 2024). In the future, in terms of research objects, methods, and objectives in this field, it will develop in the direction of refinement research, global sensitivity research, multi-objective optimisation research, BIM-based "general integration," "whole life cycle connection," and so on (Wang et al., 2024).

Above all, the development and talent training of QS specialty are also inseparable from BIM technology; therefore, for future research on QS talent training, it is also necessary to be with BIM.

CONCLUSION

In conclusion, this paper adopts the bibliometric analysis with Cite Space method to analyse the research literature on QS personnel training in China over the past two decades and, to some extent, in order to sort out the research context, research status, research hotspots, and subsequent research trends.

First, consider the trend in the number of pieces of literature. Overall, the study literature on QS personnel training is increasing. Prior to 2013, the average yearly number of papers published did not surpass two digits; however, after 2014, the average annual number of articles published has reached three digits. From 2018 to 2021, the average yearly number of publications published remained relatively stable at around 200, indicating that the academic community is more interested in relevant research on QS personnel training.

Second, from the study of the distribution of forces. The top 25 institutions in terms of the number of publications are all general undergraduate and higher vocational colleges. According to further analysis of the literature, it is found that most of the research of these two types of universities in this field is independent research, and there is little crossover research or collaborative research. Therefore, this research direction is cross-comparison research, which can be regarded as a research supplement in this academic research field.

Third, from the analysis of research hotspots. The important information about high-frequency keywords with frequencies greater than 15 was filtered out using the keyword clustering map analysis. At the same time, through an in-depth study of relevant research direction literature, the four research hotspots of quantity surveying talent training are finally concluded as connotation definition, higher vocational education, talent training, and innovation and development.

Fourth, consider the growth of research trends. Cite Space is used in this article to extract mutation phrases from literature data and map the field's research boundaries and development patterns. Through the analysis, it was found that the early research hotspots focused on the QS major itself, focusing on professional construction. With the development of the QS industry and the university itself, the research is more focused on "new engineering," vocational education, reform, and other aspects. Especially since 2019, the research has focused on "new engineering," vocational education, ideological and political curriculum, BIM, and other topics, and the research on these topics will continue.

However, in the following two aspects, this study is not enough and needs further research. First, it does not thoroughly extract the common characteristics of this research literature based on in-depth reading of highly cited literature. Second, on the basis of a full analysis of research hotspots and research frontier trends, it is not enough to provide constructive suggestions for the highly quality development of quantity surveying in higher vocational colleges and universities. As a result, the next research will concentrate on the aforementioned weaknesses and delve deeper into material with greater theoretical and practical value in order to serve as a reference for the innovation and development of QS personnel training.

Abbreviations

QS Quantity Surveying
WTO World Trade Organization
BIM Building Information Modelling
CNKI China National Knowledge Internet
CSSCI Chinese Social Sciences Citation Index
CSCD Chinese Science Citation Database

REFERENCES

Akin, S., Ergun, O., Surer, E., & Dino, I. G. (2022). An immersive performative architectural design tool with daylighting sim-ulations: A building information modeling-based approach. *Engineering, Construction, and Architectural Management*, 28(4), 1319–1344. 10.1108/ECAM-07-2020-0562

Clyde, Z. L., Yu, Z., Hengqin, W., Zhe, C., Bing, X., & Vivian, W. Y. T. (2023). The application of BIM in the AECO industry. *Journal of Civil Engineering and Management*, 29(3), 202–222. 10.3846/jcem.2023.18076

Guo R., et al. (2019). Job-Education mismatches in degrees and majors and employment quality of college graduates. *Labor economics research,7*(02), 78-100.

He L. L., et al. (2022). Research hotspots and development trends of BIM in the field of underground space. *Journal of Underground Space and Engineering,18*(S1),14-25+35.

Hou, H. (2022). Study on the reform of training mode of Engineering Cost Professionals under the background of new Engineering. *Science & Technology and Innovation*, 142-143(23), 148.

Hou, J. H., & Hu, Z. G. (2013). Review and prospect of Cite Space software application research. *Modern Information*, 33(04), 99–103.

Li, S., Cao, L., & Shao, Y. (2020). Research on the professional curriculum system of adult education based on vocational abilities - taking engineering cost major of the National Open University as an example. J*ournal of the Open University of Broadcasting (Philosophy and Social Sciences Edition)*, (03), 91-100+111. .10.16161/j.issn.1008-0597.2020.03.016

Lin, X. B., Yu, J. B., & Liu, M. Y. (2020). Study on the integration of Professional teaching standards and Vocational skill level Standards in Higher Vocational Colleges under the 1+X Certificate System. V*ocational. Education Research*, (01), 4–11.

Liu, Y. F., Yan, L., & Yin, Y. L. (2005). Innovation and practice of training system of characteristic talents of engineering cost. *China Higher Education Research*, (11), 85–86. 10.16298/j.cnki.1004-3667.2005.11.033

Lyu, F., & Liu, Y. D. (2023). The research on the response and development of vocational education under the background of artificial intelligence. *Continuing Education Research*, 2023(02), 59–63.

She, Y. (2021). Research on the training system of Engineering cost Professionals in vocational colleges under the background of "Three Education" reform. *Education Observation*, 10(10), 58–60. 10.16070/j.cnki.cn45-1388/g4s.2021.10.017

Shen Z. Y., & Sun D. (2022). Research hotspots and trend prospects of high quality development of Vocational education in our country. *Education and profession, 20.* carroll nki. 1004-3985.2022.03.002.10.13615 / j.carol

Su, A., Xia, M. W., & Tao, X. X. (2024). Research on the Relationship between Education policy and the development of higher vocational Colleges from the perspective of system Theory. *Journal of system science*, 1-8. http://kns.cnki.net/kcms/detail/14.1333.N.20240305.1849.002.html

The Ministry of Housing and Urban-Rural Development. (2022). *13th Five-Year Plan for the Development of Project Cost Undertakings*. https://www.mohurd.gov.cn/gongkai/fdzdgknr/tzgg/201708/20170804_232859.html

Wang J. et al. (2024). Nearly 20 years in the study of Chinese building energy saving process, hot spot and trend analysis. *Building science,40* (02), 179-188. The DOI: carroll nki. 11-1962 / tu 2024.02.21.10.13614 / j.carol

Wang, J. Q. (2023). Construction and analysis of knowledge graph of BIM research in railway field. *Journal of Railway Engineering*, 40(11), 87–93.

Wang, X. M. (2018). Research on Teaching Quality in Chinese Universities: Trajectory, hotspots and Future trend: CiteSpace visualization analysis based on 14 journals of higher Education. *Journal of Education Academic*, (1), 91–103.

Wang, Y. S., & Fu, Y. Y. (2020). The current situation, hotspots, and trends of international higher education evaluation research in the past decade. *Higher Education Research in Heilongjiang Province*, 38(03), 6–11. 10.19903/j.cnki.cn23-1074/g.2020.03.002

Wei S. M., & He Z. (2024). Bottleneck Restriction and Practice Direction of Higher Vocational Education Promoting skill Transformation in our country. *China's higher education research,2024* (02), 94-100. DOI: carroll nki. 1004-3667.2024.02.13.10.16298 / j.carol

Wu, L., & Li, Z. (2020). Study on local market demand and occupational ability composition of engineering cost professionals. *Times Finance*, 161-162.

Xie, M., & Chen, W. J. (2021). Knowledge graph of Mooc research in China: Hot spots, current situation and trend analysis based on Cite Space. *Journal of Southwest University for Nationalities*, 42(01), 229–235.

Ye, M., Kong, H. B., & Zhang, W. (2018). New Engineering: From Concept to Action. *Higher Engineering Education Research*, 2018(01), 24–31.

Yin, Y. L., & Bai, J. (2015). Exploration and practice of the training mode of applied engineering cost professionals - taking Tianjin University of Technology as an example. *China Engineering Science*, 17(01), 114–119.

Zhang, H. Q. (2016). Study on the Enrollment Ratio of Higher Vocational Majors Based on the Adjustment of Industrial Structure in Shandong Province. *China Vocational and Technical Education*, (35), 14–19.

Zhang, H. Q. (2017). Study on the enrollment ratio of higher vocational majors based on the adjustment of industrial structure in Shandong Province. *China Vocational and Technical Education*, (35), 14–19.

Zhang, M., & Wang, C. Y. (2020). *Research on the transformation of training mode of Engineering cost professionals under the background of new engineering*. International Public Relations.

Zhao, J., & Luo, Y. C. (2019). Some thoughts and preliminary practice on promoting the pilot work of 1+X certificate system. *China Vocational and Technical Education*, (27), 5–8.

Zhuang, X. Z. (2014). Transformation of ordinary undergraduate colleges: Why and What and How to change. *China Vocational and Technical Education*, (21), 84–89.

Chapter 14
The Implementation of BIM Software in Different Construction Stages

Tze Shwan Lim
https://orcid.org/0000-0002-6291-5287
Taylor's University, Malaysia

Zhao Yang Low
Taylor's University, Malaysia

Lam Tatt Soon
Taylor's University, Malaysia

Habizah Sheikh Ilmi
https://orcid.org/0009-0009-3382-5213
Taylor's University, Malaysia

Kenn Jhun Kam
https://orcid.org/0000-0003-4927-4611
Taylor's University, Malaysia

Myzatul Aishah Kamarazaly
Taylor's University, Malaysia

ABSTRACT

BIM enhances construction industry outcomes by connecting people, technology, and processes by incorporating extensive additional information and facilitating information exchange with project stakeholders. However, BIM software has limitations that have led to reluctance in its adoption. This research aims to explore the positive impacts and limitations of BIM software in various construction stages (pre-construction, design, construction, handover, and closeout), and seeks to improve understanding and find solutions for improvement. The research employed quantitative method by survey approach using questionnaires distributed to consultant parties in the industry. Results show that BIM has brought different unique functions that help assist the respondents. Despite that, limitations of BIM implementation found from the results are due to the lack of knowledge, high cost, and software capabilities to be further improved on. Respondents believed that the most effective solutions to tackle the limitations were mainly involved with education, enhancing BIM tools, and training.

DOI: 10.4018/979-8-3693-1650-4.ch014

INTRODUCTION

In the construction industry, digitalization and evolving technology have revolutionized work processes. To efficiently handle data in design, construction, and operation stages, the industry has adopted new technology (McPartland, 2017). Building Information Modeling (BIM) software, a key tool in integrating these technologies, is crucial for the industry (GeoSLAM, 2021). BIM aids the construction industry in all stages of infrastructure development. It offers benefits like material analysis, predictive maintenance, and improved efficiency (The Geospatial, 2019). BIM software ensures accuracy, cost-effectiveness, and standardized data throughout the project (Proforma Construction Inc, 2019).

Computer Aided Design (CAD) revolutionized engineering and architecture by replacing manual drafting with faster design creation. However, BIM as an advanced software, offers AI-driven generative design and superior features compared to CAD. BIM utilizes AI to generate multiple innovative projects starting points by manipulating design tools and parameters, leading to a significant increase in quality data (Storrier, 2021). Czmoch & Pekala (2014) described BIM as a 3D virtual model with project information that has expanded to include 7D dimensions, outperforming competitors. BIM provides enhanced scheduling, estimating, sustainability, and facility management. Despite limited adoption, BIM software showcases its advanced capabilities and benefits for construction parties.

Limited BIM understanding persists in the construction industry (Latiffi et al., 2013). Architects and engineers use BIM for accurate 3D modeling, error prevention, and cost control during design (Reyes, 2019). BIM software aids engineers in advanced analysis, simulations, and risk assessment (Aung, 2022). BIM engineers combine software and engineering expertise for improved project safety and efficiency (Latiffi et al., 2013; Aung, 2022). BIM software benefits all construction stages, improving design insight, ROI, cost control, and reducing waste (Singh & Gajjar, 2022). It minimizes rework, streamlines processes, and simplifies data transition, enhancing efficiency (Vinzero, n.d.). BIM enables collaborative project management for successful execution (TeamSystem, 2021).

This research has 3 main objectives. The first objective is to discover the positive impacts that BIM software brings towards different construction stages. The second objective of this research is to identify the limitations that BIM software has during different construction stages. The third objective is to examine the possible solutions that can be done to overcome the limitations.

BACKGROUND

The use of BIM software in the construction industry has greatly impacted all parties involved in construction projects (Liu, 2021). It provides construction engineers with increased protection during site visits and analysis accidents and offers a more intuitive and convenient experience for architectural design and construction. BIM software also performs various calculations, including 3D modeling, rendering, load computation (e.g., wind, snow, self-weight), and analysis of energy use and daylighting.

During the pre-construction stage of a project, initial ideas are considered, and the location and feasibility of the project are assessed. The design phase involves collaboration between architects, engineers, and other stakeholders to develop contemporary concepts. A cost estimate is prepared by the project team, considering factors such as labor, materials, and legal requirements, with input from Quantity Surveyors. The construction stage is led by the main contractor, who oversees resources and project updates, while the design team ensures quality control and reviews documentation. The construction closeout stage involves a final inspection and the compilation of project data and lessons learned (Gerardi, 2021).

BIM SOFTWARE

There are various definitions for the term 'BIM software' in the construction industry across the world, but in general they share a very similar meaning and fundamental idea. BIM, or Building Information Modelling, is defined as the foundation of digital transformation in the architecture, engineering, and construction (AEC) industry. BIM creates a digital representation of an asset throughout its existence, from planning and design to building and operations, based on an intelligent model and made possible by a cloud platform. Lindblad (2013) stated that BIM is a tool that facilitates the preservation and reuse of data and domain knowledge over the course of a project's lifecycle. Therefore, a key function of BIM is to coordinate and integrate the flow of data and knowledge throughout various project disciplines and phases. This idea of BIM software has been consistent throughout the decade and is still relevant in today's world. Dakhil (2017) defined BIM is an approach to designing, constructing, and operating buildings that uses modelling technology, related sets of procedures, and people to create, share, and analyse building information models. BIM software according to Naim (2018), is described as technology, or a process, or a mix of the two in a new approach that aims to cause a paradigm shift in how the building industry operates.

In the construction industry, BIM software has had a huge impact on all the parties involved in construction projects. With the use of BIM software, construction engineers are given greater protection in the event of a site visit and analysis accident, in addition to allowing construction industry practitioners to have a more intuitive and convenient experience with architectural design and construction (Liu, 2021). In addition to 3D modelling and rendering, BIM performs several calculations, such as the computation of loads like wind, snow, and self-weight, as well as the analysis of energy use and daylighting.

CONSTRUCTION STAGES

Before beginning a comprehensive design phase, a project's initial ideas are considered during the project's pre-construction stage. The location of the project should be known and the details of it as well as whether it would be able to be done. Later, there would be checking on whether the proposed solutions are in line with the primary objectives. After that, it would be the start of the design phase. Anything related to the project's design shall be discussed and started during this stage which would require the cooperation of the architects and engineers, etc. Later, while engineers guarantee the project's safety, architects would evaluate the project's viability and develop contemporary concepts. Although they are heavily involved in this stage, the architect and engineer would mostly finish the bulk of the design work. Moving on to the construction stage would then take place, at first it is to determine whether the project is feasible, and the project team would develop a reasonable cost estimate. The cost estimate for the project considers labour, schedule, site conditions, materials, and legal requirements which is mainly done by Quantity Surveyors. They would also consider the costs associated with similar projects as well as which contractor would be able to finish the job within realistic financial limits. During the construction stage, the project is under the general direction of the main contractor. They would oversee resources, keep track of paperwork, and share project updates. The design team, on the other hand, oversees quality control to make sure the project follows the approved plans. The design team also looks over submittals, change orders, and replacement requests. At this point, every team has its schedule. Lastly, it would the construction closeout stage, a final inspection would be done at the site and make sure everything is finished completely without any mistakes and all the project data would be given. Aside from that, the project manager also must compile a list of instructive lessons learned which can benefit the future (Gerardi, 2021).

POSITIVE IMPACTS OF BIM IMPLEMENTATION IN CONSTRUCTION STAGES

BIM implementation in construction helps in the smooth procedures and activities in all the phases of a project, including construction management, pre-construction planning, construction processes and post-construction as well as empower the act of facilities management. All project valuations are considered during the pre-construction stage to predict future financial benefits for the customer and the contractor. BIM software is useful for providing a framework for evaluating the project's viability as well as for providing an assessment of the building's future performance in terms of regulations pertaining to functionality and constructability. Studies on energy, acoustics, and illumination are now being conducted. They will eventually have a significant impact on the outcome if properly examined. With the conventional method, it was not possible to conduct this kind of sustainability analysis. During the pre-construction stage, BIM software is used to assess project viability and future performance. It enables sustainability analysis and reduces delays and costs in the early stage of the project planning (Singh & Gajjar, 2022). BIM can start communication and visualization in the pre-construction period of the project among various stakeholders and the implementation give the positive impact in building a solid communication channel between all the stakeholders associated with the project. Furthermore, according to Soon et al., (2019), BIM assists quantity surveyors in creating initial cost estimates by extracting data from 3D models. It also facilitates seamless construction and decreases request for information and project queries coordination.

During the conceptual design stage, following the creation of design drawings by the design team, quantity surveying consultants will prepare a cost comparison of the developed design to the prior cost plan. Traditionally, the quantity surveying consultant would manually measure the most recent design drawings or utilize CAD measurement to make incremental revisions to the cost targets for each group element and sub-element. However, with the utilization of BIM, all consulting team members can better comprehend the project design with the help of the generated 3D model, which also helps to prevent miscommunication issues (Singh & Gajjar, 2022). One of the impacts of BIM on the consultants' design is its ability to enhance simulation and visualization. With the application of BIM, architects and the other consultants can create detailed 3D models that allow a comprehensive view of the entire building in its context, which impacts to facilitates more informed decision-making and innovative solutions to design challenges. With that, it helps to facilitates quantity surveyor consultants to set cost targets for each element based on the client's budget. Implementing BIM software allows consultants to create a 3D model, aiding project comprehension and reducing miscommunication. In the technical design stage, design consultants use BIM software to identify design conflicts, automate quantity calculations, and facilitate collaboration between structural engineers and architects (Singh & Gajjar, 2022; Latiffi et al., 2013). Furthermore, BIM implementation allows for the visual detection of design conflicts between different professionals (Latiffi et al, 2013). Finding drawing inconsistencies earlier in the building process helps to cut down on variant orders. Additionally, the BIM software can automatically quantify every quantity in the 3D model with measurement information and specifics. If BIM software is used, cost estimating processes in this technical design stage will improve (Singh & Gajjar, 2022).

Other positive impacts of BIM implementation include offers cost savings in both construction and pre-construction stages. It helps to reduce rework and is quite alluring, especially for larger organizations. The use of the BIM system before the commencement of work was credited with a 40 to 90% reduction in rework in some of the construction projects that were completed in the past. During construction stage, BIM allows for advanced inspection, planning, and inventory management which resulting in lower expenses throughout the project. Besides, one advantage of BIM during this stage is the ability to inspect all project components in advance, managing and planning the construction process beforehand, minimizing the need for on-site inventory, and lowering expenses. BIM's virtual monitoring and simulation capabilities help with logistics and identifying flaws early on, minimizing conflicts in the construction cycle (TeamSystem, 2021). Quantity Surveyors overseas payment, claims, and cost validation, while BIM software detects clashes resulting the reduces of variations and expenses. It also enables easy data access and documentation for handling of expenses, site visits, and re-measurement (Soon et al., 2019). During the handover and closeout stage, BIM facilitates facility managers in eliminating on-site surveys and streamlines information gathering (TeamSystem, 2021).

During the handover and closeout stage, important data was gathered and kept in the BIM system throughout the entire construction process which allows for easy access and documentation even after the construction. For facility managers, most of the issues and time-consuming tasks are the time spent gathering information which can now easily be faced by the usage of BIM software. In fact, if they needed to make improvements to the building through renovations or maintenance work, they had to first speak with the direct person with on-site surveys, but this is no longer required. (TeamSystem, 2021).

LIMITATION OF BIM IN CONSTRUCTION STAGES

BIM software's limitations include limited capability, interoperability issues, and inadequate cost information (Kuenzel et al., 2019; Sullivan & Pisoni, 2019). It also faces data quality problems, leading to design flaws, incomplete designs, inaccurate cost estimates, and project delays (Abu-Zeid et al., 2020). Additional limitations include limited scope, lack of standardization, a steep learning curve, and integration challenges.

According to Kuenzel et al. (2019), BIM software has limited capability and frequently requires the use of other programmes and plugins to do certain tasks. The process of integrating data from diverse sources might also be complicated by BIM software's interoperability issues with other software products. The limited access to cost information is also a BIM limitation during the pre-construction stage. Cost information is usually not readily available in BIM models, requiring more work to gather and input, according to Sullivan and Pisoni (2019). This may lead to erroneous cost estimates and late project completion. BIM's multidimensional nature poses challenges due to a steep learning curve and specialized knowledge requirements (Miettinen & Paavola, 2014). According to Kuenzel et al. (2019), BIM software has limited capability and frequently requires the use of other programmes and plugins to do certain tasks. The process of integrating data from diverse sources might also be complicated by BIM software's interoperability issues with other software products. The limited access to cost information is also a BIM limitation during the pre-construction stage. Cost information is usually not readily available in BIM models, requiring more work to gather and input, according to Sullivan and Pisoni (2019). This may lead to erroneous cost estimates and late project completion. Inconsistent or inaccurate data input into BIM software may result in significant mistakes in the outcome of the project (Succar, 2009). Additionally, limitations in scope, detailing, time consumption, and modeling are other drawbacks in the design stage.

High implementation costs hinder widespread adoption of BIM in Malaysia (Baharum et al., 2019). In implementing BIM, limited financial resources and expenses for hardware, software, skilled labor, and training pose challenges for construction businesses. Furthermore, Proper training, collaboration skills, and qualified employees are necessary but add to the implementation difficulties. Lack of shared understanding and regulatory structure further impedes adoption. Other limitations include incomplete/inaccurate data, technology dependence, and time-consuming processes. BIM models' accuracy relies on input data quality and stakeholder assumptions. Incomplete or incorrect data hampers problem recognition and site information precision. BIM may overlook important aspects, limiting defect detection and site information accuracy (Azhar et al., 2012). Challenges arise in detecting problems and providing current information due to revisions and lack of real-time updates (Eadie et al., 2013).

Limitation of BIM during the handover and closeout stage is the accuracy of BIM models is heavily dependent on the precision and thoroughness of the input data and assumptions made by project stakeholders, such as designers and engineers. According to Eadie et al. (2013), the model's capacity to recognize problems and provide exact information on site and environmental conditions might be hampered by incomplete or incorrect input data. In addition, Azhar et al. (2012) claim that BIM models might not consider all the important physical and environmental aspects related to the site and its surrounds, its ability to detect defects and provide exact information about a building site may be constrained. Additionally, due to the possibility of alterations and revisions, it may be challenging for BIM models to detect problems or offer current information regarding project status during the handover and closeout stage. According to Eadie et al. (2013), this is because BIM models might not be updated in real-time at this point, thereby reducing their usefulness.

POTENTIAL SOLUTIONS TO OVERCOME THE LIMITATINS OF BIM SOFTWARE

Dynamic 3D BIM models improve understanding of architectural designs and facilitate easier design procedures (Witt & Kähkönen, 2019). Integrating BIM software into university education can address its limitations and promote AEC education. This integration can also address the shortage of skilled BIM practitioners. Offering BIM learning practice programs in companies enhances staff learning experiences and visualization of BIM models (Tanko & Mbugua, 2022). Familiarity with BIM software reduces errors and increases productivity. Increased BIM adoption by companies can lead to clearer rules and guidelines in the future (Zhao et al., 2015).

PROBLEM STATEMENT

The usage of CAD in the construction industry has been widespread for decades, but it has been overtaken by the rapid adoption of BIM software, which offers real-time visualization and model analysis (TheCADRoom, 2022). BIM allows professionals to collaborate using a shared database and building model. However, the dawning digital transformation will require people and organisations to embrace digitalisation to keep pace with productivity and innovation. It is due to the limitations and variations in results among different BIM software due to the dynamic nature of the construction industry (Czmoch & Pekala, 2014) (Staub-French & Fischer, 2000). The adoption of BIM throughout the whole construction lifecycle needs a collaborative and integrated platform and support from the industry players. Many architectural firms and construction companies are hesitant to adopt BIM software, despite its capabilities (Langar & Pearce, 2014). In Malaysia, awareness of BIM software is still relatively low, and those who are aware often only know its advantages and limitations without fully understanding its impact on different construction stages (MyBIMMalaysia, 2021). The constant updates to BIM software have improved its efficiency but also introduced more limitations. In measuring the BIM implementation in the Malaysian construction industry, it is important to assess different perspectives and views from the different professions that are currently practicing BIM in the construction cycle. Therefore, this research aims to identify the positive impacts and limitations of BIM software in different construction stages to inform improvements and overcome its limitations.

RESEARCH METHODOLOGY

Research Approach

The study uses a quantitative research method, which involves representing data numerically and using statistical analysis (Goertzen, 2017). This method focuses on gathering numerical data to generalize and characterize occurrences (Babbie, 2010). Questionnaire surveys were used to collect data from Registered Architects, Engineers, and Quantity Surveyors in Malaysia via Google Forms. The study focuses on registered Architects, Engineers, and Quantity Surveyors working in Malaysia. The questionnaire is distributed via email using official website addresses of relevant firms. Multiple follow-up emails are sent to increase the response rate. The calculation of the sample size is based on the technique introduced by Krejcie and Morgan (1970). The formula is: $n = N / (1 + (N * e^2) / (N * (e^2) – 1))$. By using the formula, a total of 378 respondents are required. Descriptive analysis is used to summarize statistical data using tables, diagrams, and graphs (Trochim, 2020). It helps describe the data set by providing statistics such as frequency, variance, means, and ranking (Rawat, 2021). This analysis determines variables based on frequencies, importance levels, and significance, enabling the identification of important subjects and addressing relevant issues (Rawat, 2021; Trochim, 2020).

FINDINGS

Respondent Profile

Analysis profile of the respondents to provide an overview of the demographic details including educational background and working experience in construction industry.

Table 1 showed the frequency table for professional field of respondents. The respondents were made up of Quantity Surveyors (42%), Architects (35%), Engineers (23%).

Table 1. Frequency table for professional field of respondents

Professional Field	Valid Percent	Cumulative Percent
Quantity Surveyor	42	42
Architect	35	77
Engineer	23	100
Total	100.00	

The working experience of the respondents in the construction industry is crucial in assessing the average population experience of the industry players. Table 2 shows the frequency for the working experience of respondents in the construction industry. The experience of all respondent's experience is grouped into four (4) different ranges of years. 7% of the Quantity Surveyors respondents have less than 5 years of experience while 11% of them have 6 to 10 years of experience. Followed by 20% which have 11 to 15 years of experience and lastly 62% with more than 15 years of experience. On the other hand, the respondents from the Architects that have less than 5 years of experience is 0%, followed by 11% with 6 to 10 years of experience, then 21% with 11 to 15 years of experience and 68% with more than 15 years of experience. Respondents for Engineers are quite like Quantity Surveyors and Architects where they are 2% have less than 5 years of experience, 50% with 6 to 10 years of experience, 30% with 11 to 15 years of experience, and 63% with more than 15 years of experience. The respondents with more than 15 years of working experience amounted to 62% to 68% of the total population of each group, which is considerably more than half. It shows that most respondents in this study have high working experience in the industry to provide insightful views.

Table 2. Working experience in the construction industry

Years of Working Experience	Professional Field		
	Quantity Surveyor	Architect	Engineer
Less than 5 years	7%	0%	2%
6-10 years	11%	11%	5%
11-15 years	20%	21%	30%
More than 15 years	62%	68%	63%

Respondent's Opinion Towards the Importance of BIM Software Application

To keep up with the BIM working environment, industry players must be aware of the importance of BIM application. The following descriptive statistics indicate the opinion of the respondents towards the importance of BIM software application. Result from Table 3 shows the perception of the respondents on the overall importance that BIM would be needed for all construction stages. 48% of the respondents considered that the importance of BIM application to be important while 27% of the respondents think that BIM application is moderate for all construction stages. It is followed by 23% of respondents agreed that BIM application is very important throughout the construction stages. This result shows that BIM software is now much more developed to be able to be used for almost all stages of construction but still yet to be perfected. Eventually, the number of BIM players will increase in the future.

Table 3. Importance of BIM applications

Level of Importance	Percentage
Very Important	23%
Important	48%
Moderately Important	27%
Low Important	2%
Not At All Important	0%

Application and Practice of BIM

The industry players play a vital role in transforming their workflow or working environment to apply BIM practices. Dodge Data and Analytics (2015) stated that the success of implementing BIM comprises several factors including visualization, behaviour, cultural and technological changes by adopting new processes, and standards throughout the workflow. This section provides an overview of BIM application in different construction stages. This would provide a more in-depth idea on when is BIM mostly being applied by Quantity Surveyors, Architects, and Engineers.

Table 4. BIM application in pre-construction stage

Frequency of BIM used	Quantity Surveyor	Architect	Engineer
Always	28%	52%	30%
Often	36%	21%	32%
Sometimes	19%	20%	36%
Rarely	13&	7%	2%
Never	4%	0%	0%

The above table reflects the respondent's views towards the usage of BIM during the pre-construction stage. 28% of Quantity Surveyor respondents were of the view that BIM is "always" being used by the industry during the pre-construction stage, where 52% from Architects, and 30% from Engineers have the same view. While 36% of Quantity Surveyors thought that BIM is used “often”, which is higher compared to Architects with 21% and 32% from Engineers.

Table 5. BIM application in design stage

Frequency of BIM used	Quantity Surveyor	Architect	Engineer
Always	7%	58%	42%
Often	10%	31&	28%
Sometimes	24%	7%	24%
Rarely	38%	4%	6%
Never	21%	0%	0%

The above table presents the data of respondents with the view towards the usage of BIM during the design stage. 7% of Quantity Surveyor believed that BIM is “always” used during the design stage, with 58% from Architects, and 42% from Engineers. While most Quantity Surveyor viewed that BIM is “rarely” used (48%). This result indicates that BIM software is still being used quite frequently during the design stage which might be because it helps provide great benefits to most professions in the construction industry during this construction stage.

Table 6. BIM application in construction stage

Frequency of BIM used	Quantity Surveyor	Architect	Engineer
Always	26%	46%	20%
Often	33%	35%	28%
Sometimes	21%	13%	13%
Rarely	14%	6%	8%
Never	4%	0%	2%

The table above shows the data of the respondents’ application of BIM during the construction stage. Quantity Surveyors with 33% considered BIM to be used “often” during the construction stage, followed by 26% for “always” used. Architects has the highest percentage for “always”, followed by 35% for “often” used. Like Quantity Surveyors, 28% of Engineers selected “often’ used and followed by 20% for “always” used. The result implies that BIM is used more frequently for Architects as compared to Quantity Surveyors and Engineers.

Table 7. BIM application in handover and closeout stage

Frequency of BIM used	Quantity Surveyor	Architect	Engineer
Always	2%	5%	3%
Often	9%	13%	7%
Sometimes	30%	36%	37%
Rarely	53%	42%	50%
Never	6%	4%	2%

The table above provides the data of the views from respondents towards the BIM usage during the handover and closeout stage. The majority of the Quantity Surveyor's respondents (53%) leaned towards "rarely" option. Followed by 42% from Architects and 50% from Engineers. The results indicate that BIM is not used as much during this construction stage by all the parties including Quantity Surveyors, Architects and Engineers as compared to earlier stages.

Positive Impacts of BIM Implementation

The following descriptive statistics indicate the impacts of BIM implementation in different construction stages from the perspective of Quantity Surveyors, Architects and Engineers.

Table 8. Impacts of BIM implementation (quantity surveyors' perspective)

Positive impacts of BIM implementation	Rank	Mean
Pre-construction stage		
Better cost plan & estimating	1	0.908
Better preparation in tender documentation	2	0.682
Better value engineering	3	0.600
Design stage		
Better cost planning	1	0.838
Better cost control	2	0.786
Better risk management	3	0.722
Construction stage		
Improved visualisation	1	0.832
Better tracking of project cost	2	0.800
More accurate quantity take-off	3	0.756
Handover & closeout stage		
Better efficiency & accuracy of contract documents review	1	0.722
Improved verification of final account	2	0.558
Better conduct in final inspection	3	0.538

From Table 8 above, BIM provides valuable functions for Quantity Surveyors. During the pre-construction and design stage, it improves cost planning and estimating, reducing estimating time by 30% and improving accuracy by 6% (Arayici et al., 2011). In the construction stage, BIM enhances visualization. Additionally, it increases efficiency and accuracy in reviewing contract documents by generating schedules of works and BQ (Wong and Zhou, 2016).

continued on following page

Table 9. Continued

Table 9. Impacts of BIM implementation (architects' perspective)

Positive impacts of BIM implementation	Rank	Mean
Pre-construction stage		
Better conduction on site analysis	1	0.930
Better development in design concepts	2	0.916
Better in creating construction documents	3	0.778
Design stage		
Better conceptual design	1	0.926
Better schematic design	2	0.886
Better design development	3	0.834
Construction stage		
Better construction observation	1	0.800
Better design review	2	0.780
More building analysis	3	0.750
Handover & closeout stage		
Better conduct in final inspection	1	0.634
Better handover to client	2	0.608
Better certifying completion of works	3	0.480

Table 9 shows that during pre-construction, BIM excels in site analysis with digital simulations (BIM Strategy, 2017). During design, it aids in better conceptual design and issue identification (Kuen-Yang et al., 2021). For construction stage, BIM enables improved observation through virtual inspections (Succar et al., 2018). In the final stage, BIM provides better conduct in final inspection.

Table 10. Impacts of BIM implementation (engineers' perspective)

Positive impacts of BIM implementation	Rank	Mean
Pre-construction stage		
Better visualisation	1	0.838
Better sustainability and efficiency	2	0.800
Enhanced accuracy and precision	3	0.774
Design stage		
Better visualisation	1	0.900
Better at preparing engineering drawings & specifications	2	0.768
Better simulation	3	0.754
Construction stage		
Improved visualisation	1	0.818
Improved clash detection	2	0.790
Better building analysis	3	0.702
Handover & closeout stage		

continued on following page

Table 10. Continued

Positive impacts of BIM implementation	Rank	Mean
Improved defect management	1	0.600
Improved asset management	2	0.580
Improved documentation management	3	0.550

The data in Table 10 shows that the highest benefits of BIM implementation for Engineers are better visualization during pre-construction, design, and construction, and improved defect management during handover and closeout. However, 21% of Engineers disagree with the improved defect management during the last stage of construction, indicating room for improvement in implementing BIM in handover and closeout stage for Architects and Engineers.

Limitation of BIM Implementation in Construction Stages

Table 11 indicates that the biggest limitation of BIM implementation towards Quantity Surveyors during the pre-construction stage would be the lack of standardization since Differences may occur while transferring BIM data between programs (Zhang et al., 2021). Next would be the limited scope in providing all aspects of construction project as there might be Absence of accurate and up-to-date cost and quantity information (Chan and Kumaraswamy, 2016). During the construction stage, the biggest limitation from BIM would be its high cost since the costs of purchasing and maintaining BIM may be a significant barrier for QS (Rezgui et al., 2018). While the main limitation during the handover and closeout stage would be its limited ability to detect defects as there might be a lack access to the most recent information on changes made during construction (Tah and Carr's, 2017).

Table 11. Limitations of BIM implementation (quantity surveyors' perspective)

Limitations of BIM implementation	Rank	Mean
Pre-construction stage		
Lack of standardisation	1	0.772
Limited availability of cost data	2	0.676
Limited scope	3	0.662
Design stage		
Limited scope in providing all aspects of construction project	1	0.900
Difficulty in extracting accurate quantities for complex projects	2	0.768
High learning curve	3	0.754
Construction stage		
High cost	1	0.818
Limited adoption	2	0.790
Incomplete/inaccurate data quality	3	0.702
Handover & closeout stage		
Limited ability to detect defects	1	0.600

continued on following page

Table 11. Continued

Limitations of BIM implementation	Rank	Mean
Difficulty in visualization real-world conditions	2	0.580
Lack of human judgement	3	0.550

Table 12 shows that the high learning curve of BIM is the biggest limitation during the pre-construction stage towards Architects as it requires a significant time & effort to become skilled in (Azhar et al., 2012). While during the design stage, it would be the fact that BIM is time-consuming to use, particularly when multiple revisions need to be integrated into the design (Succar, 2018). The high cost is once again the biggest limitation during the construction stage while the limited ability to detect defects during the last construction stage is once again the biggest limitation.

Table 12. Limitations of BIM implementation (architects' perspective)

Limitations of BIM implementation	Rank	Mean
Pre-construction stage		
High learning curve	1	0.720
Software limitations	2	0.700
Incomplete/outdated/inaccurate data quality	3	0.690
Design stage		
Time-consuming	1	0.738
Potential errors	2	0.732
Limitations in detailing	3	0.692
Construction stage		
High cost	1	0.738
Require visualization training/support	2	0.708
Too dependant on technology	3	0.644
Handover & closeout stage		
Limited ability to detect defects	1	0.712
Limited information on site conditions	2	0.684
Difficulty in visualization the real-world conditions	3	0.636

Table 13 shows that the main limitation experienced by Engineer respondents when utilizing BIM during the pre-construction stage is the data quality issues since it might be caused by the fact that people are reluctant to share their data (Li et al., 2020). The potential errors would be the biggest limitation during the design stage since sometimes there are software bugs, inappropriate program usage, or errors in data entry (Sacks et al., 2018). The highest RII for the construction stage is also the high cost while the limited real-time updates are BIM's main limitation during the handover and closeout stage.

Table 13. Limitations of BIM implementation (engineers' perspective)

Limitations of BIM implementation	Rank	Mean
Pre-construction stage		
Incomplete/outdated/inaccurate data quality	1	0.720
Integration issues	2	0.700
Software limitations	3	0.664
Design stage		
Potential errors	1	0.736
High learning curve	2	0.720
Limitations in modelling	3	0.704
Construction stage		
High cost	1	0.744
Time-consuming	2	0.688
Too dependant on technology	3	0.676
Handover & closeout stage		
Limited-real time updates	1	0.716
Highly complex	2	0.674
Lack of specificity	3	0.622

Solution to Overcome Limitations of BIM Implementation

The result findings in table 14 show that the main solution to overcome the limitations of BIM that all respondents agreed on is through tertiary education. Sacks et al. (2018) suggest that to surmount the restrictions of BIM, tertiary education institutions can furnish students with practical exposure to BIM software and tools. The incorporation of real-world projects and datasets into BIM education can enable students to cultivate crucial skills in critical thinking and problem-solving that may aid them in addressing limitations in BIM during the construction process.

From the Quantity Surveyor respondents' side, the data shows that specialized Quantity Surveying BIM software could incorporate enhanced cost estimation tools and improved automated quantity take-offs as these two points have been ranked the second and third most important solutions for BIM to overcome its limitations. With these two functions improved, BIM would then be more accurate and efficient while reducing errors and costs as studied by Zhang et al. (2021).

Moreover, at the Architect respondents' side, the data shows that specialized Architect BIM software could incorporate better visualization tools and enhanced modelling tools as these two points have been ranked the second and third most important solutions for BIM to overcome its limitations. According to Eadie et al. (2013), the use of advanced visualization tools may encourage stakeholder collaboration and idea sharing, allowing them to identify possible design issues before the building process begins. By reducing the need for design revisions during the building phase, this might result in considerable time and cost savings. However, Abdellatif et al. (2018) have proposed that the use of sophisticated modelling approaches can make it easier to combine various data sources, such as laser scanning and photogrammetry, to build more accurate BIM models. This can reduce the chance of conflicts and errors in BIM data, which will help stakeholders make decisions more effectively.

Lastly, at the Engineer respondents' side, the data shows that firms should provide training to its employees. This way the employees would be able to further understand the usage of BIM and use it more efficiently. The visualization Engineer BIM software could also enhance the modelling tools as previously analyzed.

Table 14. Potential solutions to overcome limitations of BIM implementation

Rank	Potential solutions to overcome limitations of BIM implementation		
	Quantity Surveyors' perspective	Architects' perspective	Engineers' perspective
1	Tertiary education	Tertiary education	Tertiary education
2	Enhanced cost estimation tools	Improved visualization tools	Firms giving training to employees
3	Improved automated quantity take-off	Enhanced modelling tools	Enhanced modelling tools
4	Firms giving training to employees	Firms giving training to employees	Improved sustainability analysis
5	Improved visualization tools	Real-time collaboration	Real-time collaboration
6	Real-time collaboration	Enhanced cost estimation tools	Enhanced cost estimation tools
7	Enhanced modelling tools	Improved automated quantity take-off	Improved automated quantity take-off
8	Improved sustainability analysis	Improved sustainability analysis	Improved visualization tools

CONCLUSION

As shown by the data, BIM has greatly impacted the pre-construction, design, and construction stage by providing various specific functions to assist Quantity Surveyors, Architects, and Engineers. As far as the benefits BIM provides are concerned, BIM has ultimately excelled in providing detailed visualizations of buildings which in return greatly enhanced the design process and analysis to be done as well. Furthermore, the cost estimation tools that BIM provides have also been much more efficient with most respondents agreeing with the benefits that it brings towards the costing aspects of buildings. However, the handover and closeout stage are where BIM still requires more improvements since it still is not yet capable of providing clearer real world conditions.

The learning curve and high cost of BIM are generally the main factors of limitation that it has considering that it can be quite complicated which makes the learning process of BIM to be much complex. The difficulty of using BIM to visualize real-world conditions during the handover and closeout stage has also limited the adoption of BIM being used towards that construction stage. Overall, it can be observed that the more complex a particular project is, the more limitations there would be arise from BIM.

Solutions that are related to training and education are deemed to be exceptionally effective. Many respondents have opined that tertiary education should include the learning and use of BIM while many respondents have also agreed that firms to include training on BIM. This trend of responses goes in line with the results for Objective 2 as many respondents have thought that BIM is complex and has a high learning curve. Nevertheless, software improvements on real-time collaboration as well as enhanced cost estimation tools and modelling tools are also a few main solutions that can be improved in the software area of BIM.

This study mainly focuses on the views from Quantity Surveyors, Architects, and Engineers in the Malaysian construction industry regarding to the effects of BIM software during different construction stages. It is recommended for future researchers can perhaps try to expand the target population to other countries and explore what the ultimate solutions that can help overcome the limitations of BIM. Additionally, it is advised for future researchers to use a qualitative or even mixed method to their research to obtain a more in-depth discussion and subsequently improve the overall quality and validity of the research findings.

ACKNOWLEDGMENT

The researchers of this article would like to appreciate the journal's editors and the anonymous reviewers for their positive feedback and suggestions for the improvement of this journal.

This research received no specific grant from any funding agency in the public, commercial, or not-for-profit sectors.

REFERENCES

Abu-Zeid, M., El-Tawil, S., & Kandil, A. (2020). Analyzing the impact of data quality on the effectiveness of building information 6odelling for construction projects. *Journal of Construction Engineering and Management*, 146(3), 04019110.

Aung, T. H. S. (2022). BIM for Engineering: How and why engineers should use BIM for better project delivery. *Oneistox*. https://www.oneistox.com/blog/bim-for- engineering

Azhar, S., Khalfan, M., & Maqsood, T. (2012). Building information 6odelling (BIM): A new paradigm for visual interactive 6odelling and simulation for construction projects. In *Handbook of research on building information 6odelling and construction informatics* (pp. 135-159). IGI Global.

Baharum, M. R., Kasim, N., & Taib, M. R. (2019). Barriers to Building Information Modeling (BIM) Implementation in Malaysia. *Journal of Construction in Developing Countries*, 24(1), 67–84.

Czmoch, I., & Pekala, A. (2014). Traditional Design versus BIM Based Design. *Traditional Design versus BIM Based Design*, 210–215. 10.1016/j.proeng.2014.12.048

Eadie, R., Browne, M., Odeyinka, H., McKeown, C., & McNiff, S. (2013). BIM implementation throughout the UK construction project lifecycle: *An analysis.Automation in Construction*, 36, 145–151. 10.1016/j.autcon.2013.09.001

Eastman, C. M., Teicholz, P., Sacks, R., & Liston, K. (2011). *BIM handbook: A guide to building information 6odelling for owners, managers, designers, engineers, and contractors*. John Wiley & Sons.

GeoSLAM. (2021). What is BIM and how does it impact the construction industry? – *GeoSLAM*. https://geoslam.com/blog/2021/12/23/what-is-bim- and-how- does-it-impact-the-construction- industry/

Gerardi, J. (2021). *The 6 Stages Of Construction – Key Phases Explained*. ProEst. https://proest.com/construction/process/stages-of- construction/

Kuenzel, R., Schevers, H., & Fischer, M. (2019). The limitations of BIM for project management. *Procedia Engineering*, 196, 693–700.

Latiffi, A. A., Mohd, S., Kasim, N., & Fathi, M. S. (2013). Building Information Modeling (BIM) Application in Malaysian Construction Industry. *International Journal of Construction Engineering and Management*, 2(4A), 1–6. 10.5923/s.ijcem.201309.01

Liu, Y. (2021). Introfuction of Different BIM Software in the Construction Industry. Bachelor's thesis Visamäki Campus, *Degree Programme in Construction Engineering Spring semester 2021*, 1–28.

McPartland, R. (2017). Top 10 BIM terms explained. *NBS*. https://www.thenbs.com/knowledge/top-10 - bim-terms- explained#:~:text=What%20is%20BIM?aspect%20o f%20the%20built%20asset

Miettinen, R., & Paavola, S. (2014). The concept of BIM maturity: What is it, and how to measure it? *Journal of Information Technology in Construction*, 19, 28–44.

Proforma Construction Inc. (2019). *Why BIM is Important for the Construction Industry*. Proforma Construction. General Contractor Pleasanton CA. https://www.proformaco.com/blog/why-bim-is- importan t-for-the-construction- industry/

Reyes, N. (2019). *Building Information Modeling Benefits for Architecture and Construction | Thought Leadership*. HMC Architects. https://hmcarchitects.com/news/building- information-modeling-benefit s-for-architecture-and-construction-2019-06-05/#:~:text=Building%20Information%20Modeling%20 (BIM)%20is,and%20how%2 0it%20will%20operate.

Singh, H., & Gajjar, K. (2022). 5 Benefits of BIM for Contractors During the Pre-construction Stage. *CAD Services: Outsource CAD Drafting, 3D Design & BIM Services*.https://www.hitechcaddservices .com/news/bim-for- contractors-during-pre- construction-stage/

Soon, L. T. S., Hassan, H., & Abidin, N. Z. (2019). Measuring the Effects and Efficiency of 3D modelling in Quantity Surveying Profession. *MCRJ*, *2*(7), 58–64. https://expert.taylors.edu.my/file/rems/ publication/10 5958_6709_1.

Staub-French, S., & Fischer, M. A. (2000). *Formalisms and Mechanisms Needed to Maintain Cost Estimates Based on an IFC Product Model*. American Society of Civil Engineers. 10.1061/40513(279)94

Storrier, A. (2021). *How CAD and BIM Changed the Construction Industry*. Resource Center. https:// resources.hexagonppm.com/articles- blogs/how-cad-and-bim-changed-the-construction-industry

Succar, B. (2009). Building information modelling framework: A research and delivery foundation for industry stakeholders. *Automation in Construction*, 18(3), 357–375. 10.1016/j.autcon.2008.10.003

Sullivan, K. T., & Pisoni, R. (2019). BIM and cost estimating: Perceptions, issues, and strategies. *Journal of Management Engineering*, 35(6), 05019008.

Tanko, B. l., & Mbugua, L. (2022). BIM Education in Higher Learning Institutions: A Scientometric Review and the Malaysia Perspective. *International Journal of Built Environment and Sustainability*, *9*(1), 23–37. https://ijbes.utm.my/index.php/ijbes/article/view/864/ 250

TeamSystem. (2021). *Advantages of using BIM in Construction*. TeamSystem Construction. https://www .teamsystemconstruction.com/magazine/advantages-of- bim/#:~:text=Pre%20-%20Construction%20 Stage&text=BIM%20is%20helpful%20not%20only,regulations%20regarding%20functi onality%20 and%20constructability.

The Geospatial. (2019). What is BIM? Why is it Important? *Medium*. https://medium.com/@ thegeospatialnews/what-is- bim-why-is-it-important- a662df34b60b

TheCADRoom. (2022, July 28). CAD Vs BIM: What is the Difference? | The CAD Room. *The CAD Room*. https://www.thecadroom.com/difference-between-cad-and- bim/#:~:text=CAD%20has%20rapidly%20been%20overtaken,same%20database%20and%20building%20model

Vinzero. (2019). *What can BIM do after construction is complete?*https://www.a2kstore.com/what-can-bim- do-after-construction-is-complete

Witt, E., & Kähkönen, K. (2019). BIM-Enabled Education: a Systematic Literature Review. *10th Nordic Conference on Construction Economics and Organization (Emerald Reach Proceedings Series, Vol. 2)*. Emerald. 10.1108/S2516-285320190000002042

Zhao, D., Mccoy, A. P., Bulbul, T., Fiori, C., & Nikkhoo, P. (2015). Building Collaborative Construction Skills through BIM-integrated Learning Environment. *International Journal of Construction Education and Research*, 11(2), 97–120. 10.1080/15578771.2014.986251

KEY TERMS AND DEFINITIONS

BIM: Building Information Modeling (BIM) is a digital representation of physical and functional characteristics of a facility.

Construction: The action of building something or structure.

Impact: A marked effect or influence.

Limitation: Something that controls or reduces something.

Software: A set of programs, procedures, and routines associated with the operation of a computer system.

Stages: A part of an activity or a period of development.

Chapter 15
The Provision of Elderly Care Centre and User Perception Towards Its Built Environment in Kuala Lumpur

Myzatul Aishah Kamarazaly
Taylor's University, Malaysia

Chiam Winnie
https://orcid.org/0009-0007-7825-2661
Taylor's University, Malaysia

Filzani Illia Ibrahim
Taylor's University, Malaysia

Nurul Aini Osman
Taylor's University, Malaysia

Siti Norzaini Zainal Abidin
https://orcid.org/0000-0002-2637-4794
Taylor's University, Malaysia

ABSTRACT

The population growth in Malaysia since Independence Day is attributed to improved healthcare and longer lifespans. However, as the older generation ages, some are forced to live independently, while others lack suitable institutions for their daily activities. Neglected conditions in elderly care institutions have led to concern about the physical and mental health of these individuals. The research aimed to analyze the existing built environment of elderly care centers and the needs of users for an improved future environment that prioritizes the comfort and safety of the elderly. The objective of this research is to determine the strategies for the future development of elderly care centers. A qualitative method adopted by recruiting five voluntary participants from Pusat Jagaan Rumah Love and Care Kajang Home. Key findings revealed diverse needs and provisions influenced by social factors and living infrastructure.

DOI: 10.4018/979-8-3693-1650-4.ch015

Ultimately, accommodating the changing behaviors and characteristics of the elderly was crucial for developing infrastructure that promoted positive and healthy aging.

INTRODUCTION

The population in Malaysia has been growing since Independence Day (Department of Statistics Malaysia, 2020), leading to the necessity for proper institutional facilities and a comfortable environment. However, developmental progress in Malaysia often prioritizes the younger generation rather than the elderly. Statistics indicate that the elderly constitute 9.6% of the total population, with certain states having a higher concentration of the elderly population (World Health Organization, 2020). The research specifically focused on aging individuals residing in Kuala Lumpur.

The Ministry of Urban Wellbeing, Housing, and Local Government identified five issues related to aging, including medical illness, physical changes, financial constraints, low social interaction, and non-elderly friendly building designs (Mafauzy, 2000). While some of these issues are unavoidable, there is still potential for improvement to enhance the wellbeing of the elderly.

PROBLEM STATEMENT

The aging population in Malaysia raised significant concerns with various predictions, suggesting that Malaysia might become an aging nation by 2035 (Nuradzimmah, 2016). Factors such as improved health and access to subsidized healthcare have contributed to an increased lifespan in the elderly. However, the built environment in Malaysia is suitable to accommodate the growing aging population (Aini, Murni & Abdul Aziz, 2016). Elderly individuals living in rural areas prefer staying in their hometown due to comfort, while those in busy cities often choose between living with their children or residing in a care center (Nurul Aini et al., 2018). Existing care institutions were lacking in focusing on the cognitive and active development of the elderly, as well as a lack of awareness of their specific needs (Elderly in the Rural vs Urban Settings, 2020). Therefore, a comprehensive study of existing care centers was necessary to identify areas for improvement. By understanding the needs of the end-users and addressing the issues in the built environment, this study aimed to provide recommendations and design strategies that could promote active development for elderlies in Malaysia.

LITERATURE REVIEW

Aging Population

The definition of "Elderly" varies globally, but the World Health Organization (WHO) has classified individuals over 60 as elderly or as senior citizens (World Health Organization, 2020). Further classifications included the oldest old (over 80), centenarians (over 100), and super-centenarians (over 110). The longevity of the elderly has been increasing, attributed to improvements in healthcare and technology (World Health Organization, 2020).

The government of Malaysia has made efforts to enhance healthcare services, resulting in subsidized consultations and cheaper services for senior citizens (Khoo, 2018). This has contributed to better health and wellbeing for the elderly. Additionally, a medical study highlighted an increase in the number of elderly citizens and a significant drop in the percentage of deaths over the years (Harvard Business Review, 2018).

Malaysia is projected to have a rapidly aging population, with an estimate suggesting a 269% increase by 2040 (Kinsella and Wan, 2018). The elderly population accounted for 9.96% of Malaysia's total population, with the highest distribution in Selangor, Johor, and Perak (Department of Statistics Malaysia Official Portal, 2015). There have also been significant differences in the rates of aging based on gender, ethnicity, and geographical location. The proportion of elderly people living in urban areas has been rising, partly due to the rapid development of rural areas, transforming them into urban settings and elderly migrating back to their hometowns (Seamon & Sowers, 2008).

Overall, Malaysia's aging population has posed various challenges and implications, necessitating the development of policies and strategies to address the needs of the elderly in healthcare, social welfare, and infrastructure.

The Changing in Characteristics of Malaysian Elderly

The aging process and its effects on individuals were highlighted, noting that it varied based on genetics, environmental factors, and lifestyle choices (Besdine, 2019). Physical changes during the aging process, such as skin wrinkling, changes in body weight, and deterioration of the senses, digestive system, bones, joints, muscles, and cardiovascular system, were observed (Mayo Clinic, 2019). The importance of maintaining muscle mass, strength, and balance through physical activity was emphasized (The Centre for Aging, 2011).

Health conditions commonly found during aging include Alzheimer's, arthritis, lung disease, stroke, osteoporosis, malnutrition, eye problems, back and neck pain, diabetes, and depression (Lunenfeld & Stratton, 2013). Geriatric syndromes and an inactive lifestyle could lead to depression and stress among the elderly (McMaster University, 2018).

The social and physical needs of the elderly are essential for their wellbeing (Lim, 2012). Different living arrangements for the elderly, such as living alone, with a spouse, or with extended family, were emphasized (Parker et al., 2004). The importance of providing comfortable and safe living spaces for the elderly was emphasized. The living typology affects residents' preferences, and it is essential to consider individual needs and physical abilities in design criteria (Fon & Philips, 2007).

Elderly Care Centre Design Typology

Worldwide, discussions have focused on the impact of an aging population on housing typology and the needs of the elderly (Fon & Philips, 2007). Older individuals, particularly those residing in cities, emerge as discerning consumers compared to younger generations, driving the demand for improved housing options tailored to their specific needs and preferences (Fon & Philips, 2007). The wellbeing and quality of life of the elderly become primary concerns as they approach old age, emphasizing the critical role of housing typology in creating a comfortable environment for them.

However, designing housing that meets the needs and preferences of the elderly has been a challenge to planners, designers, builders, and even end users (Sungur & Polatoglu, 2012). Different experiences and needs exist among the elderly, and in the Malaysian context, issues related to the built environment for the elderly have often been overlooked (Lim, 2012). Therefore, understanding their needs and desires was crucial before developing design strategies for the built environment.

Common types of elderly housing globally include Independent Living, Assisted Living, Skilled Nursing Communities, and Continuous Care Retirement Communities. However, these typologies and terms are not commonly used in the Malaysian guidelines (Nan et al., 2006). Instead, Malaysian guidelines mentioned three types of living settlements, for example "Aging-In-Place Housing Developments," "Retirement Villages", and "Care Centers." These settlements required harmonious layouts, compactness, mixed-use facilities, and a universal building design to cater to the elderly's needs. Accessibility to public amenities, clear labelling, and safety considerations were also important (Shyam & Zarren, 2021). The "aging-in-place" concept allows the elderly to live independently in a familiar environment, either through new elderly-friendly developments or retrofitting existing houses. Essential features of barrier-free design, such as non-slip flooring, grab bars, proper placement of switches, wider doors, ramps, and sufficient lighting, are emphasized (Jabatan Perancangan Bandar Dan Desa, 2018).

Retirement villages or special housing developments offered temporary or permanent residence options. Active settlements included landed housing, while special care settlements consisted of free-standing buildings with integrated healthcare facilities (Hashim, 2020).

Respite care is available in active communities, providing temporary assistance to elderly individuals and their families. Day Care Centers, classified as both Day Care and Residential Care Centers, offer continuous care and include Senior Citizen Activity Centers (Johannes, Thomas, and James, 2016). Residential care communities are categorized into low, medium, and high care based on the level of assistance required by the residents. These communities must provide appropriate facilities and support for the elderly (Norani Hj. Mohd Hashim, 2020).

The overarching objective is to establish living environments that prioritize the wellbeing, safety, and independence of the elderly, ensuring their needs are met through considerate design and comprehensive care services (Norani Hj. Mohd Hashim, 2020).

Physiological Aspects

When designing living spaces for the elderly, several critical factors must be taken into account, including lighting, safety requirements, and space analysis. To enhance the wellbeing of older individuals, lighting levels must be increased while minimizing extreme contrasts to reduce glare. Switches and electrical outlets should be strategically placed for easy accessibility, and natural lighting needs to be maximized whenever possible. Safety considerations were paramount, recognizing the increased vulnerability of the elderly to accidents and fires. Falling risks were minimized by eliminating stairs, incorporating ramps, and using skid-resistant floor surfaces. Consideration of bathroom safety through the installation of handrails and seats in tubs. By addressing these factors comprehensively, living spaces can be customized to be more comfortable and secure for the elderly (Johannes et al., 2016).

Psychological Aspects

Psychological factors exert a significant influence on the wellbeing of the elderly. Challenges such as the loss of financial resources, retirement, and loneliness can contribute to heightened stress. Addressing the psychological needs of aging individuals involves providing privacy, opportunities for social interaction, access to knowledge about physical and mental health, and recognition of their individuality.

Meeting these needs includes ensuring adequate spaces for outdoor recreation, such as gardens, which promote exercise and psychological wellbeing. Additionally, thoughtful choices in lighting and color in living spaces can positively impact mood and mental activity. Safety measures, including the use of color codes and specialized devices to facilitate movement, are considered to prevent accidents and improve overall comfort (Creative Safety Supply, 2022).

Design Consideration

Consideration in housing design is essential for the elderly; it provides them with a healthy and congenial environment. Such designs could allow seniors to participate in community activities and contribute actively. The design has to ensure that elderly individuals can live comfortably and maintain their health. The elderly spend more time at home, relying on its facilities for social activities and pursuing their interests (Pekalee & Gray, 2023). Design considerations include analyzing basic standards for different areas of the house, emphasizing convenience, safety, and privacy, and avoiding an institutional atmosphere (Clegg et al.,2013).

Essential design and planning principles for elderly housing encompass several aspects, including a focus on small size and compactness, fireproof construction for safety, efforts to minimize housekeeping tasks, and the creation of a pleasant and spacious environment while avoiding an institutional appearance (Farage et al., 2012). Leisure spaces are designed for comfort, featuring interesting views and ample space for living, dining, and storage. Bedroom areas are customized for single or two-person occupancy, ensuring privacy, sufficient bed space, and convenient access to windows and outlets. Bathrooms incorporate safety features such as tubs with built-in seats, grab bars, non-slip tile floors, and well-arranged fixtures. Kitchens are equipped with proper enclosures, sinks with drainboards, mechanical ventilation, and ample storage capacity.

Additional considerations encompass windows designed for comfortable viewing (Constructororghq., n.d), doors and hardware for easy accessibility, non-slip flooring throughout, and appropriate vertical circulation with ramps and handrails to facilitate safe movement.

Overall, the design of housing for the elderly prioritized their comfort, safety, and independence while maintaining a pleasant and homely atmosphere.

METHODOLOGY

The qualitative research method was identified as the most suitable approach for discovering and exploring a new area and developing hypotheses (Creswell, 2003). The study employed an exploratory research design, utilizing a qualitative method for data collection through open-ended questionnaires. Various researchers have suggested that exploratory research is valuable for defining a set of investigative questions that can guide detailed research design (Hall, Chai and Albrecht, 2016). The present study adopted an interpretive philosophical stance, specifically interpretivism, which is associated with people's experiences and focuses on individuals' interactions with society (Advance Care Planning, 2018). The research design allowed the researcher to address research questions and align with the overall research scheme (Advance Care Planning, 2018). To meet the objectives of the research, an explanatory research design was deemed more appropriate, given the exploratory nature of the study.

A total of 365 registered aging care centers were operational in various states of Malaysia (Bryman, and Bell, 2007; Saldana, and Omasta, 2016). It is important to note that all respondents were recruited from the same care center, named Pusat Jagaan Rumah Love and Care Kajang Home. The study employed purposive sampling, selecting two caretakers and three elderly individuals from selected aging care centers. Interviews were conducted in both English and Malay languages, with each session lasting forty-five minutes to one hour. Semi-structured interview questions were categorized into Caretakers and Elderly Individuals. Caretakers, as employees providing services to the elderly, were provided with questions tailored to address the research questions and objectives outlined in the literature review.

A previous study by Merriam (2009) emphasized avoiding lengthy and negative questions in an interview guide and recommended using clear, non-technical language to ensure participant understanding. Another study [42] introduced a seven-step technique, adapted from a method for qualitative analysis, which utilized semi-structured questionnaires and audio-recorded interviews for data gathering. This systematic and organized approach to analyzing collected data has been employed in numerous studies (Advance Care Planning, 2018; Meriam Syed Akil and Abdullah, 2014).

FINDINGS AND DISCUSSION

The study adopted the qualitative research method; interviews were conducted with the elderlies and caretakers to determine the strategies needed for enhancing the future development of elderly care centers that were favoured by the current and future elderly populations. Hence, the coded data from each of the interviewees was cross-referenced with other interviewees to locate the similarities and dissimilarities among the participants' views. Table 1 explains the generated themes from the transcribed interview under interview questions asked to the centres' operators.

Table 1. Coding legend on category A

Category	Themes	Sub-Themes	Codes	Group
A. Strategies needed for enhancing the future development of elderly care center that were favored by the current and future elderly population.	1. Indoor living space	a. Living condition	A1a	1
		b. Space quality	A1b	
		c. Surrounding ambience	A1c	
	2. Outdoor recreational space	-	A2	2
	3. Travel distance and wayfinding	-	A3	3

Indoor Living Space - The responses from the survey were categorized under living conditions, space quality, and the surrounding ambience. The majority of the elderly revealed that they wish to have more privacy and extra spaces for their movements. These findings are in line with previous findings (Merriam, 2009). On the other hand, the majority of the caretakers reported that the challenge in managing the elderly care centre is the shortage of well-trained staff, and these findings concurred with previous findings by Murphy et al. (2006). The responses from the participants are as follows.

> "There is a lack of privacy and security. We wish to have more privacy space rather than having a shared area in an open dormitory concept. Too many people occupying the same bedroom and this bedroom is an open area concept; lack of privacy." - Chinese Lady (Elderly R1)
>
> "I am facing a lack of trained staff. This environment is systematic and easier to manage, especially for those elderly who require full attention. Also, this will become more practical, especially during a period when we are facing a shortage in the number of caretakers." - Caretaker R4.

Spatial Quality: Both Elderly R1 and R3 expressed satisfaction with the quality of the space during the interview. It is worth noting that Elderly R1 initially faced difficulties upon her first arrival at the elderly care center, as supported by a previous study (Meriam and Abdullah, 2014). One of the feedback items from a respondent is as follows.

> "Yes ... upon my arrival or at the beginning, it was quite hard for me to adapt to this environment, but after some time, I am getting used to it. All of us living here have already been able to adjust our lifestyle in the care center. I always believe that whatever possession we have, we will contend with it. That is how and what I feel to brighten up my life ..." - Chinese Lady (Elderly R1)

Nevertheless, Respondent R2 expressed dissatisfaction with the quality of space in the elderly care center, citing issues with compactness. Some elderly individuals faced challenges finding their way to designated spaces, a sentiment supported by previous research (ScienceDaily, 2018; Peninsular, 2021). One of the feedback items from R2 encapsulates this concern.

> "When moving around in an unfamiliar environment, it is perfectly normal to get lost. However, this tends to happen more often to certain elderly individuals diagnosed with Alzheimer's or Parkinson's disease. You may use me as an example, too." - Chinese Lady (Elderly R2)

In terms of the Surrounding Ambience/Environment, psychologically, the elements present in the surroundings of the elderly significantly impacted their mental health. Factors such as color coordination, furniture placement, and appropriate lighting in the main living space were elements that subconsciously brought peace to the users of the space. These findings align with a previous research study (Ishihara, 2001).

> "The ambience in the existing care center needed to be improved. The dark, dimmed, and stagnant ambience in the bedroom of the care center did not help uplift the emotions" – Elderly 1 and 2.

Outdoor Recreational Activities - Most of the elderlies mentioned their preference to have more outdoor activities. These were in line with past research studies (Sugiyama & Thompson, 2006). Some of the feedback from the participants is stated below:

"Yes, I love being outside. I enjoy the theatre and cinema. If someday, somewhere, I have the opportunity to go out; I think it's essential as it is just a simple pleasure of life." - Chinese Lady (Elderly R1)

"I love engaging in activities that bring productivity to my mind, body, and soul. Through these activities, I can enjoy a different environment, allowing me to feel refreshed."- Chinese lady (Elderly R2)

"If I can walk, I prefer walking and doing some gardening. I can do it either in the morning or the late evening, especially when the sun is not too harsh. Unfortunately, I cannot because I am in a wheelchair. Regardless, I noticed that the outdoor space in this care center is very limited, and there are restrictions for us to take a walk in the small garden." - Chinese Lady (Elderly R3)

Travel Distance and Wayfinding - Many of the respondents revealed difficulties walking long distances, and wayfinding is deemed essential to guide the elderly to the end point within the built environment. This finding aligns with the findings from the past researchers Mnea and Zairul (2023). The following are feedback from the respondents.

"I prefer having facilities that allow us to mark down reference points when maneuvering around. For example, having the location of my bed with the largest mural beside it serves as a landmark." - Chinese Lady (Elderly R2)

"The location for elderly activity space needs to be properly positioned within reachable distances that would not tire the elderly." - Elderly R1.

All the responses are summarized in the following tables.

Table 2. Summary of responses on category A - Coding group 1

Category						
Coding Group 1						
Themes	***Sub-Themes***	***R1***	***R2***	***R3***	***R4***	***R5***
Indoor Living Space	A1a– Living Condition	P	P	P	P	
	A1b – Space Quality	P	P	P	P	P
	A1c – Surrounding Ambience	P	P			

Table 3. Summary of responses on category A - Coding group 2

Category					
Coding Group 2					
Themes	**R1**	**R2**	**R3**	**R4**	**R5**
A2 – Outdoor Recreational Space	P	P	P	P	

Table 4. Summary of responses on category A - Coding group 3

Category					
Coding Group 3					
Themes	**R1**	**R2**	**R3**	**R4**	**R5**
A3 – Travel distance and wayfinding	P	P	P	P	

CONCLUSION

To ensure the wellbeing of the aging populations, one of the objectives is to determine future strategies for the development of elderly care center institutions. Feedback and reviews from published sources were analyzed to inform these strategies. The literature reviews consistently emphasized the importance of promoting active aging, which has the potential to improve the quality of life in care centers in Malaysia. The analysis and synthesis conducted concluded that incorporating design language into local standards and guidelines is crucial.

Various suggestions are provided to enhance the activeness of the elderly within care centers. One key recommendation is to incorporate architectural elements that facilitate communication and interaction, thereby stimulating the elderly's sense of active development. The literature review presented examples where interactive architectural languages are successfully integrated with physical planning. These examples included the application of sensory design, the creation of diverse community spaces, and the implementation of spatial planning strategies to encourage the active participation of the elderly. Such measures can help prevent the deterioration of social and mental health among the occupants.

RECOMMENDATION

The study has the potential to delve further into broader aspects related to the characteristics of the elderly, such as details of their financial background, diverse cultural upbringing, or distinct health environments. In future research, the survey may consider Malaysian elderly individuals staying in private care to understand their needs and perceptions toward the built environment of the care center. This could contribute to constructing a new development scheme, fostering a consciously designed care center according to the needs of the new generation of elderly.

REFERENCES

Aged Care. (2018). Conversation About Your Future Self and Care Management and Finance. AgedCare. www.agedcare.com.my/acgconcept/0.2017/july/caregiverpreneurs

Aini, A. M., Murni, N., & Abd Aziz, W. N. A. W. (2016). Housing Asipirations of the Elderly in Malaysia: A Comparison of Urban and Rural Areas. *Journal of Design and Built Environment*, 16(2).

Alcañiz, M., Riera-Prunera, M.-C., & Solé-Auró, A. (2020, April). Elderly in the Rural vs Urban Settings. *International Journal of Environmental Research and Public Health*, 17(7), 2442. 10.3390/ijerph1707244232260213

Asli Sungur, E. (2012). Design for the Elderly: Housing and Re-arrangements. Journal of Civil Engineering and Architecture. Yildiz Technical University, Turnkey.

Bryman, A., & Bell, E. (2007). *Business Research Methods* (2nd ed.). Oxford University Press.

Clegg, A., Young, J., Iliffe, S., Rikkert, M. O., & Rockwood, K. (2013). Frailty in elderly people. *Lancet*, 381(9868), 752–762. 10.1016/S0140-6736(12)62167-923395245

Constructororghq. (n.d). *How to Achieve the Perfect Window Height from Floor*. The Constructor. https://theconstructor.org/building/building-tips/how-to-achieve-the-perfect-window-height-from-floor/569392/?amp=1#top

Creative Safety Supply. (2022). *The meaning behind OSHA's safety colors*. CSS. https://www.creativesafetysupply.com/articles/safety-colors/

Creswell, J. W. (2003). *Research Design: Qualitative, Quantitative, and Mixed Method Approaches* (3rd ed.). Sage Publibacations, Inc.

Department of Statistic Malaysia Official Portal. (2015). *Total Fertility Rate*. Malaysia Economic Statistic Time Series.

Department of Statistics Malaysia. (2020). *Current Population Estimates, Malaysia, 2020*. DOSM. https://www.dosm.gov.my/portal-main/release-content/current-population-estimates-malaysia-2020

Easterby-Smith, M., Thorpe, R., & Lowe, A. (1991). Qualitative methods. Management Research: An Introduction.

Farage, M., Miller, K. W., Ajayi, F., & Hutchins, D. (2012). Design principles to accommodate older adults. *Global Journal of Health Science*, 4(2), 2–25. 10.5539/gjhs.v4n2p222980147

Hall, E., Chai, W. and Albrecht, J.A. (2016), "A qualitative phenomenological exploration of teacher's experience with nutrition education". American Journal of Health Education, Vol. 47 No. 3, pp. 136-148, dou: .10.1080/19325037.2016.1157532

Ishihara, Keiko & Ishihara, Shigekazu & Nagamachi, Mitsuo & Hiramatsu, Sugaru & Osaki, Hirokazu. (2001). Age-related decline in color perception and difficulties with daily activities–measurement, questionnaire, optical and computer-graphics simulation studies. International Journal of Industrial Ergonomics - INT J IND ERGONOMIC. 28. 153-163. .10.1016/S0169-8141(01)00028-2

Khoo, D. (2018). Medicine Suppliers to Benefit from higher healthcare location. *The Star*.

Kinsella, K., & Wan, H. (2009). *An Aging World. US Census Bureau, International Population Reports, P95/09-1*. Washington DC: US Government Printing Office..

Lim, C. K. (2012). The Changing Needs of Older Malaysians: A Selangor Case Study. [Doctoral Dissertation, Adelaide: University of Adelaide].

Lunenfeld, B., & Stratton, P. (2013). The Clinical Consequence of An Aging World and Preventive Strategies. *Best Practice & Research. Clinical Obstetrics & Gynaecology*, 27(5), 643–659. 10.1016/j.bpobgyn.2013.02.00523541823

Mafauzy. (2000). The Problems and Challenges of the Aging Population of Malaysia. *Malaysian Journal of Medical Sciences 7*(1), 1-3.

Mayo Clinic. (2019). *Aging: What to Expect*. Mayo Clinic. https://www.mayoclinic.org/healthy-lifestyle/healthy-aging /in-depth/aging/art-29946979.

McMaster University. (2018). *Addressing The Social Needs of Older Adults: A Contributing Factor to Their Health and Wellbeing*. Optimal Aging Portal.

Meher, S., & Singh, Z. (2021). Universal Design and All-Age Friendly Planning. 10.13140/RG.2.2.13889.07523

Meriam Syed Akil, S., & Abdullah, S. (2014). Challenges in managing elderly care centres in Malaysia. *The International Journal of the Arts in Society*, 07(3), 1944–6934.

Merriam, S. B. (2009). *Qualitative Research: A Guide to Design and Implementation*. Jossey-Bass.

Mnea, A., & Zairul, M. (2023). Evaluating the Impact of Housing Interior Design on Elderly Independence and Activity: A Thematic Review. *Buildings*, 13(4), 1099. 10.3390/buildings13041099

Murphy, K., O'Shea, E., Cooney, A., Shiel, A., & Hodgins, M. (2006). *Improving quality of life for older people in long-stay care settings in Ireland Dublin*. National Council on Aging and Older People.

Norani, H. (2020). Age Friendly – Cities and Communities. *12th Asean & Japan High Level Officials Meeting on Caring Societies*. Ministry of Woman, Family and Community Development. https://www .mhlw.go.jp/file/06-Sesiakujouhou-10500000- Daijinkanboukokusiaka/01-12_Malaysia.pdf

Nuradzimmah, D. (2016). *Malaysia to become "aging notion" by 2035*. NST. https://www.nst.com.my/news/2016/03/134562/malaysia-become-aging nation-2035.

Nurul Aini B. (2018). Assessment of Health Status of The Elderly and Pre-elderly at A Malaysia Elderly Care Centre. *International Journal of Integrated Engineering 10*(7).

Park, N. S., Zimmerman, S., Sloane, P. D., Gruber-Baldini, A. L., & Kevein Eckert, J. (2006, May). "An Empirical Typology of Residential Care/Assited Livng Based on A Four-State-Study," The Gerontological Society of America. *The Gerontologist*, 46(2), 238–248. 10.1093/geront/46.2.23816581888

Parker, C., Barnes, S., McKee, K., Morgan, K., Torrington, J., & Tregenza, P. (2004). Quality of Life and Building Design in Residential and Nursing Homes for Older People. *Ageing and Society*, 24(6), 941–962. 10.1017/S0144686X04002387

Pekalee, A., & Gray, R. S. (2023). In-home environment and happiness among older adults in Thailand. *Frontiers in Public Health*, 11, 1159350. 10.3389/fpubh.2023.115935037077188

Phua, K. H., Yap, M. T., & Goh, L. G. (2019). *Aging in Asia: Contemporary Trends and Policy Issues.* ebcHost. http://search.ebscohost.com/login.aspx?direct=true&scope+site&db=nlebk&db=nlabk&AN=2158050

Richard, W. (2019). *Physical Changes with Aging.* Warren Alpert Medical School of Brown University.

Saldana, J., & Omasta, M. (2016). *Qualitative Research: Analyzing Life*. Sage.

ScienceDaily. (2018). *Getting lost: Why older people might lose their way*. ScienceDaily. https://www.sciencedaily.com/releases/2018/03/180315140715.htm

Seamon, D., & Sowers, J. (2008). Place and Placelessness, Edward Relph. *Key Texts in Human Geography, London.*, (January), 43–51. 10.4135/9781446213742.n6

Sugiyama, T., & Thompson, C. W. (2006). Environmental support for outdoor activities and older people's quality of life. *Journal of Housing for the Elderly*, 19(3-4), 167–185. 10.1300/J081v19n03_09

The Centre for Aging. (2011). *Aging: Age-Related Physical Changes.* College of Nursing, University of Cincinnati. https://nursing.uc.edu/content/dam/nursing/docs/CFAWD/Aging %20Series/Part%202%Aging%20 Changes.pdf.

The Peninsular. (2021). *Why do older adults struggle to adapt to new environments?* The Peninsular. https://thepeninsulaqatar.com/article/27/04/2016/Why-do-older-adults-struggle-to-adapt-to-new-environments

Van Beek, J. H. G. M., Kirkwood, T. B. L., & Bassingthwaighte, J. B. (2016). Understanding the Physiology of the Aging Individual: Computational Modelling of Changes in Metabolism and Endurance. *Interface Focus*, 6(2), 20150079. 10.1098/rsfs.2015.007927051508

World Health Organisation. (2020). *Health Statistics and Information Systems*. WHO. https://www.int/healthinfo/survey/aging defnolder/en/. Accessed on October 6 2020.

Compilation of References

Abanda, F. H., Vidalakis, C., Oti, A. H., & Tah, J. H. M. (2015). A critical analysis of Building Information Modelling systems used in construction projects. *Advances in Engineering Software*, 90, 183–201. 10.1016/j.advengsoft.2015.08.009

Abbasi, S., Taghizade, K., & Noorzai, E. (2020). BIM-based combination of takt time and discrete event simulation for implementing just in time in construction scheduling under constraints. *Journal of Construction Engineering and Management*, 146(12), 04020143. 10.1061/(ASCE)CO.1943-7862.0001940

Abdollahi, K. K., Ning, Z. H., & Appeaning, A. (2000). Gulf Coast Regional Climate Change Council. In *Global Climate Change & the Urban Forest*. Franklin Press.

Abdrabo, K. I., Kantosh, S. A., Saber, M., Sumi, T., Elleithy, D., Habiba, O. M., & Alboshy, B. (2022). The Role of Urban Planning and Landscape Tools Concerning Flash Flood Risk Reduction Within Arid and Semiarid Regions. In Sumi, T., Kantoush, S. A., & Saber, M. (Eds.), *Wadi Flash Floods. Natural Disaster Science and Mitigation Engineering: DPRI reports*. Springer. 10.1007/978-981-16-2904-4_11

Abdullah, N. R. W., Rosli, N. A., & Ramasamy, R. (2021). Private Higher Education Institutions (PHEIs) in Malaysia: An Assessment of Service Quality and Students' Satisfaction. *International Journal of Management, Accounting, Governance and Education, 1*(2).

Abergel, T., Dean, B., & Dulac, J. (2017). *Towards a Zero-Emission, Efficient, and Resilient Buildings and Construction Sector: Global Status Report 2017*. UN Environment and International Energy Agency.

Aboulnaga, M., Puma, P., Eletrby, D., Bayomi, M., & Farid, M. (2022). Sustainability Assessment of the National Museum of Egyptian Civilization (NMEC): Environmental, Social, Economic, and Cultural Analysis. *Sustainability, 14*(20), 13080. Q3. 10.3390/su142013080

Abu-Zeid, M., El-Tawil, S., & Kandil, A. (2020). Analyzing the impact of data quality on the effectiveness of building information 6odelling for construction projects. *Journal of Construction Engineering and Management*, 146(3), 04019110.

Accordino, J., & Johnson, G. T. (2000). Addressing the vacant and abandoned property problem. *Journal of Urban Affairs*, 22(3), 301–315. 10.1111/0735-2166.00058

Ackerman, K. (2012). *The Potential for Urban Agriculture in New York City: Growing Capacity, Food Security, & Green Infrastructure*. Urban Design Lab at the Columbia Earth Institute.

Acquah, I. S. K., Naude, M. J., & Soni, S. (2021). How the dimensions of culture influence supply chain collaboration: An explanatory sequential mixed-methods investigation. *Revista de Gestao*, 28(3), 241–262. 10.1108/REGE-11-2020-0105

Adams, E. (n.d.). Towards Sustainability Indicators For Museums In Australia.

Aged Care. (2018). Conversation About Your Future Self and Care Management and Finance. AgedCare. www.agedcare .com.my/acgconcept/0.2017/july/caregiverpreneurs

Aini, A. M., Murni, N., & Abd Aziz, W. N. A. W. (2016). Housing Asipirations of the Elderly in Malaysia: A Comparison of Urban and Rural Areas. *Journal of Design and Built Environment*, 16(2).

Akbarian, H., Gheibi, M., Hajiaghaei-Keshteli, M., & Rahamani, M. (2022). A hybrid novel framework for flood disaster risk control in developing countries based on smart prediction systems and prioritize scenarios. *Journal of Environmental Management*, 312, 114939. 10.1016/j.jenvman.2022.11493935338986

Akboy-İlk, S. (2017). Crafting the architectural measured drawings. *The Plant Journal*, 2(1), 39–61.

Akin, S., Ergun, O., Surer, E., & Dino, I. G. (2022). An immersive performative architectural design tool with daylighting sim-ulations: A building information modeling-based approach. *Engineering, Construction, and Architectural Management*, 28(4), 1319–1344. 10.1108/ECAM-07-2020-0562

Al Husban, S. A. M., Al Husban, A. A. S., & Al Betawi, Y. (2021). The Impact of the Cultural Beliefs on Forming and Designing Spatial Organizations, Spaces Hierarchy, and Privacy of Detached Houses and Apartments in Jordan. *Space and Culture*, 24(1), 66–82. 10.1177/1206331218791934

Al Mehedi, M.-A., Smith, V., Hosseiny, H., & Jiao, X. (2022). Unravelling the complexities of urban fluvial flood hydraulics through AI. *Scientific Reports*, 12(18738), 1–18. 10.1038/s41598-022-23214-9

Al Nasiri, N., Al-Awadhi, T., Hereher, M., Ahsan, R., & AlRubkhi, A. G. (2020). Changing Urban Ecology a Challenge for Coastal Urban Resilience: A Study on Muscat. *Environment & Urbanization Asia*, 11(1), 10–28. 10.1177/0975425320906307

Al-Ashmori, Y. Y., Othman, I., Rahmawati, Y., Amran, Y. M., Sabah, S. A., Rafindadi, A. D. U., & Mikić, M. (2020). BIM benefits and its influence on the BIM implementation in Malaysia. *Ain Shams Engineering Journal*, 11(4), 1013–1019. 10.1016/j.asej.2020.02.002

Al-Badi, A., Tarhini, A., & Al-Kaaf, W. (2017). Financial Incentives for Adopting Cloud Computing in Higher Educational Institutions. *Asian Social Science*, 13(4), 162–174. 10.5539/ass.v13n4p162

Alberti, M., Booth, D., Hill, K., Coburn, B., Avolio, C., Coe, S., & Spirandelli, D. (2003). The impact of urban patterns on aquatic ecosystems: An empirical analysis in Puget lowland sub-basins. *Landscape and Urban Planning*, 64(3), 175–198.

Alcañiz, M., Riera-Prunera, M.-C., & Solé-Auró, A. (2020, April). Elderly in the Rural vs Urban Settings. *International Journal of Environmental Research and Public Health*, 17(7), 2442. 10.3390/ijerph1707244232260213

Alexander, F. S. (2005). *Land Bank Authorities: A Guide for the Creation and Operation of Local Land Banks*. Local Initiatives Support Corporation. http://www.nhc.org/media/files/Land_Bank_Authorities-Alexander.pdf

Alexandra, C., Daniell, K. A., Guillaume, J., Saraswat, C., & Feldman, H. R. (2023). Cyber-physical systems in water management and governance. *Current Opinion in Environmental Sustainability*, 62(101290), 1–11. 10.1016/j.cosust.2023.101290

Alhassan, N. U., Aede, H. M., Hassan, A. M., & Idris, A. M. (2020). *Effects of place-based and activity-based approaches in technical education, interest and retention.*

Ali, M., Ismail, K. M., Hashim, K. S. H. Y., Suhaimi, S., & Mustafa, M. H. (2018). Heritage Building Preservation through Building Information Modelling: Reviving Cultural Values through Level of Development Exploration. *Planning Malaysia Journal, 16*(6).

Aliasgar, K. (2012). *Developing a geoinformatics-based early warning system for floods in the Caribbean, Trinidad, and Tobago*. Southern Cross University.

Ali, M. M., Abas, N. H., Affandi, H. M., & Abas, N. A. (2018). Factors impeding the industrialized building system (IBS) implementation of building construction in malaysia. *International Journal of Engineering and Technology(UAE)*, 7(4), 2209–2212. 10.14419/ijet.v7i4.17863

Alkhater, N., Wills, G., & Walters, R. (2015, August). Factors affecting an organization's decision to adopt cloud services in Saudi Arabia. In 2015 3rd International Conference on Future Internet of Things and Cloud (pp. 553–557). IEEE, Rome, pp. 553–557.

Alkhater, N., Walters, R., & Wills, G. (2018). An empirical study of factors influencing cloud adoption among private sector organizations. *Telematics and Informatics*, 35(1), 38–54. 10.1016/j.tele.2017.09.017

Almeida, S. L. G. D. (2014). *Retrofitting and Refurbishment Processes of Heritage Buildings: Application to Three Case Studies*. [Doctoral Dissertation].

Al-Obaidi, K. M., Wei, S. L., Ismail, M. A., & Kam, K. J. (2017). Sustainable Building Assessment of Colonial Shop-houses After Adaptive Reuse in Kuala Lumpur. *Buildings*, 7(4), 87. 10.3390/buildings7040087

Alok, S., & Mishra, S. B. (2017). *Handbook of Research Methodology*. Educreation Publishing. 10.1097/0000346 5-199001000-00018

Alshamaila, Y., Papagiannidis, S., & Li, F. (2013). Cloud computing adoption by SMEs in the north east of England: A multi-perspective framework. *Journal of Enterprise Information Management*, 26(3), 250–275. 10.1108/17410391311325225

Alsumayt, A., El-haggar, N., Amouri, L., Alfawaer, Z. M., & Aljameel, S. S. (2023). Smart Flood Detection with AI and Blockchain Integration in Saudi Arabia Using Drones. *Sensors (Basel)*, 23(5148), 1–30. 10.3390/s2311514837299876

Alzarrad, M. A., Moynihan, G. P., Parajuli, A., & Mehra, M. (2021). 4D BIM Simulation Guideline for Construction Visualization and Analysis of Renovation Projects: A Case Study. *Frontiers in Built Environment*, 7, 617031. 10.3389/fbuil.2021.617031

Ancheh, K. S. B., Krishnan, A., & Nurtjahja, O. (2007). Evaluative criteria for selection of private universities and colleges in Malaysia. *Journal of International Management Studies*, 2(1), 1–11.

Andrews, D., Bedford, J., Blake, B., Cromwell, T., & Lea, R. (2010). *Measured and Drawn: Techniques and practice for the metric survey of historic buildings | Historic England* (J. Bedford & H. Papworth, Eds.). English Heritage.

Angel, S., Parent, J., Civco, D. L., Blei, A., & Potere, D. (2011). The dimensions of global urban expansion: Estimates and projections for all countries, 2000-2050. *Progress in Planning*, 75(2), 53–107. 10.1016/j.progress.2011.04.001

Angel, S., Sheppard, S., & Civco, D. (2005). *The dynamics of global urban expansion*. Transport and Urban Development Department, World Bank.

Arakpogun, E., Elsahn, Z., Olan, F., & Elsahn, F. (2021). Artificial intelligence in Africa: challenges and opportunities. In *The Fourth Industrial Revolution: Implementation of Artificial Intelligence for Growing Business Success. Studies in Computational Intelligence* (pp. 375–388). Springer., 10.1007/978-3-030-62796-6_22

Ariffin, H. L. B. T., Lynn, B. Y. H., Shukery, N. B. M., Rahiman, N. B. A., Mahmud, S. H. B., & Raslim, F. B. M. (2018). Innovative procurement adoption for industrialised building system (IBS) projects. *International Journal of Engineering and Technology(UAE), 7*(29), 887–892. 10.14419/ijet.v7i2.29.14277

Ariyanto, S. R., & Muslim, S. (2019). Collaborative Problem-Based Learning Models Implementation in Vocational High Schools. *1st Vocational Education International Conference (VEIC 2019)*, (pp. 238–245). IEEE. 10.2991/as-sehr.k.191217.039

Arkle, J., Owen, L., & Weber, J. (2017). Trinidad and Tobago. In Allen, C. (Ed.), *Landscapes and Landforms of the Lesser Antilles. World Geomorphological Landscapes* (pp. 267–291). Springer., 10.1007/978-3-319-55787-8_17

Armayuda, E.-, Lin, P. H., & Lin, R. (2020). Sustainability in Transforming Cultural Feature Into Cultural Creative Product Design (Case Study of Malangan Mask). *Journal of Arts and Humanities*, 9(3), 3. 10.18533/journal.v9i3.1842

Arnaud, L. (2021). Trinidad & Tobago's Mangrove Ecosystems suffocating in plastic waste: Recommendations moving forward. *Academic Letters*, 3745, 1–9. 10.20935/AL3745

Arnold, M. (2022). Problem-based Learning and Community Engagement: A Service-Learning Project with Social Pedagogues. In Fahr, U., Alessandra, K., Angenent, H., & Eßer-Lüghausen, A. (Eds.), *Hochschullehre erforschen* (pp. 199–214). Springer Fachmedien Wiesbaden., 10.1007/978-3-658-34185-5_11

ASHRAE. (2019). *ASHRAE terminology. ASHRAE searchable terminology tool. American Association of Heating, Refrigeration and Air-conditioning Engineers*. ASHRAE.

Ashworth, A., & Perera, S. (2015). Cost Modelling. In *Cost Studies of Buildings* (6th ed.). Routledge. 10.4324/9781315708867

Asli Sungur, E. (2012). Design for the Elderly: Housing and Re-arrangements. Journal of Civil Engineering and Architecture. Yildiz Technical University, Turnkey.

Attwood, S., & Hajat, C. (2020). How will the COVID-19 pandemic shape the future of meat consumption? *Public Health Nutrition*, 23(17), 3116–3120. 10.1017/S136898002000316X32782062

Audit Commission. (1993). *Unfinished Business. Fulltime education courses for 16-19 year olds*. HMSO.

Aung, T. H. S. (2022). BIM for Engineering: How and why engineers should use BIM for better project delivery. *Oneistox*. https://www.oneistox.com/blog/bim-for- engineering

Awang, Y. A. (2015, January). *Students' Factors Preference Of Choosing Private University In Sarawak, Malaysia. Swinburne University Technology, Sarawak, Malaysia. Proceedings of the Asia Pacific Conference on Business and Social Sciences 2015*, Kuala Lumpur. ISBN 978-0-9925622-2-9.

Azhar, S., Khalfan, M., & Maqsood, T. (2012). Building information 6odelling (BIM): A new paradigm for visual interactive 6odelling and simulation for construction projects. In *Handbook of research on building information 6odelling and construction informatics* (pp. 135-159). IGI Global.

Azis, I. A. A., Khaidzir, K. A. M., & Rashid, A. M. (2022). Identifying Guidelines Limitation of Historic Interior in Adaptive Re-Use of Heritage Building in Malaysia. *Journal of Tourism Hospitality and Environment Management*, 7(27), 69–76. 10.35631/JTHEM.727006

B, J. C., & B, M. Z. (2023). Application of Aging-Appropriate Smart Home. In *Proceedings of the 2023 2nd International Conference on Social Sciences and Humanities and Arts (SSHA 2023)*. Atlantis Press SARL. 10.2991/978-2-38476-062-6

Baharum, M. R., Kasim, N., & Taib, M. R. (2019). Barriers to Building Information Modeling (BIM) Implementation in Malaysia. *Journal of Construction in Developing Countries*, 24(1), 67–84.

Bakhtiari, V., Piadeh, F., Behzadian, K., & Kapelan, Z. (2023). A critical review of the application of cutting-edge digital visualisation technologies for effective urban flood risk management. *Sustainable Cities and Society*, 99(104958), 1–14. 10.1016/j.scs.2023.104958

Balaras, C. A., Dascalaki, E., & Kontoyiannidis, S. (2004). Decision support software for sustainable building refurbishment. *Transactions-American Society of Heating Refrigerating and air Conditioning Engineers, 110*(1), 592-601.

Balasbaneh, A. T., & Bin Marsono, A. K. (2017). Strategies for reducing greenhouse gas emissions from residential sector by proposing new building structures in hot and humid climatic conditions. *Building and Environment*, 124, 357–368. 10.1016/j.buildenv.2017.08.025

Banerjee, H. K., & Graaff, E. D. (1996). Problem-based learning in architecture: Problems of integration of technical disciplines. *European Journal of Engineering Education*, 21(2), 185–195. 10.1080/03043799608923402

Baró, F., Chaparro, L., Gómez-Baggethun, E., Langemeyer, J., Nowak, D. J., & Terradas, J. (2014). *Contribution of ecosystem services to air quality and climate change mitigation policies: The case of urban forests in Barcelona*, Spain. *Ambio*, 43(4), 466–479. 10.1007/s13280-014-0507-x24740618

Barrett, J., Birch, R., Cherrett, N., & Wiedmann, T. (2005). Exploring the application of the ecological footprint to sustainable consumption policy. *Journal of Environmental Policy and Planning*, 7(4), 303–316. 10.1080/15239080500441095

Barringer, H. R., Gardner, R. W., & Levin, M. J. (1993). *Asians and Pacific Islanders in the United States.* Russell Sage Foundation.

Bataglin, F. S., Viana, D. D., Formoso, C. T., & Bulhões, I. R. (2020). Model for planning and controlling the delivery and assembly of engineer-to-order prefabricated building systems: Exploring synergies between lean and BIM. *Canadian Journal of Civil Engineering*, 47(2), 165–177. 10.1139/cjce-2018-0462

Batty, M. (2013). *The new science of cities.* MIT Press. 10.7551/mitpress/9399.001.0001

Batty, M., Besussi, E., & Chin, N. (2012). Traffic evolution: Exploring the historical morphology of London's street network. *Environment and Planning. B, Planning & Design*, 39(5), 838–860.

Batty, M., & Marshall, S. (2019). *Cities, Planning and Simulation.* Routledge.

Bazan, G. (1997). Our Ecological Footprint: Reducing human impact on the earth. *Electronic Green Journal*, 1(7). Advance online publication. 10.5070/G31710273

Beijing Municipal Housing and Urban-Rural Development Commission. (2021). *Notice on the issuance of the "Beijing green building creation action implementation plan (2020-2022)".* Beijing Municipal Housing and Urban-Rural Development Commission. https://zjw.beijing.gov.cn/bjjs/gcjs/jzjnyjcjg/tzgg/tzgg98/325777657/index.shtml

Benenson, I., & Torrens, P. M. (2004). *Geosimulation: Automata-based modeling of urban phenomena.* John Wiley & Sons. 10.1002/0470020997

Bergh, J. C. J. M. Van Den, & Verbruggen, H. (1999). *Spatial Sustainability, Trade and Indicators: An Evaluation of the 'Ecological Footprint.'* EconPapers.

Bergmark, U., & Westman, S. (2018). Student participation within teacher education: Emphasising democratic values, engagement and learning for a future profession. *Higher Education Research & Development*, 37(7), 1352–1365. 10.1080/07294360.2018.1484708

Berke, P. R., Cooper, J., Aminto, M., Reams, M., & Lee, J. S. (2015). Local mitigation planning to reduce social vulnerability: Southeast Florida's regional climate change compact. *Mitigation and Adaptation Strategies for Global Change*, 20(6), 991–1013.

Bhattacharjee, S., & Sattar, S. (2021). Pandemic And Urban Sustainability: Analyzing The Covid-19 Scenario In Mumbai, India. Geography, Environment. *Sustainability*, 14(1), 17–24. 10.24057/2071-9388-2020-189

Bigelow, J. D. (2004). Using problem-based learning to develop skills in solving unstructured problems. *Journal of Management Education*, 28(5), 591–609. 10.1177/1052562903257310

Blakeman, C., Brown, B., Fitzpatrick, B. T., Shaw, I., & Williamson, A. (2008). City-Wide Brownfield Redevelopment Plan; *Report Prepared for the City of Roanoke*. City Council: Roanoke, VA.

Blakeman, C., Brown, B., Fitzpatrick, B. T., Shaw, I., & Williamson, A. (2008). *City-Wide Brownfield Redevelopment Plan*. City Council.

Blanco, H., McCarney, P., Parnell, S., Schmidt, M., & Seto, K. (2011). The role of urban land in climate change. In Rozensweig, C., Solecki, W., & Hammer, S. (Eds.), *First Urban Climate Change Research Network (UCCRN) assessment report on climate change in cities*. Cambridge University Press. 10.1017/CBO9780511783142.014

Bongaarts, J. (2001). Household size and composition in the developing world in the 1990s. *Population Studies*, 55(3), 263–279. 10.1080/00324720127697

Bonham, J. B., Spilka, G., & Rastorfer, D. (2002). *Old Cities/Green Cities: Planning Advisory Service Report Number 506/507*. American Planning Association: Chicago, IL, USA.

Bonham, J. B.Jr, & Spilka, G. (1995). *Urban Vacant Land: Issues and Recommendations*. Pennsylvania Horticultural Society.

Boon, M., Orozco, M., & Sivakumar, K. (2022). Epistemological and educational issues in teaching practice-oriented scientific research: Roles for philosophers of science. *European Journal for Philosophy of Science*, 12(1), 16. 10.1007/s13194-022-00447-z

Borucke, M., Moore, D., Cranston, G., Gracey, K., Iha, K., Larson, J., Lazarus, E., Morales, J. C., Wackernagel, M., & Galli, A. (2013). Accounting for demand and supply of the biosphere's regenerative capacity: The National Footprint Accounts' underlying methodology and framework. *Ecological Indicators*, 24, 518–533. 10.1016/j.ecolind.2012.08.005

Bosupeng, M. (2017). Is China's target of a 40-45% reduction in carbon dioxide emission plausible? *Environmental & Socio-Economic Studies*, 5(1), 46–50. 10.1515/environ-2017-0005

Bowman, A. O. M., & Pagano, M. A. (2004). *Terra Incognita: Vacant Land and Urban Strategies*. Georgetown University Press.

Branas, C. C., Cheney, R. A., MacDonald, J. M., Tam, V. W., Jackson, T. D., & Ten Have, T. R. (2011). A difference in differences analysis of health, safety, and greening vacant urban space. *American Journal of Epidemiology*, 174(11), 1296–1306. 10.1093/aje/kwr27322079788

Brandes, D. (1995). The flora of old town centers in Europe. In Sukoop, H., Numata, M., & Huber, A. (Eds.), *Urban Ecology as the Basis for Urban Planning*. SPB Academic Publishing.

Brinkman, J., & Mangum, K. (2020). *The Geography of Travel Behavior in the Early Phase of the Covid-19 Pandemic*. SSRN. https://papers.ssrn.com/abstract=3717188

Brookhuis, B., & Hein, L. (2015). The value of the flood control service of tropical forests: A case study for Trinidad. *Forest Policy and Economics*, 62, 118–124. 10.1016/j.forpol.2015.10.002

Brown, S., & Vacca, F. (2022). Cultural sustainability in fashion: Reflections on craft and sustainable development models. *Sustainability: Science, Practice and Policy, 18,* 590–600. N/A. 10.1080/15487733.2022.2100102

Bryman, A., & Bell, E. (2007). *Business Research Methods* (2nd ed.). Oxford University Press.

BSI. (2017). *Conservation of Cultural Heritage*. Guidelines for Improving the Energy Performance of Historic Buildings. BSI Standards Publication.

Büchmann-Slorup, R., & Andersson, N. (2010). BIM-based scheduling of Construction–A comparative analysis of prevailing and BIM-based scheduling processes. In *Proc., 27 th Int. Conf. of the CIB W78* (pp. 113-123). Research Gate.

Burhanuddin, M. A., Halawani, S. M., & Ahmad, A. R. (2011). A costing analysis for Decision Making Grid model in failure-based maintenance. *Advances in Decision Sciences*, 2011. 10.1155/2011/205039

Burkholder, S. (2012). The new ecology of vacancy: Rethinking land use in shrinking cities. *Sustainability (Basel)*, 4(6), 1154–1172. 10.3390/su4061154

Buyan & Cao, L. (2020). The road of R&D and thinking of educational cultural creativity in museums--Taking Shaanxi History Museum as an example. *Museum of China, 3,* 117-121.

Cabrera, A. F., & Nasa, S. M. L. (2000). Understanding the college-choice process. In A. F. Cabrera & S. M. L. Nasa (Eds.), *Understanding the college choice of disadvantaged students* (pp. 5-22). San Francisco: Jossey-Bass Publishers.

Camilleri, M. A., Cricelli, L., Mauriello, R., & Strazzullo, S. (2023). Consumer Perceptions of Sustainable Products: A Systematic Literature Review. *Sustainability, 15*, 8923. Q3. 10.3390/su15118923

Campagna, L. M., & Fiorito, F. (2022). On the Impact of Climate Change on Building Energy Consumptions: A Meta-Analysis. *Energies*, 15(1), 354. 10.3390/en15010354

Canisius, F., & Nancy, C. (2009). GeoFIS flood insurance system for Trinidad: A case study for San Juan downstream. *Urban and Regional Information Systems Association (URISA). Journal*, 21(1), 1–11.19331084

Cao, P., & Lei, X. (2023). Evaluating risk in prefabricated building construction under epc contracting using structural equation modeling: A case study of Shaanxi Province, china. *Buildings*, 13(6), 1465. 10.3390/buildings13061465

Carter, J. G., Cavan, G., Connelly, A., Guy, S., Handley, J., & Kazmierczak, A. (2015). Climate change and the city: Building capacity for urban adaptation. *Progress in Planning*, 95, 1–66. 10.1016/j.progress.2013.08.001

Cascini, G., O'Hare, J., Dekoninck, E., Becattini, N., Boujut, J.-F., Ben Guefrache, F., Carli, I., Caruso, G., Giunta, L., & Morosi, F. (2020). Exploring the use of AR technology for co-creative product and packaging design. *Computers in Industry*, 123, 103308. 10.1016/j.compind.2020.103308

Catalina, I. S. (2014). *Cloud computing - Impact on business.* [Master thesis, Aalborg University Copenhagen Denmark].

Central Committee of the Communist Party of China and the State Council. (2021). *Opinions on complete, accurate and comprehensive implementation of the new development concept and carbon implementation to peak carbon neutrality.* Central Committee of the Communist Party of China and the State Council. https://www.gov.cn/zhengce/202110/24/content_5644613.htm

Chand, S. (2014). Meetings: Types, Purpose, Advantages and Disadvantages. Your Article Library. https://www.yourarticlelibrary.com/businesscommunication/meetings-types-purpose-advantages-and-disadvantages/27671

Chandrasekaran, K. (2014). *Essentials of cloud computing.* CRC Press. 10.1201/b17805

Chang, C., Wu, X., & Yan, X. (2020). Multiobjective optimization of safety risk of prefabricated building construction considering risk correlation. *Mathematical Problems in Engineering*, 2020, 1–11. 10.1155/2020/3923486

Chang, Y., Li, X., Masanet, E., Zhang, L., Huang, Z., & Ries, R. (2018). Unlocking the green opportunity for prefabricated buildings and construction in china. *Resources, Conservation and Recycling*, 139, 259–261. 10.1016/j.resconrec.2018.08.025

Chan, N., Go, T. F., Moey, L. K., & Chia, C. M. (2022). Short Review on Renewable Energy Policy and Energy Consumption of Buildings in Malaysia. *Journal of Engineering & Technological Advances*, 7(1), 64–82.

Charlesraj, V. P. C., & Dinesh, T. (2020). Status of 4D BIM implementation in Indian construction. In *ISARC. Proceedings of the International Symposium on Automation and Robotics in Construction* (Vol. 37, pp. 199-206). IAARC Publications.

Che Husin, S. M. (2017). *Retrofitting Existing Building to Improve Energy Efficiency.*

Che Maznah, M. I., Fatin Najwa, M. N., Nur Kamaliah, M., Jeffery, L., Sahithi, A. S., & Preece, C. N. (2021). Sustainable Township and Sustainable Home: Public Perceptions. *Journal of the Society of Automotive Engineers Malaysia*, 5(3), 331–347. 10.56381/jsaem.v5i3.176

Che, M. R. (2022). *Research on incentive effect and optimization simulation of government policy on prefabricated buildings* [Master's thesis, Chongqing Jiaotong University]. CNKI. https://link.cnki.net/doi/10.27671/d.cnki.gcjtc.2022.000779doi:10.27671/d.cnki.gcjtc.2022.000779

Cheema, S. (2020, November 11). *Everything you need to know about Malaysia's Education Blueprint 2013-2025.* Mashable SE Asia. https://sea.mashable.com/culture/13175/everything-you-need-to-know-about-Malaysias-education-blueprint-2013-2025

Chen, B.-S. (2021). Research on cultural industry fund to support the development of cultural and creative industries in museums. *Cultural Industry*, 26, 159–161.

Chen, D., Deng, J., Cheng, B., Wang, Q., & Zhao, B. (2021). New anticracking glass-fiber-reinforced cement material and integrated composite technology with lightweight concrete panels. *Advances in Civil Engineering*, 2021, 1–17. 10.1155/2021/7447066

Cheng, X. D., & Wang, D. (2023). Research on the high-level development path of prefabricated buildings: csQCA analysis based on TOE framework. *Journal of Engineering Management*, (6), 1–6. 10.13991/j.cnki.jem.2023.06.005

Cheng, X., & Guo, Y. (2022). *A Quantitative Evaluation System of Yan'an Red Cultural and Creative Product Design Based on AHP-TOPSIS Method. Proceedings of the International Conference on Art Design and Digital Technology, ADDT 2022*, 16-18 September 2022, Nanjing, China. 10.4108/eai.16-9-2022.2324890

Chen, J., Li, Y., Zhang, C., Tian, Y., & Guo, Z. (2023). Urban Flooding Prediction Method Based on the Combination of LSTM Neural Network and Numerical Model. *International Journal of Environmental Research and Public Health*, 20(1043), 1–12. 10.3390/ijerph2002104336673799

Chen, S., Wang, S., & Xu, H. (2017). Influence of Place Identity on Residents' Attitudes to Dark Tourism. *Journal of China Tourism Research*, 13(4), 338–356. 10.1080/19388160.2017.1401023

Chen, X., & Yang, S. (2022). Research on sustainable design strategy of creative products from the perspective of cultural consumption. *Packaging Engineering*, 43(14), 320–325. 10.19554/j.cnki.1001-3563.2022.14.039

Chen, Y., Jiao, J., Bai, S., & Lindquist, J. (2020). Modeling the spatial factors of COVID-19 in New York City. https://ssrn.com/abstract=360671910.2139/ssrn.3606719

China's State Council. (2021). *Notice of the state council on issuing the action plan for carbon peaking before 2030.* CSC. https://www.gov.cn/zhengce/content/2021-10/26/content_5644984.htm

Chitwatkulsiri, D., & Miyamoto, H. (2023). Real-Time Urban Flood Forecasting Systems for Southeast Asia- A Review of Present Modelling and Its Future Prospects. *Water (Basel)*, 15(178), 1–18. 10.3390/w15010178

Cho, H. M., Yun, B. Y., Kim, Y. U., Yuk, H., & Kim, S. (2022). Integrated retrofit solutions for improving the energy performance of historic buildings through energy technology suitability analyses: Retrofit plan of wooden truss and masonry composite structure in Korea in the 1920s. *Renewable & Sustainable Energy Reviews*, 168, 11280. 10.1016/j.rser.2022.112800

Chris Corker and Sarah Holland (2005). Introducing students to employability, skills and reflection: A case study from history. *Student Engagement and Experience Journal, 4*(1). 10.7190/seej.v4i1.113

Chua, B. L., Al-Ansi, A., Lee, M. J., & Han, H. (2020). Tourists' outbound travel behavior in the aftermath of the COVID-19: Role of corporate social responsibility, response effort, and health prevention. *Journal of Sustainable Tourism*. 10.1080/09669582.2020.1849236

Chung, E. Y. (2019). Facilitating learning of community-based rehabilitation through problem-based learning in higher education. *BMC Medical Education*, 19(1), 433. 10.1186/s12909-019-1868-431752842

Ciriaci, D., & Muscio, A. (2011). University choice, research quality and graduates' employability: Evidence from Italian national survey data. *AlmaLaurea Working Papers, 49*, 1-14.

City of Pittsburgh. (2012). *Greening Resources*. https://www.pittsburghpa.gov/neighborhoodinitiatives/greenup/resources.htm

City of Portland. (2021). GIS data. https://www.portland.gov/bps/maps-and-gis/gis-data

Clarke, N., Kuipers, M., & Stroux, S. (2020). Embedding built heritage values in architectural design education. *International Journal of Technology and Design Education*, 30(5), 867–883. 10.1007/s10798-019-09534-4

Clegg, A., Young, J., Iliffe, S., Rikkert, M. O., & Rockwood, K. (2013). Frailty in elderly people. *Lancet*, 381(9868), 752–762. 10.1016/S0140-6736(12)62167-923395245

Climate Watch. (2022). *Historical GHG Emissions*. Climate Watch. https://www.climatewatchdata.org/ghg

Clyde, Z. L., Yu, Z., Hengqin, W., Zhe, C., Bing, X., & Vivian, W. Y. T. (2023). The application of BIM in the AECO industry. *Journal of Civil Engineering and Management*, 29(3), 202–222. 10.3846/jcem.2023.18076

Colding, J., & Barthel, S. (2019). Urban Greening. In Newton, R. W. (Ed.), *Water and Urban Development Paradigms: Towards an Integration of Engineering, Design and Management Approaches* (pp. 77–98). Springer.

Coleman, A. (1982). Dead space in the dying inner city. *The International Journal of Environmental Studies*, 19(2), 103–107. 10.1080/00207238208709976

Constructororghq. (n.d). *How to Achieve the Perfect Window Height from Floor*. The Constructor. https://theconstructor.org/building/building-tips/how-to-achieve-the-perfect-window-height-from-floor/569392/?amp=1#top

Corlett, R. T., Primack, R. B., Devictor, V., Maas, B., Goswami, V. R., Bates, A. E., Koh, L. P., Regan, T. J., Loyola, R., Pakeman, R. J., Cumming, G. S., Pidgeon, A., Johns, D., & Roth, R. (2020). Impacts of the coronavirus pandemic on biodiversity conservation. *Biological Conservation*, 246, 108571. 10.1016/j.biocon.2020.10857132292203

Costanza, R. (2000). The dynamics of the ecological footprint concept. *Ecological Economics*, 32(3), 341–345. 10.1016/S0921-8009(99)00150-0

Crauderueff, R., Margolis, S., & Tanikawa, S. (2012). *Greening Vacant Lots: Planning and Implementation Strategies*. The Nature Conservancy.

Creaney, R., Reid, L., & Currie, M. (2021). The Contribution of Healthcare Smart Homes to Older Peoples' Wellbeing: A New Conceptual Framework. *Wellbeing, Space and Society, 2*(June 2020), 100031. 10.1016/j.wss.2021.100031

Creative Safety Supply. (2022). *The meaning behind OSHA's safety colors*. CSS. https://www.creativesafetysupply.com/articles/safety-colors/

Creswell, J. W. (2014). The Selection of a Research Approach. *Research Design*, 3–23. https://doi.org/45593:01

Creswell, J. W. (2003). *Research Design: Qualitative, Quantitative, and Mixed Method Approaches* (3rd ed.). Sage Publibacations, Inc.

Cucinotta, D., & Vanelli, M. (2020). WHO declares COVID-19 a pandemic. In *Acta Biomedica, 91*(1). 10.23750/abm.v91i1.9397

Cui, J., Li, F., & Shi, Z. L. (2019). Origin and evolution of pathogenic coronaviruses. *Nature Reviews. Microbiology*, 17(3), 181–192. 10.1038/s41579-018-0118-930531947

Curto, R., Barreca, A., Coscia, C., Ferrando, D. G., Fregonara, E., & Rolando, D. (2021). The Active Role of Students, Teachers, and Stakeholders in Managing Economic and Cultural Value, Urban and Built Heritage. *The Interdisciplinary Journal of Problem-Based Learning*, 15(1), 1. 10.14434/ijpbl.v15i1.29626

Czmoch, I., & Pekala, A. (2014). Traditional Design versus BIM Based Design. *Traditional Design versus BIM Based Design*, 210–215. 10.1016/j.proeng.2014.12.048

Dabbour, E., & Sliuzas, R. V. (2013). Integrating system dynamics and cellular automata for urban simulation: Prospects and challenges. In *Proceedings of the 7th International Conference on Geographic Information Science (GIScience 2012)* (pp. 63-77). IEEE.

Daga, L. (2022). *4D BIM - Collaboration of Schedule with 3D Model.*

Dam, R. F., & Siang, T. Y. (2020, June 8). *10 Insightful Design Thinking Frameworks: A Quick Overview.* The Interaction Design Foundation. https://www.interaction-design.org/literature/article/design-thinking-a-quick-overview

Daniel, T. (2022). Rankings released! QS World University Rankings: Asia 2023. https://www.qs.com/rankings-released-qs-world-university-rankings-asia-2023/\

Darlington, A. (1969). *Ecology of Refuse Tips*. Heinemann Educational.

Dawadi, S., Shrestha, S., & Giri, R. A. (2021). Mixed-Methods Research: A Discussion on its Types, Challenges, and Criticisms. *Journal of Practical Studies in Education*, 2(2), 25–36. 10.46809/jpse.v2i2.20

De Merode, E., Smeets, R., & Westrik, C. (2004). Linking universal and local values: Managing a sustainable future for world heritage. *World Heritage Paper*, (13).

Deng, Q. J. (2024). Research on the application of BIM technology in the quality chain system of prefabricated buildings. *Journal of Beijing Industrial Vocational and Technical College*, 23(1), 1–5.

Department of Statistic Malaysia Official Portal. (2015). *Total Fertility Rate*. Malaysia Economic Statistic Time Series.

Department of Statistics Malaysia. (2020). *Current Population Estimates, Malaysia, 2020.* DOSM. https://www.dosm.gov.my/portal-main/release-content/current-population-estimates-malaysia-2020

Dewberry, E., & Sherwin, C. (2002). E., & Sherwin, C. (2002). Visioning Sustainability through Design. *Greener Management International*, 2002(37), 125–139. 10.9774/GLEAF.3062.2002.sp.00011

Dewiyana, E., Ibrahim, N., & Hajar, N. H. (2016). The Green Aspects of Adaptive Reuse of Hotel Penaga. *Procedia: Social and Behavioral Sciences*, 222, 631–643. 10.1016/j.sbspro.2016.05.220

Dey, A. K. (2001). Understanding and using context. *Personal and Ubiquitous Computing*, 5(1), 4–7. 10.1007/s007790170019

Di Renzo, L., Gualtieri, P., Cinelli, G., Bigioni, G., Soldati, L., Attinà, A., Bianco, F. F., Caparello, G., Camodeca, V., Carrano, E., Ferraro, S., Giannattasio, S., Leggeri, C., Rampello, T., Lo Presti, L., Tarsitano, M. G., & De Lorenzo, A. (2020). Psychological aspects and eating habits during covid-19 home confinement: Results of ehlc-covid-19 italian online survey. *Nutrients*, 12(7), 1–14. 10.3390/nu1207215232707724

Di Renzo, L., Gualtieri, P., Pivari, F., Soldati, L., Attinà, A., Cinelli, G., Cinelli, G., Leggeri, C., Caparello, G., Barrea, L., Scerbo, F., Esposito, E., & De Lorenzo, A. (2020). Eating habits and lifestyle changes during COVID-19 lockdown: An Italian survey. *Journal of Translational Medicine*, 18(1), 1–15. 10.1186/s12967-020-02399-532513197

Dias Pereira, L., Tavares, V., & Soares, N. (2021). Up-to-date challenges for the conservation, rehabilitation and energy retrofitting of higher education cultural heritage buildings. *Sustainability (Basel)*, 13(4), 2061. 10.3390/su13042061

Dietersdorfer, L., Efremova, V., Fernandez Agueda, B., Fleschurz, R., Mangialardi, G., Piscitelli, C., & Wolff, M. (2012). *Urban shrinkage and chances for adaptation to climate change.* Final report EU-COST action: TU0803.

Djabarouti, J., & O'Flaherty, C. (2019). Experiential learning with building craft in the architectural design studio: A pilot study exploring its implications for built heritage in the UK. *Thinking Skills and Creativity*, 32, 102–113. 10.1016/j.tsc.2019.05.003

Doberstein, B., Fitzgibbons, J., & Mitchell, C. (2019). Protect, accommodate, retreat, or avoid (PARA): Canadian community options for flood disaster risk reduction and flood resilience. *Natural Hazards*, 98(1), 31–50. 10.1007/s11069-018-3529-z

Douglas, J. (2006). *Building adaptation.* Routledge. 10.4324/9780080458519

Duan, Y., & Li, G. (2020). Analysis on the quality problems and preventive measures of prefabricated building construction. *Journal of Physics: Conference Series*, 1648(3), 032141. 10.1088/1742-6596/1648/3/032141

Du, Q., Hao, T., Huang, Y., & Yan, Y. (2022). Prefabrication decisions of the construction supply chain under government subsidies. *Environmental Science and Pollution Research International*, 29(39), 59127–59144. 10.1007/s11356-022-19861-035380327

Eadie, R., Browne, M., Odeyinka, H., McKeown, C., & McNiff, S. (2013). BIM implementation throughout the UK construction project lifecycle: *An analysis.Automation in Construction*, 36, 145–151. 10.1016/j.autcon.2013.09.001

Eames, M., Dixon, T., Lannon, S., Hunt, M., Laurentis, C., Marvin, S., Hodson, M., Guthrie, P., Georgiadou, M. (2014). *Retrofit 2050: Critical Challenges for Urban Transitions*. Springer. 978-1-899895-12-0.

Easterby-Smith, M., Thorpe, R., & Lowe, A. (1991). Qualitative methods. Management Research: An Introduction.

Eastman, C. M., Teicholz, P., Sacks, R., & Liston, K. (2011). *BIM handbook: A guide to building information 6odelling for owners, managers, designers, engineers, and contractors*. John Wiley & Sons.

Echegaray, F. (2021). What POST-COVID-19 lifestyles may look like? Identifying scenarios and their implications for sustainability. *Sustainable Production and Consumption*, 27, 567–574. 10.1016/j.spc.2021.01.02534722841

Echenique, M. H., Mingardo, G., & Jara-Diaz, S. R. (2012). On the income elasticity of car travel demand. *PLoS One*, 7(9), e43515.

Ekung, S., & Okonkwo, E. (2018). Trends Influences on Career Choice In Quantity Surveying And Its Implications. *International Journal of Education (IJE), 3*(1), March 2015.

Embaby, M. E. (2014). Heritage conservation and architectural education: "An educational methodology for design studios". *HBRC Journal*, 10(3), 339–350. 10.1016/j.hbrcj.2013.12.007

equityXdesign. (2019, January 30). Racism and inequity are products of design. They can be redesigned. *Medium*. https://medium.com/@equityXdesign/racism-and-inequity-are-products-of-design-they-can-be-redesigned-12188363cc6a

Eroğlu, H. (2020). Effects of Covid-19 outbreak on environment and renewable energy sector. *Environment, Development and Sustainability*, 0123456789. 10.1007/s10668-020-00837-432837274

European Commission. (2013). *Building a green infrastructure for Europe*. European Commission.

European Commission. (2016). *Digitizing European Industry—Reaping the Full Benefits of a Digital Single Market COM (2016) 180 Final*. EC.

European Environment Agency (EEA). (2012). *Urban adaptation to climate change in Europe: Challenges and opportunities for cities together with supportive national and European policies*. Office for Official Publications of the European Union.

Farage, M., Miller, K. W., Ajayi, F., & Hutchins, D. (2012). Design principles to accommodate older adults. *Global Journal of Health Science*, 4(2), 2–25. 10.5539/gjhs.v4n2p222980147

Federal Emergency Management Agency (FEMA). (2017). *Homeland Security Exercise and Evaluation Program (HSEEP) overview*. FEMA. https://www.fema.gov/sites/default/files/2020-07/2020_hseep_overview.pdf

Feng, Y. (2021). Research on visualization of prefabricated building process based on BIM technology. *ACM International Conference Proceeding Series*, (pp. 307–312). ACM. 10.1145/3510858.3510955

Ferdous, W., Bai, Y., Ngo, T. D., Manalo, A., & Mendis, P. (2019). New advancements, challenges and opportunities of multi-storey modular buildings – a state-of-the-art review. *Engineering Structures*, 183, 883–893. 10.1016/j.engstruct.2019.01.061

Fernandez, J. L. (2010). An Exploratory Study Of Factors Influencing The Decision Of Students To Study At Universiti Sains Malaysia. Universiti Sains Malaysia, Penang. *Kajian Malaysia*, 28(2), 2010.

Ferry and Brandon. (2007). *Ferry and Brandon's Cost Planning of Buildings, 8th Edition*. Wiley. https://www.wiley.com/en-us/Ferry+and+Brandon%27s+Cost+Planning+of+Buildings%2C+8th+Edition-p-9781405130707

Fichinelli, A. C., Yigitcanlar, T., Sabtini-Marques, J., Cortese, T. T. P., Sotto, D., & Libardi, B. (2023). Urban Smartness and City Performance: Identifying Brazilian Smart Cities through a Novel Approach. *Sustainability (Basel)*, 15(10323), 1–24. 10.3390/su151310323

Filippi, M. (2015). Remarks on the Green Retrofitting of Historic Buildings in Italy. *Energy and Building*, 95, 15–22. 10.1016/j.enbuild.2014.11.001

Fohrer, N., Haverkamp, S., Eckhardt, K., & Frede, H. G. (2001). Hydrologic Response to Land Use Changes on the Watershed Scale. *Hydrology. Oceans and Atmosphere*, 26(7-8), 577–582.

Fontes de Meira, L., & Phillips, W. (2019). An economic analysis of flooding in the Caribbean. The case of Jamaica and Trinidad and Tobago, *Studies, and Perspectives series-ECLAC subregional headquarters for the Caribbean, 78*. Santiago: Economic Commission for Latin America and the Caribbean (ECLAC).

Foo, C. H. (2018). The Impact of Capital Contributions and Compliance Costs on Housing Affordability. *Building and Investment*, (August), 34–38.

Frenette, E., Bahn, O., & Vaillancourt, K. (2017). Meat, Dairy and Climate Change: Assessing the Long-Term Mitigation Potential of Alternative Agri-Food Consumption Patterns in Canada. *Environmental Modeling and Assessment*, 22(1), 1–16. 10.1007/s10666-016-9522-6

Fu, G., Jin, Y., Sun, S., Yuan, Z., & Butler, D. (2022). The role of deep learning in urban water management: A critical review. *Water Research*, 223(118973), 1–16. 10.1016/j.watres.2022.11897335988335

Gao, Y. (2023). Research progress on digitalization of prefabricated construction industry. *Chinese Building Metal Structure*, 22(11), 140–142. 10.20080/j.cnki.ISSN1671-3362.2023.11.046

Geertman, S., & Stillwell, J. (2009). Planning support systems: An inventory of current practice. *Computers, Environment and Urban Systems*, 33(3), 189–199.

Gehl, J. (2010). *Cities for people.* Island Press.

Gehner, E. (2009). Real Estate Development Strategies and Their Impact on the Risk Profile of a Project. *CIB Joint International Symposium*, (pp. 853–863). IEEE.

General Office of the CPC Central Committee and the General Office of the State Council. (2021). *Opinions on promoting green development of urban and rural construction.* CPC. https://www.gov.cn/zhengce/2021-10/21/content_5644083.htm

Georgetown. (2021). Rancangan Kawasan Khas Tapak Warisan Dunia UNESCO: Jilid 2 Kawalan Pembangunan.

GeoSLAM. (2021). What is BIM and how does it impact the construction industry? – *GeoSLAM.* https://geoslam.com/blog/2021/12/23/what-is-bim- and-how- does-it-impact-the-construction- industry/

Gerardi, J. (2021). *The 6 Stages Of Construction – Key Phases Explained.* ProEst. https://proest.com/construction/process/stages-of- construction/

Ghaffarianhoseini, A., Tookey, J., Ghaffarianhoseini, A., Naismith, N., Azhar, S., Efimova, O., & Raahemifar, K. (2017). Building Information Modelling (BIM) uptake: Clear benefits, understanding its implementation, risks and challenges. *Renewable & Sustainable Energy Reviews*, 75, 1046–1053. 10.1016/j.rser.2016.11.083

Ghobakhloo, M. (2020). Industry 4.0, digitization, and opportunities for sustainability. *Journal of Cleaner Production*, 252(119869), 1–21. 10.1016/j.jclepro.2019.119869

Giffinger, R., & Gudrun, H. (2010). Smart Cities Ranking: An Effective Instrument for the Positioning of the Cities? *Journal of the Centre of Land Policy and Valuation*, 4(12), 7–25. 10.5821/ace.v4i12.2483

Gill, S., Handley, J., Ennos, A., & Pauleit, S. (2007). Adapting cities for climate change: The role of the green infrastructure. *Built Environment*, 33(1), 115–133. 10.2148/benv.33.1.115

Giretti, A., Carbonari, A., Novembri, G., & Robuffo, F. (2012). Estimation of job-site work progress through on-site monitoring. In *Proceedings of the 29th International Symposium of Automation and Robotics in Construction, ISARC 2012*, Kochi,India 10.22260/ISARC2012/0089

Gledson, B. J., & Greenwood, D. J. (2016). Surveying the extent and use of 4D BIM in the UK. *Journal of Information Technology in Construction*, 21, 57–71.

Global Forest Watch (GFW). (2023). *Global Forestry Watch.* GFW. https://www.globalforestwatch.org/

Goode, S., & Stevens, K. (2000). An analysis of the business characteristics of adopters and non-adopters of World Wide Web technology. *Information Technology and Management*, 1(1), 129–154. 10.1023/A:1019112722593

Grant, M. J., & Booth, A. (2009). A typology of reviews: An analysis of 14 review types and associated methodologies. *Health Information and Libraries Journal*, 26(2), 91–108. 10.1111/j.1471-1842.2009.00848.x19490148

Grazi, F., & van den Bergh, J. C. J. M. (2008). Spatial organization, transport, and climate change: Comparing instruments of spatial planning and policy. *Ecological Economics*, 67(4), 630–639. https://www.sciencedirect.com/science/article/pii/S0921800908000426. 10.1016/j.ecolecon.2008.01.014

Greene, D. L., & Schafer, A. (2003). *Reducing greenhouse gas emissions from U.S transportation.* Pew Center on Global Climate Change.

Greene, J. H., & D'Oliveira, M. (1982). *Learning to use statistical tests in psychology: A student's guide*. Open University Press.

Grewal, S. S., Cheng, Z., Masih, S., Wolboldt, M., Huda, N., Knight, A., & Grewal, P. S. (2011). An assessment of soil nematode food webs and nutrient pools in community gardens and vacant lands in two post-industrial American cities. *Urban Ecosystems*, 14(2), 181–194. 10.1007/s11252-010-0146-3

Grimmer, A. E. (2011). *The Secretary of the Interior's Standards for Rehabilitation & Illustrated Guidelines on Sustainability for Rehabilitating Historic Buildings*. US Government Printing Office.

Grimm, N. B., Faeth, S. H., Golubiewski, N. E., Redman, C. L., Wu, J., Bai, X., & Briggs, J. M. (2008). Global change and the ecology of cities. *Science*, 319(5864), 756–760. 10.1126/science.115019518258902

Guba, E. G., & Lincoln, Y. S. (1994). Competing Paradigms in Qualitative Research. In *Handbook of qualitative research* (pp. 105–117). Sage Publications, Inc.

Guerra, A., & Kolmos, A. (2011). Comparing problem based learning models: suggestions for their implementation. *PBL across the disciplines: Research into best practice*, 3-16.

Gunewardene, N. (2016). Documentation of vernacular heritage in architectural education: "Measured Drawing of Devala Buildings" with Level 1 students of architecture, University of Moratuwa, Sri Lanka. *The International Journal of Environmental Studies*, 73(4), 502–511. 10.1080/00207233.2016.1178983

Guo R., et al. (2019). Job-Education mismatches in degrees and majors and employment quality of college graduates. *Labor economics research,7*(02), 78-100.

Guo, J., & Li, B. (2018). The application of medical artificial intelligence technology in rural areas of developing countries. *Health Equity*, 2(1), 174–181. 10.1089/heq.2018.003730283865

Guo, Y. (2021). Research on design method and construction key points of prefabricated building structure combined with machinery. *Journal of Physics: Conference Series*, 1744(2), 022091. 10.1088/1742-6596/1744/2/022091

Guo, Y. (2021). The establishment and evaluation method of prefabricated building construction quality evaluation index system taking into account multimedia sensor network nodes. *Advances in Multimedia*, 2021, 1–6. 10.1155/2021/9613100

Gutiérrez, A., Miravet, D., & Domènech, A. (2020). COVID-19 and urban public transport services: Emerging challenges and research agenda. *Cities & Health*, 00(00), 1–4. 10.1080/23748834.2020.1804291

Ha, C. Y., Radzi, I., & Khoo, T. J. (2020). The Barriers of Implementing Green Building in Penang Construction Industry. *Progress in Energy and Environment, 12*(March), 1–10. https://www.akademiabaru.com/doc/progeeV12_P1_10.pdf

Hadavi, P., & Tavakolan, M. (2018) 4D modeling and BIM-based Project planning using constraint-based simulation for implementation in Iran considering an actual case study. *Construction Research Congress, CRC 2018*, (pp. 542–551). ASCE. 10.1061/9780784481264.053

Haddad, B. (2005). Ranking the adaptive capacity of nations to climate change when socio-political goals are explicit. *Global Environmental Change*, 15(2), 165–176. 10.1016/j.gloenvcha.2004.10.002

Hagel, P., & Shaw, R. (2008). The influence of delivery mode on consumer choice of university. *European Advances in Consumer Research*, 8, 531–536.

Halim, A. H. A., Harun, S. N., & Hamid, M. Y. (2012). Peranan Sains dan Teknologi Dalam Kerja-Kerja Konservasi Bangunan Bersejarah.

Hall, E., Chai, W. and Albrecht, J.A. (2016), "A qualitative phenomenological exploration of teacher's experience with nutrition education". American Journal of Health Education, Vol. 47 No. 3, pp. 136-148, dou: .10.1080/19325037.2016.1157532

Hall, P. (1993). Forces shaping urban Europe. *Urban Studies (Edinburgh, Scotland)*, 30(6), 883–898. 10.1080/00420989320080831

Halpin, M., & Richard, N. (2021). An invitation to analytic abduction. *Methods in Psychology (Online)*, 5, 100052. 10.1016/j.metip.2021.100052

Hamzah, N., Lou, E., Kamaruzzaman, S. N., Wong, P. F., Che-Ain, A. I., & Khoiry, M. A. (2019). An Environmental Study on Development of Refurbishment Assessment Themes for Heritage Non-domestic Buildings in Malaysia. *Ekoloji*, 28(107), 3–15.

Han, J., Wang, C., Deng, S., & Lichtfouse, E. (2023). China's sponge cities alleviate urban flooding and water shortage: A review. *Environmental Chemistry Letters*, 21(3), 1297–1314. 10.1007/s10311-022-01559-x

Han, T. (2023). Comprehensive benefit analysis of prefabricated buildings under the full life cycle. *Stone*, (12), 96–98. 10.14030/j.cnki.scaa.2023.0535

Hariry, N. N. N. (2021). National Heritage Act 2005 (Act 2005): Problems and Solutions from the Perspective betweenanother Countries. [TURCOMAT]. *Turkish Journal of Computer and Mathematics Education*, 12(3), 2920–2925.

Härkönen, E., & Stöckell, A. (2019). Cultural Sustainability in Art-Based Interdisciplinary Dialogue. *International Journal of Art & Design Education, 38*(3), 639–648. Q2. 10.1111/jade.12246

Harris, F., McCaffer, R., Baldwin, A., & Edum-Fotwe, F. (2021). *Modern construction management.* John Wiley & Sons.

Harun, S. N. (2021). Pemuliharaan Bangunan Warisan. Dewan Bahasa dan Pustaka.

Harun, S. N. (2011). Heritage Building Conservation in Malaysia: Experience and Challenges. *Procedia Engineering*, 20, 41–53. 10.1016/j.proeng.2011.11.137

Hasan, H. S. M., Ab Wahab, L., & Ismail, D. (2019). Authenticity in Adaptive Reuse of Heritage Building in Malaysia. *International Journal of Cultural Heritage*, 4, 33–39.

Hassan, A. M., & Awad, A. I. (2018). Urban Transition in the Era of the Internet of Things: Social Implications and Privacy Challenges. *IEEE Access : Practical Innovations, Open Solutions*, 6, 36428–36440. 10.1109/ACCESS.2018.2838339

Hassan, M. I., & Elhassan, S. M. M. (2020). Modelling of urban growth and planning: A critical review. *Journal of Building Construction and Planning Research*, 8(4), 245–262. 10.4236/jbcpr.2020.84016

He L. L., et al. (2022). Research hotspots and development trends of BIM in the field of underground space. *Journal of Underground Space and Engineering,18*(S1),14-25+35.

He, F., Yang, J., Zhang, Y., Sun, D., Wang, L., Xiao, X., & Xia, J. (2022). Offshore Island Connection Line: A new perspective of coastal urban development boundary simulation and multi-scenario prediction. *GIScience & Remote Sensing*, 59(1), 801–821. 10.1080/15481603.2022.2071056

Hendarwati, E., Nurlaela, L., Bachri, B., & Sa'ida, N. (2021). Collaborative problem based learning integrated with online learning. [iJET]. *International Journal of Emerging Technologies in Learning*, 16(13), 29–39. 10.3991/ijet.v16i13.24159

Hendrix, J. S. (2012). Architecture as the Psyche of a Culture. In Emmons, P., Lomholt, J., & Hendrix, J. S. (Eds.), *The cultural role of architecture: Contemporary and historical perspectives.* Routledge.

Hessam, G. K., & Sotoue, E.-D. (2016, January 01). Evaluation of the relationship between culture and traditional architecture and its effects on design quality improvement. *International Journal of Applied Engineering Research: IJAER*, 11(3), 2120–2123.

Historic England. (2008). *Conservation principles: Policies and guidance for the sustainable management of the historic environment.* English Heritage.

Historic England. (2018). *Energy Efficiency and Historic Buildings: How to Improve Energy Efficiency.* Historic England.

Historic Scotland. (2020). *Historic Environment Scotland Refurbishment Case Study 37. Holyrood Park Lodge, Edinburgh: Thermal upgrade works to a 19th century lodge house.* Historic Scotland.

Historic Scotland. (2021). *Guide to Energy Retrofit of Traditional Buildings.* Historic Scotland. https://www.historicenvironment.scot/archives-and research/publications/publication/?publicationId=47c9f2eb-1ade-4a76-a775add0008972f3

Hjorth, T., Huseinovic, E., Hallström, E., Strid, A., Johansson, I., Lindahl, B., Sonesson, U., & Winkvist, A. (2020). Changes in dietary carbon footprint over ten years relative to individual characteristics and food intake in the Västerbotten Intervention Programme. *Scientific Reports*, 10(1), 1–14. 10.1038/s41598-019-56924-831913331

Hodgson, K., Campbell, M. C., & Bailkey, M. (2011). *Urban Agriculture: Growing Healthy, Sustainable Places.* American Planning Association.

Holmberg, J., Lundqvist, U., Robèrt, K. H., & Wackernagel, M. (1999). The ecological footprint from a systems perspective of sustainability. *International Journal of Sustainable Development and World Ecology*, 6(1), 17–33. 10.1080/13504509.1999.9728469

Hong, P., & Qianli, S. (2019). Researches on Museum Cultural Creative Products Based on Domains Crossing. *IOP Conference Series. Materials Science and Engineering*, 573(1), 012032. 10.1088/1757-899X/573/1/012032

Hooimeijer, F. L., Yoshida, Y., Bortolotti, A., & Iuorio, L. (2022). Integrated urban flood design in the United States and the Netherlands. In Brody, S., Lee, Y., & Kothuis, B. (Eds.), *Coastal Flood Risk Reduction: The Netherlands and the U.S. Upper Texas Coast* (pp. 241–254). Elsevier. 10.1016/B978-0-323-85251-7.00018-4

Hope, D., Gries, C., Zhu, W., Fagan, W. F., Redman, C. L., Grimm, N. B., & Kinzig, A. (2003). Socioeconomics drive urban plant diversity. *Proceedings of the National Academy of Sciences of the United States of America*, 100(15), 8788–8792. 10.1073/pnas.153755710012847293

Hossler, D. E. F. (1995). *D Wallers, and M.K.Moss (EDS), New Directions Institutional Research, N.88: Evaluating and responding to college guidebooks and rankings.* Jossey-Bass, Inc.

Hou, H. (2022). Study on the reform of training mode of Engineering Cost Professionals under the background of new Engineering. *Science & Technology and Innovation*, 142-143(23), 148.

Hou, J. H., & Hu, Z. G. (2013). Review and prospect of Cite Space software application research. *Modern Information*, 33(04), 99–103.

Hridoy, R. M., Parvez, M. S., & Mohsin, N. (2020). Joining Methods of Analytic Hierarchy Process (AHP), Kano Model and Quality Function Deployment (QFD) to Improve the Tractor's Seat Design for Tractor Drivers in Bangladesh. *American Journal of Industrial and Business Management*, 10(05), 05. Advance online publication. 10.4236/ajibm.2020.105071

Hsueh, S.-L., Zhou, B., Chen, Y.-L., & Yan, M.-R. (2022). Supporting technology-enabled design education and practices by DFuzzy decision model: Applications of cultural and creative product design. *International Journal of Technology and Design Education, 32*(4), 2239–2256. Q4. 10.1007/s10798-021-09681-7

Hsu, H., & Lachenbruch, P. A. (2014). Paired t Test. Wiley StatsRef. *Statistics Reference Online*, 1–3. 10.1002/9781118445112.stat05929

Huang, C., Wang, Y., Li, X., Ren, L., Zhao, J., Hu, Y., Zhang, L., Fan, G., Xu, J., Gu, X., Cheng, Z., Yu, T., Xia, J., Wei, Y., Wu, W., Xie, X., Yin, W., Li, H., Liu, M., & Cao, B. (2020). Clinical features of patients infected with 2019 novel coronavirus in Wuhan, China. *Lancet*, 395(10223), 497–506. 10.1016/S0140-6736(20)30183-531986264

Huhnt, W., Richter, S., Wallner, S., Habashi, T., & Krämer, T. (2010). Data management for animation of construction processes. *Advanced Engineering Informatics*, 24(4), 404–416. 10.1016/j.aei.2010.07.009

Hui, T. L., & Khoon, N. C. (2020). Comparative study on precast building construction and conventional building construction for housing project in sarawak. *Jurnal Teknologi*, 82(1), 75–84. 10.11113/jt.v82.13776

Hunter, C., & Shaw, J. (2007). The ecological footprint as a key indicator of sustainable tourism. *Tourism Management*, 28(1), 46–57. 10.1016/j.tourman.2005.07.016

Huo, T., Ye, C., & Zhang, X. (2021). Research on construction technology of external hanging scaffold for prefabricated concrete shear wall. *Journal of Physics: Conference Series*, 1952(3), 032073. 10.1088/1742-6596/1952/3/032073

Husain, S., Safiei, H., Ghazali, A., Nazir, N. N. M., Redzuan, N. I. N., & Ghani, F. A. (2018). Factors Affecting Students' Preferences Towards Private University College in Malaysia. *Universiti Tenaga Nasional, Malaysia. Global Business and Management Research: An International Journal, 10*(3).

Husin, S., Zaki, N., & Husain, M. (2019). Implementing Sustainability in Existing Building through Retrofitting Measures. [IJCIET]. *International Journal of Civil Engineering and Technology*, 10(01), 1450–1471.

IEA. (2006). *Energy technology perspectives: Scenario and strategies to 2050, International Energy Agency*. IEA.

Ifatimehin, O. O. (2007). An Assessment of Urban Heat Island of Lokoja Town and Surroundings Using Landsat ETM Data. *FUTY Journal of the Environment*, 2(1), 100–108.

INDC. (2015). (Submitted to). Intended Nationally Determined Contribution (INDC) of the Government of Malaysia. *the United Nations Framework.*

Inter-American Development Bank (IDB). (2010a). *Indicators of Disaster Risk and Risk Management Program for Latin America and the Caribbean Summary Report.* IDB. https://publications.iadb.org/publications/english/viewer/Indicators-of-Disaster-Risk-and-Risk-Management-Program-for-Latin-America-and-the-Caribbean-Summary-Report.pdf

Inter-American Development Bank (IDB). (2010b). *Indicators of Disaster Risk and Risk Management Program for Latin America and the Caribbean, Barbados.* IDB. https://www.imf.org/external/np/seminars/eng/2013/caribbean/pdf/barbados.pdf

International Energy Agency. (2017). Towards a Zero-Emission, Efficient, and Resilient Buildings and Construction Sector. *UN Environment and International Energy Agency*, 2017, 2017.

International Federation of Red Cross and Red Crescent Societies (IFRC). (2022). *DREF Application-Trinidad and Tobago Floods-November 2022.* IFRC. https://reliefweb.int/attachments/4fb5022e-a0e1-4ea3-ab80-0415ba23ee2e/MDRTT002do.pdf

International Monetary Fund (IMF). (2013). *Measuring Disaster Risk.* IMF. https://www.imf.org/external/np/seminars/eng/2013/caribbean/pdf/measuring.pdf

IPCC. (2001). *Climate Change 2001: The Scientific Basis* (Albritton, D. L., Eds.). Cambridge University Press.

IPCC. (2014). *Climate change 2014: Impacts, adaptation, and vulnerability. Part A: Global and sectoral aspects. Intergovernmental Panel on Climate Change. Contribution of working group II to the fifth assessment report of the intergovernmental panel on climate change*. Cambridge: Cambridge University Press.

Ishihara, Keiko & Ishihara, Shigekazu & Nagamachi, Mitsuo & Hiramatsu, Sugaru & Osaki, Hirokazu. (2001). Age-related decline in color perception and difficulties with daily activities–measurement, questionnaire, optical and computer-graphics simulation studies. International Journal of Industrial Ergonomics - INT J IND ERGONOMIC. 28. 153-163. .10.1016/S0169-8141(01)00028-2

Ismail, L. C., Osaili, T. M., Mohamad, M. N., Al Marzouqi, A., Jarrar, A. H., Jamous, D. O. A., Magriplis, E., Ali, H. I., Al Sabbah, H., Hasan, H., Almarzooqi, L. M. R., Stojanovska, L., Hashim, M., Obaid, R. R. S., Saleh, S. T., & Dhaheri, A. S. A. (2020). Eating habits and lifestyle during covid-19 lockdown in the united arab emirates: A cross-sectional study. *Nutrients*, 12(11), 1–20. 10.3390/nu1211331433137947

Ismail, Z. A. (2019). Optimising the safety of road transport workers on IBS building construction projects: A review. *Social Responsibility Journal*, 15(6), 837–851. 10.1108/SRJ-09-2018-0240

Ivy, J. (2008). A New Higher Education Marketing Mix: The 7Ps for MBA Marketing. *International Journal of Educational Management*, 22(4), 288–299. 10.1108/09513540810875635

Jabatan Warisan Negara. (2016). *Garis Panduan Pemuliharaan Bangunan Warisan*. Jabatan Warisan Negara.

Jackson, G. A. (1982). Public efficiency and private choice in higher education. *Educational Evaluation and Policy Analysis*, 4(2), 237–247. 10.3102/01623737004002237

Jadresin Milic, R., McPherson, P., McConchie, G., Reutlinger, T., & Singh, S. (2022). Architectural history and sustainable architectural heritage education: Digitalisation of heritage in New Zealand. *Sustainability (Basel)*, 14(24), 16432. 10.3390/su142416432

Jadresin Milic, R., & Nikezic, A. (2018). Communicating Heritage Through Intertwining Theory and Studio Based Course in Architectural Education. In Amoruso, G. (Ed.), *Putting Tradition into Practice: Heritage, Place and Design* (pp. 626–635). Springer International Publishing. 10.1007/978-3-319-57937-5_65

Jaeger, P. T., Lin, J., Grimes, J. M., & Simmons, S. N. (2009). Where is the cloud? Geography, economics, environment, and jurisdiction in cloud computing. *First Monday*, 14(5). 10.5210/fm.v14i5.2456

Jagarajan, R., Abdullah, M. N., Asmoni, M., Lee, J. Y., & Jaafar, M. N. (2015). An Overview of Green Retrofitting Implementation in Non-Residential Existing Buildings. *Jurnal Teknologi*, 73(5). 10.11113/jt.v73.4324

Jahed, N., Aktaş, Y. D., Rickaby, P., & Bilgin Altınöz, A. G. (2020). Policy Framework for Energy Retrofitting of Built Heritage: A Critical Comparison of UK and Turkey. *Atmosphere (Basel)*, 11(6), 674. 10.3390/atmos11060674

Jain, L., & Bhardwaj. (2010). Enterprise cloud computing: Key considerations for adoption. *International Journal of Engineering and Information Technology*, 2(2), 113–117.

Jakle, J. A., & Wilson, D. (1992). *Derelict Landscapes: The Wasting of America's Built Environment*. Rowman & Littlefield.

Janiszewski, C., & van Osselaer, S. M. J. (2022). Abductive Theory Construction. *Journal of Consumer Psychology*, 32(1), 175–193. 10.1002/jcpy.1280

Jauhari, M., & Prayudi, Y. (2023). IMPLEMENTASI METODE DOUBLE DIAMOND DALAM PERANCANGAN PROTOTIPE APLIKASI SISTEM ERP BERBASIS WEBSITE. *AKSELERASI: Jurnal Ilmiah Nasional*, 5(1), 85–98. 10.54783/jin.v5i1.699

Jehlík, V. (1986) *The Vegetation of Railways in Northern Bohemia*. Academia: Praha, Czech Republic.

Jiao, Y., Wang, Y., Zhang, S., Li, Y., Yang, B., & Yuan, L. (2013). A cloud approach to unified lifecycle data management in architecture, engineering, construction and facilities management: Integrating BIMs and SNS. *Advanced Engineering Informatics*, 27(2), 173–188. 10.1016/j.aei.2012.11.006

Jin, Y., Zhang, J., & Sun, L. (2019). Safety risk assessment of prefabricated building construction based on bayesian network. *IOP Conference Series. Earth and Environmental Science*, 371(3), 032052. 10.1088/1755-1315/371/3/032052

Jin, Y., Zhou, X., & Zhang, Z. (2023). Design and evaluation of domestic baby bathtubs based on fuzzy hierarchical analysis. *Furniture and Interior Decoration*, 30(8), 72–75. 10.16771/j.cn43-1247/ts.2023.08.012

Ji, Y., Qi, K., Qi, Y., Li, Y., Li, H. X., Lei, Z., & Liu, Y. (2020). BIM-based life-cycle environmental assessment of prefabricated buildings. *Engineering, Construction, and Architectural Management*, 27(8), 1703–1725. 10.1108/ECAM-01-2020-0017

Jokilehto, J. (2007). An international perspective to conservation education. *Built Environment*, 33(3), 275–286. 10.2148/benv.33.3.275

Jorgenson, A. K. (2003). Consumption and Environmental Degradation: A Cross-National Analysis of the Ecological Footprint. In *Social Problems, 50*(3). University of California Press. 10.1525/sp.2003.50.3.374

Joseph, M., & Joseph, B. (1998). Identifying need of potential students in tertiary education for strategy development. *Quality Assurance in Education*, 6(2), 90–96. 10.1108/09684889810205741

Kagermann, K., Helbig, J., Hellinger, A., & Wahlster, W. (2013). *Recommendations for implementing the strategic initiative Industrie 4.0: Securing the future of German manufacturing industry; final report of the Industrie 4.0 Working Group*. Forschungsunion.

Kalnay, E & Cai, M. (2003). Impact of Urbanization and Land Use change on Climate. *Nature*, *423*, 528-531. , http://dx.doi.org/10.1038/nature01675

Kassem, M., Brogden, T., & Dawood, N. (2012). BIM and 4D planning: A holistic study of the barriers and drivers to widespread adoption. *Journal of Construction Engineering and Project Management*, 2(4), 1–10. 10.6106/JCEPM.2012.2.4.001

Kaur, S. P., & Gupta, V. (2020). COVID-19 Vaccine: A comprehensive status report. In *Virus Research (Vol. 288)*. Elsevier B.V. 10.1016/j.virusres.2020.198114

Kenayathulla, H., B., Ahmad, N., A., & Idris, A., R. (2019, August 14). Gaps between competence and importance of employability skills: evidence from Malaysia. Higher Education Evaluation and Development, 13(2), 97-112.

Khairani, A. Z., & Razak, N. N. (2013). Assessing Factors Influencing Students' Choice of Malaysian Public University: A Rasch Model Analysis. Universiti Sains Malaysia. *International Journal of Applied Psychology*, 3(1), 19–24.

Khalaf, R. W. (2021). World Heritage on the Move: Abandoning the Assessment of Authenticity to Meet the Challenges of the Twenty-First Century. *Heritage*, 4(1), 371–386. 10.3390/heritage4010023

Khalil, A., Stravoravdis, S., & Backes, D. (2021). Categorisation of Building Data in the Digital Documentation of Heritage Buildings. *Applied Geomatics*, 13(1), 29–54. 10.1007/s12518-020-00322-7

Khan, K., & Malluhi, Q. (2010). Establishing trust in cloud computing. *IT Professional*, 12(5), 20–27. 10.1109/MITP.2010.128

Khatib, A. N., Carvalho, A. M., Primavesi, R., To, K., & Poirier, V. (2020). Navigating the risks of flying during COVID-19: A review for safe air travel. *Journal of Travel Medicine*, 27(8), 1–9. 10.1093/jtm/taaa21233184655

Khazanah Research Institute. (2021). *Building Resilience: Towards inclusive social protection in Malaysia*. Khazanah Research Institute.

Khoo, D. (2018). Medicine Suppliers to Benefit from higher healthcare location. *The Star*.

Khursheed, A., Alam, S., Tyagi, V. K., Nagpure, A. S., Khan, A. A., Gaur, R. Z., Singh, S., Bhattacharya, P., Mukherjee, S., & Kumar, M. (2020). Future liasing of the lockdown during COVID-19 pandemic: The dawn is expected at hand from the darkest hour. *Groundwater for Sustainable Development*, 11(336), 100433. 10.1016/j.gsd.2020.10043332835071

Kim, C., Son, H., & Kim, C. (2013). Automated construction progress measurement using a 4D building information model and 3D data. *Automation in Construction*, 31, 75–82. 10.1016/j.autcon.2012.11.041

Kim, G. (2016). The Public Value of Urban Vacant Land: Social Responses and Ecological Value. *Sustainability (Basel)*, 8(5), 486. 10.3390/su8050486

Kim, G., Miller, P. A., & Nowak, D. J. (2015). Assessing urban vacant land ecosystem services: Urban vacant land as green infrastructure in the City of Roanoke, Virginia. *Urban Forestry & Urban Greening*, 14, 519–526. 10.1016/j.ufug.2015.05.003

Kim, G., Miller, P. A., & Nowak, D. J. (2016). The value of green infrastructure on vacant and residential land in Roanoke, Virginia. *Sustainability (Basel)*, 8(4), 296. 10.3390/su8040296

Kim, W. (2009). Cloud computing: Today and tomorrow. *Journal of Object Technology*, 8(1), 65–72. 10.5381/jot.2009.8.1.c4

Kinsella, K., & Wan, H. (2009). *An Aging World. US Census Bureau, International Population Reports, P95/09-1*. Washington DC: US Government Printing Office..

Kitzes, J., & Wackernagel, M. (2009). Answers to common questions in Ecological Footprint accounting. *Ecological Indicators*, 9(4), 812–817. 10.1016/j.ecolind.2008.09.014

KLCP. (2020). *Kuala Lumpur City Plan 2020*. KLCP. http://klcityplan2020.dbkl.gov.my/sme/

Klimeš, L. (1987). Succession in road bank vegetation. *Folia Geobotanica et Phytotaxonomica*, 22(4), 435–440. 10.1007/BF02853242

Koh, J. H., & Ahamad, N. A. (2023). A Phenomenological Study of Learner's Virtual Place Experience. In *Handbook of Research on Inclusive and Innovative Architecture and the Built Environment* (pp. 368–386). IGI Global. 10.4018/978-1-6684-8253-7.ch020

Kremer, P., Hamstead, Z. A., & McPhearson, T. A. (2013). Social-ecological assessment of vacant lots in New York City. *Landscape and Urban Planning*, 2013(120), 218–233. 10.1016/j.landurbplan.2013.05.003

Kudryavtseva, O. V., Malikova, O. I., & Egorov, E. G. (2021). Sustainable Urban Development And Ecological Externalities: Russian Case. Geography, Environment. *Sustainability*, 14(1), 81–90. 10.24057/2071-9388-2020-151

Kuenzel, R., Schevers, H., & Fischer, M. (2019). The limitations of BIM for project management. *Procedia Engineering*, 196, 693–700.

Kulatunga, K., Amaratunga, D., & Haigh, R. (2007). Researching construction client and innovation: methodological perspective. *7th International Postgraduate Research Conference in the Built and Human Environment*. Research Gate.

Kumar, P., & Morawska, L. (2019). Could fighting airborne transmission be the next line of defence against COVID-19 spread? *City and Environment Interactions*, 4, 100033. 10.1016/j.cacint.2020.10003334235420

Langeveld, L., & Das, A. K. (2021). Impact of Culture on Design Versus Design on Culture. In Chakrabarti, A., Poovaiah, R., Bokil, P., & Kant, V. (Eds.), *Design for Tomorrow—Volume 2* (pp. 455–467). Springer. 10.1007/978-981-16-0119-4_37

Lapadula, M. I., & Quiroga, C. (2012). Heritage as a pedagogical resource and platform for exploration in architectural design education. *Journal of Architecture (London)*, 17(4), 591–607. 10.1080/13602365.2012.709028

Latiffi, A. A., Mohd, S., Kasim, N., & Fathi, M. S. (2013). Building Information Modeling (BIM) Application in Malaysian Construction Industry. *International Journal of Construction Engineering and Management*, 2(4A), 1–6. 10.5923/s.ijcem.201309.01

Lawson, L. J. (2005). *City Bountiful: A Century of Community Gardening in America*. University of California Press. 10.1525/9780520931558

LCCF. (2017). Low Carbon Cities Framework Version 2. Ministry of Energy, Green Technology and Water Malaysia (KeTTHA). Springer. ISBN 978-967-5893-10-0.

Lee, H., Song, K., Kim, G. W., & Chon, J. (2021). Flood-adaptive green infrastructure planning for urban resilience. *Landscape and Ecological Engineering*, 17(4), 1–11. 10.1007/s11355-021-00458-7

Leeraphong, A., Papasratorn, B., & Chongsuphajaisiddhi, V. (2015). A Study on Factors Influencing Elderly Intention to Use Smart Home in Thailand : A Pilot Study. *The 10th International Conference on E-Business*, 1–10.

Leidig, P., & Oakes, W. (2021). Model for project-based community engagement. *International Journal for Service Learning in Engineering. Humanitarian Engineering and Social Entrepreneurship*, 16(2), 1–13.

Lemos, M., Eakin, H., Nelson, D., & Engle, N. (2013). Building adaptive capacity to climate change in less developed countries. In *Climate science for serving society* (pp. 437–457). Springer. 10.1007/978-94-007-6692-1_16

Lennon, M., & Scott, M. (2014). Delivering ecosystems services via spatial planning: Reviewing the possibilities and implications of a green infrastructure approach. *The Town Planning Review*, 85(5), 563–587. 10.3828/tpr.2014.35

Lenzen, M., & Murray, S. A. (2001). A modified ecological footprint method and its application to Australia. *Ecological Economics*, 37(2), 229–255. 10.1016/S0921-8009(00)00275-5

Leong, F. T. L. (1985). Career development of Asian Americans. *Journal of College Student Personnel*, 26, 539–546.

Leung, K., Bhagat, R. S., Buchan, N. R., Erez, M., & Gibson, C. B. (2005). Culture and international business: Recent advances and their implications for future research. *Journal of International Business Studies*, 36(4), 357–378. 10.1057/palgrave.jibs.8400150

Li, R. M. (2016). *Jianzhu lishi kecheng jiaoxuezhong de kunhuo yu duice: Yi Qingdao Binhai xueyuan jianzhulishi kecheng weili* [Problems of Architectural History Teaching and their Solutions: A Case of the Qingdao Binhai College Architectural History Course]. *Jiancai yu zhuangshi [Construction Materials & Decoration], 1*, 176–177.

Li, S., Cao, L., & Shao, Y. (2020). Research on the professional curriculum system of adult education based on vocational abilities - taking engineering cost major of the National Open University as an example. *Journal of the Open University of Broadcasting (Philosophy and Social Sciences Edition)*, (03), 91-100+111. .10.16161/j.issn.1008-0597.2020.03.016

Liang, S. M. (2020). Research on the development, design and promotion of cultural creations of the Forbidden City in Beijing under the background of "Internet+". *Packaging Engineering, 41*(8), 309-312+316. 10.19554/j.cnki.1001-3563.2020.08.050

Lian, J. W., Yen, D. C., & Wang, Y. T. (2014). An exploratory study to understand the critical factors affecting the decision to adopt cloud computing in Taiwan hospital. *International Journal of Information Management*, 34(1), 28–36. 10.1016/j.ijinfomgt.2013.09.004

Liberati, A., Altman, D. G., Tetzlaff, J., Mulrow, C., Gotzsche, P. C., Ioannidis, J. P. A., Clarke, M., Devereaux, P. J., Kleijnen, J., & Moher, D. (2009). The PRISMA Statement for Reporting Systematic Reviews and Meta-Analyses of Studies that Evaluate Healthcare Interventions: Explanation and Elaboration. *BMJ, 339*(jul21 1), b2700–b2700. 10.1136/bmj.b2700

Lichtenstein, S., & Slovic, P. (2006). *The construction of preference*. Cambridge University Press. 10.1017/CBO9780511618031

Lim, C. K. (2012). The Changing Needs of Older Malaysians: A Selangor Case Study. [Doctoral Dissertation, Adelaide: University of Adelaide].

Li, M., Zhao, D., & Yu, Y. (2015). TOE drivers for cloud transformation: Direct or trust-mediated? *Asia Pacific Journal of Marketing and Logistics*, 27(2), 226–248. 10.1108/APJML-03-2014-0040

Lin, D., Hanscom, L., Murthy, A., Galli, A., Evans, M., Neill, E., Mancini, M. S., Martindill, J., Medouar, F. Z., Huang, S., & Wackernagel, M. (2018). Ecological footprint accounting for countries: Updates and results of the national footprint accounts, 2012-2018. *Resources*, 7(3), 58. 10.3390/resources7030058

Lin, H., Bingli, X., Yuting, C., Wenhang, L., Lan, Y., & Jie, H. (2022). VGEs as a New Platform for Urban Modeling and Simulation. *Sustainability (Basel)*, 14(13), 7980. 10.3390/su14137980

Lin, X. B., Yu, J. B., & Liu, M. Y. (2020). Study on the integration of Professional teaching standards and Vocational skill level Standards in Higher Vocational Colleges under the 1+X Certificate System. V*ocational. Education Research*, (01), 4–11.

Li, R., Chen, K., & Wu, D. (2020). Challenges and opportunities for coping with the smart divide in rural America. *Annals of the American Association of Geographers*, 110(2), 559–570. 10.1080/24694452.2019.1694402

Little, G. M. (2008). *629 Main Street—A Revisioning of an Urban Void: An Exploration into Challenging Cultural Perceptions of an Urban Vacant Lot*. The University of Manitoba.

Liu, Y. (2021). Introfuction of Different BIM Software in the Construction Industry. Bachelor's thesis Visamäki Campus, *Degree Programme in Construction Engineering Spring semester 2021*, 1–28.

Liu, F., Li, L., & Liu, C. (2022). Research on the design and evaluation model of Jing tourism souvenirs under the perspective of cultural genes. *Packaging Engineering*, 43(2), 333–340. 10.19554/j.cnki.1001-3563.2022.02.044

Liu, X., Wei, M., & Zeng, J. (2020). Simulating Urban Growth Scenarios Based on Ecological Security Pattern: A Case Study in Quanzhou, China. *International Journal of Environmental Research and Public Health*, 17(19), 7282. 10.3390/ijerph1719728233028037

Liu, Y. F., Yan, L., & Yin, Y. L. (2005). Innovation and practice of training system of characteristic talents of engineering cost. *China Higher Education Research*, (11), 85–86. 10.16298/j.cnki.1004-3667.2005.11.033

Liu, Y., & Cui, J. (2020). Identification of hazard sources in prefabricated building construction by entropy weight method. *IOP Conference Series. Earth and Environmental Science*, 560(1), 012073. 10.1088/1755-1315/560/1/012073

Liu, Z., Meng, X., Xing, Z., & Jiang, A. (2021). Digital twin-based safety risk coupling of prefabricated building hoisting. *Sensors (Basel)*, 21(11), 3583. 10.3390/s2111358334064110

Li, W. (2018). A Research on Undergraduate Architecture Teaching Approach Based on Integration of Architectural Design and Architectural History Teaching. *Creative Education*, 9(12), 12. 10.4236/ce.2018.912135

Loach, K., & Rowley, J. (2022). Cultural sustainability: A perspective from independent libraries in the United Kingdom and the United States. *Journal of Librarianship and Information Science, 54*(1), 80–94. Q3. 10.1177/0961000621992824

Lokhandwala, S., & Gautam, P. (2020). Indirect impact of COVID-19 on environment: A brief study in Indian context. *Environmental Research*, 188(June), 109807. 10.1016/j.envres.2020.10980732574854

Lokhorst, A. M., Werner, C., Staats, H., van Dijk, E., & Gale, J. L. (2013). Commitment and Behavior Change: A Meta-Analysis and Critical Review of Commitment-Making Strategies in Environmental Research. *Environment and Behavior*, 45(1), 3–34. 10.1177/0013916511411477

López-Guerrero, R. E., Vera, S., & Carpio, M. (2022). A quantitative and qualitative evaluation of the sustainability of industrialised building systems: A bibliographic review and analysis of case studies. *Renewable & Sustainable Energy Reviews*, 157, 112034. 10.1016/j.rser.2021.112034

Lorenzoni, I., Nicholson-Cole, S., & Whitmarsh, L. (2007). Barriers perceived to engaging with climate change among the UK public and their policy implications. *Global Environmental Change*, 17(3–4), 445–459. 10.1016/j.gloenvcha.2007.01.004

Loss, C., Rossi, S., & Tannert, T. (2018). In-plane stiffness of hybrid steel–cross-laminated timber floor diaphragms. *Journal of Structural Engineering*, 144(8), 04018128. 10.1061/(ASCE)ST.1943-541X.0002105

Lourenco, I. B., Beleno de Oliveira, A. K., Marques, L., Quintanilha Barbosa, A. A., Veról, A. P., Magalhães, P. C., & Miguez, M. G. (2020). A framework to support flood prevention and mitigation in the landscape and urban planning process regarding water dynamics. *Journal of Cleaner Production*, 227(122983), 1–17. 10.1016/j.jclepro.2020.122983

Lowden, K., Hall, S., Elliot, D., & Lewin, J. (2011). *Employers' perceptions of the employability skills of new graduates.* University of Glasgow. https://www.educationandemployers.org/wp-content/uploads/2014/06/employability_skills_as_pdf_-_final_online_version.pdf

Lucas, H. C., Jr. (1989). Methodological issues in information systems survey research. The Information Systems Research Challenge: Survey Research Methods, New York.

Lu, M., Wu, J., & Meng, Q. (2015). Comprehensive Development Evaluation System of Asian Infrastructure Investment Bank Based on Double Diamond Model. *American Journal of Industrial and Business Management*, 05(07), 07. 10.4236/ajibm.2015.57051

Lunenfeld, B., & Stratton, P. (2013). The Clinical Consequence of An Aging World and Preventive Strategies. *Best Practice & Research. Clinical Obstetrics & Gynaecology*, 27(5), 643–659. 10.1016/j.bpobgyn.2013.02.00523541823

Lu, Q., & Zhu, J. H. (2020). A study on the design strategy of cultural and creative products in museums: The case of Yachou-axe. *Architecture and Culture*, 5, 46–48.

Luxner, T. (2023, May 18). Cloud computing stats: Flexera 2023 State of the cloud report. *Flexera Blog*. https://www.flexera.com/blog/cloud/cloud-computing-trends-flexera-2023-state-of-the-cloud-report/

Lymath, A. (2014) *The Top Five Barriers to BIM Implementation.* NBS. https://www.thenbs.com/knowledge/the-top-five-barriers-to-bimimplementation

Lynch, K. (1995). The openness of open space. In Banerjee, T., & Southworth, M. (Eds.), *City sense and city design: Writings and projects of Kevin Lynch* (pp. 396–412). The MIT Press.

Lyu, F., & Liu, Y. D. (2023). The research on the response and development of vocational education under the background of artificial intelligence. *Continuing Education Research*, 2023(02), 59–63.

Mabrouk, M., Han, H., Fan, C., Abdrabo, K. I., Shen, G., Saber, M., Kantoush, S. A., & Sumi, T. (2023). Assessing the effectiveness of nature-based solutions-strengthened urban planning mechanisms in forming flood-resilient cities. *Journal of Environmental Management*, 344, 118260. 10.1016/j.jenvman.2023.11826037354590

MacDonald, E., Cairns, P., Cormack, D., Máté, I., & McLaren, D. (2009). *Scottish House Condition Survey: Key Findings for 2008.*

Mafauzy. (2000). The Problems and Challenges of the Aging Population of Malaysia. *Malaysian Journal of Medical Sciences 7*(1), 1-3.

Malaysia Energy Statistics Handbook. (2021). Department of Energy Management and Industrial Development Suruhanjaya Tenaga (Energy Commission), 1, 86. www.st.gov.my

Mansir, D., Muhammad, J. A., & Kasim, N. (2016). Reviewing the Need for Historic Building Information Modelling (HBIM) in the Conservation of Heritage Buildings in Melaka World Heritage City. *Social Sciences*, 11(11), 2777–2782.

Marino, C., Nucara, A., Pietrafesa, M., & Pudano, A. (2016). The Assessment of Road Traffic Air Pollution by Means of an Average Emission Parameter. *Environmental Modeling and Assessment*, 21(1), 53–69. 10.1007/s10666-015-9489-8

Mark, N. P. (2003). Book Review: Accountability in Social Research: Issues and Debates. *Sociological Methods & Research*, 31(3), 430–434. 10.1177/0049124102239084

Marston, S., Li, Z., Bandyopadhyay, S., Zhang, J., & Ghalsasi, A. (2011). Cloud computing—The business perspective. *Decision Support Systems*, 51(1), 176–189. 10.1016/j.dss.2010.12.006

Martinez-Fernandez, C., Audirac, I., Fol, S., & Cunningham-Sabot, E. (2012). Shrinking cities: Urban challenges of globalization. *International Journal of Urban and Regional Research*, 36(2), 213–225. 10.1111/j.1468-2427.2011.01092.x22518881

Mason, M. (2022). The Contribution of Design Thinking to Museum Digital Transformation in Post-Pandemic Times. Multimodal Technologies and Interaction, 6(9). *Multimodal Technologies and Interaction*, 9(9), 79. 10.3390/mti6090079

Mayo Clinic. (2019). *Aging: What to Expect.* Mayo Clinic. https://www.mayoclinic.org/healthy-lifestyle/healthy-aging/in-depth/aging/art-29946979.

Ma, Z., Badi, A., & Jørgensen, B. N. (2016). Market opportunities and barriers for smart buildings. *2016 IEEE Green Energy and Systems Conference, IGSEC 2016.* IEEE. 10.1109/IGESC.2016.7790078

Ma, Z., Cooper, P., Daly, D., & Ledo, L. (2012). Existing Building Retrofits: Methodology and State-of-the-Art. *Energy and Building*, 55, 889–902. 10.1016/j.enbuild.2012.08.018

McKeen, H. (2019). Culturally Relevant Preparedness Using STEAM Integrative Teaching in the Classroom. In Keengwe, J., & Onchwari, G. (Eds.), *Handbook of Research on Assessment Practices and Pedagogical Models for Immigrant Students* (pp. 127–138). IGI Global. 10.4018/978-1-5225-9348-5.ch008

McKinsey and Company. (2016). *Urban world: Meeting the demographic challenge.* McKinsey and Company.

McMaster University. (2018). *Addressing The Social Needs of Older Adults: A Contributing Factor to Their Health and Wellbeing.* Optimal Aging Portal.

McMINN. J., & Polo, M. (2005, September). Sustainable Architecture as a Cultural Project. In *The 2005 World Sustainable Building Conference, Tokyo* (pp. 27-29).

McPartland, R. (2017). Top 10 BIM terms explained. *NBS*. https://www.thenbs.com/knowledge/top-10- bim-terms- explained#:~:text=What%20is%20BIM?aspect%20o f%20the%20built%20asset

Meena, A. K., & Yadav, T. K. (2019). What is Ecological Footprint and Why Is It Important? *Agrobios, 18*(June), 25–26. http://www.mpict.org/ict_education_defined_importance.html

Megahed, N. A., & Ghoneim, E. M. (2020). Antivirus-built environment: Lessons learned from Covid-19 pandemic. *Sustainable Cities and Society*, 61, 102350. 10.1016/j.scs.2020.10235032834930

Meher, S., & Singh, Z. (2021). Universal Design and All-Age Friendly Planning. 10.13140/RG.2.2.13889.07523

Mehrotra, S., Bardhan, R., & Ramamritham, K. (2020, June). Diurnal thermal diversity in heterogeneous built area: Mumbai, India. *Urban Climate*, 32, 100627. 10.1016/j.uclim.2020.100627

Mell, P., & Grance, T. (2011). *The NIST Definition of Cloud Computing. Computer Security Division Information Technology Laboratory National Institute of Standards and Technology Gaithersburg*. National Institute of Standards and Technology.

Menegon, S. J., Wilson, J. L., Lam, N. T. K., & Gad, E. F. (2020). Experimental testing of innovative panel-to-panel connections for precast concrete building cores. *Engineering Structures*, 207, 110239. 10.1016/j.engstruct.2020.110239

Mengist, W., Soromessa, T., & Legese, G. (2020). Method for conducting systematic literature review and meta-analysis for environmental science research. *MethodsX*, 7, 100777. 10.1016/j.mex.2019.10077731993339

Meriam Syed Akil, S., & Abdullah, S. (2014). Challenges in managing elderly care centres in Malaysia. *The International Journal of the Arts in Society*, 07(3), 1944–6934.

Merriam, S. B. (2009). *Qualitative Research: A Guide to Design and Implementation*. Jossey-Bass.

Michele Koch. (2021). GreenPrint Survey Finds Consumers Want to Buy Eco-Friendly Products, but Don't Know How to Identify Them. *Businesswire*, 50–51.

Microsoft. (2018). *What is cloud computing? A beginner's guide*. Azure. https://azure.microsoft.com/en-us/resources/cloud-computing-dictionary/what-is-cloud-computing/

Miettinen, R., & Paavola, S. (2014). The concept of BIM maturity: What is it, and how to measure it? *Journal of Information Technology in Construction*, 19, 28–44.

Migilinskasa, D., Popovb, V., Juoceviciusc, V., & Ustinovichiu, L. (2013). The benefits, obstacles and problems of practical BIM implementation. *Proceedings of the 11th International Conference on Modern Building Materials, Structures and Techniques*. Elsevier. 10.1016/j.proeng.2013.04.097

Migin, M. W., Falahat, M., Yajid, M. S., & Khatibi, A. (2014). Impacts of Institutional Characteristics on International Students' Choice of Private Higher Education Institutions in Malaysia. *Higher Education Studies*, 5(1), 31–42.

Miguez, M. G., Raupp, I. P., & Verol, A. P. (2018). An integrated quantitative framework to support the design of resilient alternatives to manage urban flood risks. *Journal of Flood Risk Management, 12*(S2-e12514), 1-13. 10.1111/jfr3.12514

Miles, D. A. (2017). A Taxonomy of Research Gaps: Identifying and Defining the Seven Research Gaps. *Journal of Research Methods and Strategies*, 2017, 1–15.

Millennium Ecosystem Assessment. (2003). *Ecosystems and Human Well-Being: Synthesis. World Health*. Island Press.

Ministry of Higher Education Malaysia. (2022, April 22). *Higher Education Report: [MALAYSIA]. Paper presented at the UNESCO World Higher Education Conference.* UNESCO. https://whec2022.net/resources/Country%20report%20-%20Malaysia.pdf

Ministry of Housing & Local Government. (2019). Malaysia Smart City Framework. *Malaysia Smart City Framework, September*, 251. https://drive.google.com/file/d/1lgnk6rGDx7ci08-MjXNTiAyaW-PfVwWs/view%0Ahttps://www.thesundaily.my/local/ministry-launches-malaysia-smart-city-framework-BN1395377

Mnea, A., & Zairul, M. (2023). Evaluating the Impact of Housing Interior Design on Elderly Independence and Activity: A Thematic Review. *Buildings*, 13(4), 1099. 10.3390/buildings13041099

Moalosi, R., Popovic, V., & Hickling-Hudson, A. (2010). Culture-orientated product design. *International Journal of Technology and Design Education, 20*(2), 175–190. Q4. 10.1007/s10798-008-9069-1

Moey, L. K., Goh, K. S., Tong, D. L., Chong, P. L., Adam, N. M., & Ahmad, K. A. (2020). A Review on Current Energy Usage and Potential of Sustainable Energy in Southeast Asia Countries. *Journal of Sustainability Science and Management*, 15(2), 89–107.

Mofijur, M., Fattah, I. M. R., Alam, M. A., Islam, A. B. M. S., Ong, H. C., Rahman, S. M. A., Najafi, G., Ahmed, S. F., Uddin, M. A., & Mahlia, T. M. I. (2021). Impact of COVID-19 on the social, economic, environmental and energy domains: Lessons learnt from a global pandemic. *Sustainable Production and Consumption, 26*(September 2020), 343–359. 10.1016/j.spc.2020.10.016

Moher, D., Liberati, A., Tetzlaff, J., & Altman, D. G. (2009). Preferred reporting items for systematic reviews and meta-analyses: The PRISMA statement. *BMJ (Clinical Research Ed.)*, 339(7716), 332–336. 10.1136/bmj.b253519622551

Moktadir, M. A., Ali, S. M., Kusi-Sarpong, S., & Shaikh, M. A. A. (2018). Assessing challenges for implementing Industry 4.0: Implications for process safety and environmental protection. *Process Safety and Environmental Protection*, 117, 730–741. 10.1016/j.psep.2018.04.020

Molina, J. A., Ignacio Giménez-Nadal, J., & Velilla, J. (2020). Sustainable commuting: Results from a social approach and international evidence on carpooling. *Sustainability (Basel)*, 12(22), 1–12. 10.3390/su12229587

Moran, F., Blight, T., Natarajan, S., & Shea, A. (2013). The Use of Passive House Planning Package to Reduce Energy Use and CO2 Emissions in Historic Dwellings. *Energy and Building*, 75, 216–227. 10.1016/j.enbuild.2013.12.043

Morel-EdnieBrown, F. (2012). Wither Genius Loci?: The City, Urban Fabric and Identity in Perth, Western Australia. *The Role of Place Identity in the Perception, Understanding, and Design of Built Environments*, 209.

Morozova, I. A., & Yatsechko, S. S. (2022). The Risks of Smart Cities and the Perspectives of Their Management Based on Corporate Social Responsibility in the Interests of Sustainable Development. *Risks*, 10(2), 34. 10.3390/risks10020034

Munce, S. E. P., Guetterman, T. C., & Jaglal, S. B. (2021). Using the Exploratory Sequential Design for Complex Intervention Development: Example of the Development of a Self-Management Program for Spinal Cord Injury. *Journal of Mixed Methods Research*, 15(1), 37–60. 10.1177/1558689820901936

Murphy, K., O'Shea, E., Cooney, A., Shiel, A., & Hodgins, M. (2006). *Improving quality of life for older people in long-stay care settings in Ireland Dublin.* National Council on Aging and Older People.

Murray-Harvey, R., Pourshafie, T., & Reyes, W. S. (2013). What Teacher Education Students Learn about Collaboration from Problem-Based Learning. *Journal of Problem Based Learning in Higher Education*, 1(1), 114–134.

Mycoo, M. (2018). Urban Sustainability in Caribbean Small Island Developing States: A Conceptual Framework for Urban Planning using a case study of Trinidad. *International Development Planning Review*, 40(2), 143–174. 10.3828/idpr.2018.8

Nagaraj, S. S. M., Noor, I. M. J., Diana, A. W., & Mirzaei, T. (2008). *How do undergraduates choose their university? A study of first year University of Malaysia students.* (FEA Working Paper No. 2008-8) University of Malaya.

Nakajima, E. S., & Ortega, E. (2016). Carrying capacity using emergy and a new calculation of the ecological footprint. *Ecological Indicators*, 60, 1200–1207. 10.1016/j.ecolind.2015.08.054

Narinesingh, P. (2014). An Assessment of River Maintenance Practices in Trinidad and Tobago. *The Journal of the Association of Professional Engineers of Trinidad and Tobago*, 42(2), 19–28.

National Housing Department. (2016). *National Housing Policy*. NHD.

National Park Service. (2017). *The Secretary of the Interior's Strandards for the Treatment of Historic Properties with Guideline for Preserving*. Rehabilitating, Restoring and Reconstructing Historic Buildings.

Navaratnam, S., Ngo, T., Gunawardena, T., & Henderson, D. (2019). Performance review of prefabricated building systems and future research in Australia. *Buildings*, 9(2), 38. 10.3390/buildings9020038

Nawi, M. N. M., Noordin, A., Tamrin, N., Nifa, F. A. A., & Lin, C. K. (2019). An ecological study on enhancing the Malaysian construction ecosystem: Readiness implementation factors in industrialised building system (IBS) projects. *Ekoloji*, 28(107), 545–552.

Nebor, J. N. (1986). Parental Influence and Involvement on Reading Achievement. *Information Analyses (070)*. https://files.eric.ed.gov/fulltext/ED286150.pdf

Németh, J., & Langhorst, J. (2014). Rethinking urban transformation: Temporary uses for vacant land. *Cities (London, England)*, 40, 143–150. 10.1016/j.cities.2013.04.007

Neukrug, H. M. (2009). *Clean Water, Green City: Blending Interests of Land and Water*. Office of Watersheds.

Newman, P., & Kenworthy, J. (1999). *Sustainability and Cities: Overcoming Automobile Dependence*. Island Press.

Ng, F. P. (2013). Values of Learning Through 'Place-making" in the Design Studio. *Archnet-IJAR: International Journal of Architectural Research*, 7(1), 86–98.

NGTP. (2009). *National Green Technology Policy 2009*. Ministry of Energy, Green Technology and Water.

Nie, H. T., & Li, N. (2020). Protection and Creation: An Analysis of Copyright Legal Issues of Museum Cultural and Creative Products. *Chinese Museum, 1*, 13-18.

Nikezić, A., & Marković, D. (2015). Place-based education in the architectural design studio: Agrarian landscape as a resource for sustainable urban lifestyle. *Sustainability (Basel)*, 7(7), 9711–9733. 10.3390/su7079711

Nkhoma, M. Z., Dang, D. P., & De Souza-Daw, A. (2013, May). Contributing factors of cloud computing adoption: A technology-organization-environment framework approach. In *Proceedings of the European Conference on Information Management & Evaluation* (Vol. 2, No. 1, pp. 180-188). 10.24924/ijise/2013.04/v1.iss1/30.41

Norani, H. (2020). Age Friendly – Cities and Communities. *12th Asean & Japan High Level Officials Meeting on Caring Societies.* Ministry of Woman, Family and Community Development. https://www.mhlw.go.jp/file/06-Sesiakujouhou-10500000- Daijinkanboukokusiaka/01-12_Malaysia.pdf

Nowak, D. J., Crane, D. E., & Dwyer, J. F. (2002). Compensatory value of urban trees in the United States. *Arboriculture & Urban Forestry*, 28(4), 194–199. 10.48044/jauf.2002.028

NPCC. (2010). *National Policy on Climate Change 2009*. Ministry of Natural Resources and Environment Malaysia.

Nuradzimmah, D. (2016). *Malaysia to become "aging notion" by 2035*. NST. https://www.nst.com.my/news/2016/03/134562/malaysia-become-aging nation-2035.

Nurul Aini B. (2018). Assessment of Health Status of The Elderly and Pre-elderly at A Malaysia Elderly Care Centre. *International Journal of Integrated Engineering 10*(7).

Olweny, M. R. O. (2020). Architectural education in sub-Saharan Africa: An investigation into pedagogical positions and knowledge frameworks. *Journal of Architecture (London)*, 25(6), 717–735. 10.1080/13602365.2020.1800794

Oregon Climate Change Research Institute. (2020). *Oregon's Fourth Climate Assessment Report*. OCCRI. https://www.occri.net/

Oregon Department of Environmental Quality. (2019). *Data and reports*. Oregon Department of Environmental Quality. https://www.oregon.gov/deq/pages/data-and-reports.aspx

Organisation for Economic Co-operation and Development (OECD). (2022). *Achieving the SDGs in cities and regions. Promoting local and regional development*. OECD. https://www.oecd.org/about/impact/achieving-sdgs-in-cities-and-regions.htm

Othman, S., Steen, M., & Fleet, J.-A. (2020). A sequential explanatory mixed methods study design: An example of how to integrate data in a midwifery research project. *Journal of Nursing Education and Practice*, 11(2), 75. 10.5430/jnep.v11n2p75

Özsoy, H. Ö. (2020). Evaluating Competitiveness In Product Design By Using Analytic Hierarchy Process. *İstanbul Ticaret Üniversitesi Sosyal Bilimler Dergisi, 19*(37), 37.

Páez, M., & Albert, L. (2012). Cultural consciousness. In Banks, J. A. (Ed.), *Encyclopedia of diversity in education* (Vol. 1, pp. 510–510). SAGE Publications, Inc., 10.4135/9781452218533.n160

Pagano, M. A., & Bowman, A. O. M. (2000). *Vacant Land in Cities: An Urban Resource*. Brookings Institution, Center on Urban and Metropolitan Policy.

Pais, D. F., Marques, A. C., & Fuinhas, J. A. (2020). Reducing Meat Consumption to Mitigate Climate Change and Promote Health: But Is It Good for the Economy? *Environmental Modeling and Assessment*, 25(6), 793–807. 10.1007/s10666-020-09710-0

Pandey, D., Agrawal, M., & Pandey, J. S. (2011). Carbon footprint: Current methods of estimation. *Environmental Monitoring and Assessment*, 178(1–4), 135–160. 10.1007/s10661-010-1678-y20848311

Pan, Y. (2019). The British Museum's cultural and creative exploration: How do world-class museums develop cultural and creative industries? *PR World*, 22, 23–26.

Parady, G., Taniguchi, A., & Takami, K. (2020). Travel behavior changes during the COVID-19 pandemic in Japan: Analyzing the effects of risk perception and social influence on going-out self-restriction. *Transportation Research Interdisciplinary Perspectives*, 7, 100181. 10.1016/j.trip.2020.10018135782787

Parker, C., Barnes, S., McKee, K., Morgan, K., Torrington, J., & Tregenza, P. (2004). Quality of Life and Building Design in Residential and Nursing Homes for Older People. *Ageing and Society*, 24(6), 941–962. 10.1017/S0144686X04002387

Park, N. S., Zimmerman, S., Sloane, P. D., Gruber-Baldini, A. L., & Kevein Eckert, J. (2006, May). "An Empirical Typology of Residential Care/Assited Livng Based on A Four-State-Study," The Gerontological Society of America. *The Gerontologist*, 46(2), 238–248. 10.1093/geront/46.2.23816581888

Pasha, Y. N. (2020). Presence Of Culture In Architectural Education: A Case Study Of METU School of Architecture. *METU Journal of the Faculty of Architecture*, 37(1), 95–116. 10.4305/METU.JFA.2020.1.7

Patten, S. B. (1997). Performance of the Composite International Diagnostic Interview Short Form for Major Depression in Community and Clinical Samples. *Health Canada*, 18(3), 131.9375257

Patz, J. A., Gibbs, H. K., Foley, J. A., Rogers, J. V., & Smith, K. R. (2015). Climate change and global health: Quantifying a growing ethical crisis. *EcoHealth*, 12(4), 447–455.

Paulsen, M. B. (1990). ASHE-ERIC Higher Education Report: Vol. 6. *College choice: Understanding student enrollment behavior*. George Washington University.

Pekalee, A., & Gray, R. S. (2023). In-home environment and happiness among older adults in Thailand. *Frontiers in Public Health*, 11, 1159350. 10.3389/fpubh.2023.115935037077188

Peng, J., Zhou, J., Meng, F., & Yu, Y. (2021). *Analysis on the hidden cost of prefabricated buildings based on FISM-BN.* PLoS ONE. 10.1371/journal.pone.0252138

Peng, Z.-R., Lu, K.-F., Liu, Y., & Zhai, W. (2023). The Pathway of Urban Planning AI: From Planning Support to Plan-Making. *Journal of Planning Education and Research*, 0(0), 0739456X231180568. Advance online publication. 10.1177/0739456X231180568

Persson, M. (2014). *Socio-geographic Effect On Higher Education Choice.* XVIII ISA World Congress of Sociology 13-19/7 - 2014, Yokohama, Japan.

Petticrew, M., & Roberts, H. (2008). *Systematic reviews in the social sciences: A practical guide.* Blackwell Publishing., 10.1002/9780470754887

Petti, L., Trillo, C., & Makore, B. N. (2020). Cultural Heritage and Sustainable Development Targets: A Possible Harmonisation? Insights from the European Perspective. Sustainability, 12(3). *Sustainability (Basel)*, 3(3), Q3. 10.3390/su12030926

Philokyprou, M. (2011). Teaching conservation and vernacular architecture. *Journal of Architectural Conservation*, 17(2), 7–24. 10.1080/13556207.2011.10785086

Phua, K. H., Yap, M. T., & Goh, L. G. (2019). *Aging in Asia: Contemporary Trends and Policy Issues.* ebcHost. http://search.ebscohost.com/login.aspx?direct=true&scope+site&db=nlebk&db=nlabk&AN=2158050

Pickles, D., & McCaig, I. (2017). *Energy Efficiency and Historic Buildings: Application of Part L of the Building Regulations to historic and traditionally constructed buildings.* Historic England.

Pike, A., MacKinnon, D., Coombes, M., Champion, T., Bradley, D., Cumbers, A., & Wymer, C. (2016). *Uneven growth: Tackling city decline.* Joseph Rowntree Foundation.

Pop, I., & Borza, A. (2016). Factors Influencing Museum Sustainability and Indicators for Museum Sustainability Measurement. *Sustainability, 8*(1), 101. Q3. 10.3390/su8010101

Portland Bureau of Planning and Sustainability. (2016). *Comprehensive Plan.* Portland Bureau of Planning and Sustainability. https://www.portland.gov/bps/comprehensive-plan

Portland Development Commission. (2020). *Economic development.* Portland Development Commission. https://www.pdc.us/economic-development/

Portland Parks & Recreation. (2019). *Parks, gardens & natural areas.* Portland Parks & Recreation. https://www.portland.gov/parks

Postalcı, I. E., & Atay, G. F. (2018). Rethinking on Cultural Sustainability in Architecture: A Reading on Projects of Behruz Çinici.

Proforma Construction Inc. (2019). *Why BIM is Important for the Construction Industry*. Proforma Construction. General Contractor Pleasanton CA. https://www.proformaco.com/blog/why-bim-is- important-for-the-construction- industry/

Public Works Department. (2019). *Heritage Building Conservation Need Statement. Jabatan Kerja Raya*. JKR.

Qian, Q. K., Chan, E. H. W., & Khalid, A. G. (2015). Challenges in Delivering Green Building Projects: Unearthing the Transaction Costs (TCs). *Sustainability (Basel)*, 7(4), 3615–3636. 10.3390/su7043615

Qin, Z., & Ng, S. (2020). Culture as Inspiration: A Metaphorical Framework for Designing Products with Traditional Cultural Properties (TCPs). Sustainability, 12(17). *Sustainability (Basel)*, 17(17), Q3. 10.3390/su12177171

Qin, Z., Song, Y., & Tian, Y. (2019). The Impact of Product Design with Traditional Cultural Properties (TCPs) on Consumer Behavior Through Cultural Perceptions: Evidence from the Young Chinese Generation. Sustainability, 11(2). *Sustainability (Basel)*, 2(2), Q3. 10.3390/su11020426

Quan, H., Li, S., Wei, H., & Hu, J. (2019). Personalized Product Evaluation Based on GRA-TOPSIS and Kansei Engineering. *Symmetry*, 11(7), 7. 10.3390/sym11070867

Railsback, S. F., & Grimm, V. (2019). *Agent-based and individual-based modeling: A practical introduction*. Princeton University Press.

Rajgor, M., Paresh, C., Dhruv, P., Chirag, P., & Dhrmesh, B. (2016). RII & IMPI : Effective Techniques for Finding Delay in Construction Project. *International Research Journal of Engineering and Technology*, 3(1), 1173–1177.

Ramalu, J. C. D., Bakar, N. A., & Nijar, N. N. (2013). Factors That Determine Students' Preferences In Selecting Higher Learning Institution. *Infrastructure University Kuala Lumpur Research Journal, 1*(1).

Ramdani, B., Kawalek, P., & Lorenzo, O. (2009). Predicting SMEs' adoption enterprise systems. *Journal of Enterprise Information Management*, 22(1/2), 10–24. 10.1108/17410390910922796

Ramlal, B., & Baban, S. (2008). Developing a GIS-based integrated approach to flood management in Trinidad, West Indies. *Journal of Environmental Management*, 88(4), 1131–1140. 10.1016/j.jenvman.2007.06.01017714856

Rapoport, A. (2005). Culture, architecture, and design: Locke Science Pub. *Co., Chicago, US.*

Rees, W., & Wackernagel, M. (2012). Urban ecological footprints: Why cities cannot be sustainable and why they are a key to sustainability. *The Urban Sociology Reader, 9255*(96), 157–165. 10.4324/9780203103333-26

Reher, G. S. (2020). What is value? Impact assessment of cultural heritage. *Journal of Cultural Heritage Management and Sustainable Development, 10*(4), 429–436. Q4. 10.1108/JCHMSD-10-2019-0128

Reyes, N. (2019). *Building Information Modeling Benefits for Architecture and Construction | Thought Leadership*. HMC Architects. https://hmcarchitects.com/news/building- information-modeling-benefits-for-architecture-and-construct ion-2019-06-05/#:~:text=Building%20Information%20Modeling%20(BIM)%20is,and%20how%2 0it%20will%20operate.

Richard, W. (2019). *Physical Changes with Aging*. Warren Alpert Medical School of Brown University.

Richardson, J. (2022, March 27). *Museums and sustainable development goals*. MuseumNext. https://www.museumnext .com/article/museums-and-sustainable-development-goals/

Robinson, S. L., & Lundholm, J. T. (2012). Ecosystem services provided by urban spontaneous vegetation. *Urban Ecosystems*, 15(3), 545–557. 10.1007/s11252-012-0225-8

Rogerson, A., Hankins, A., Nettel, P. F., & Rahim, S. (2022). Government AI Readiness Index 2022. *Oxford Insights*. https://www.unido.org/sites/default/files/files/2023-01/Government_AI_Readiness_2022_FV.pdf

Rogner, H. H. (1997). An assessment of world hydrocarbon resources. *Annual Review of Energy and the Environment*, 22(1), 217–262. 10.1146/annurev.energy.22.1.217

Roopnarine, R., Opadeyi, J., Eudoxie, G., Thongs, G., & Edwards, E. (2018). GIS-based flood Susceptibility and Risk Mapping Trinidad Using Weight Factor Modeling. *Caribbean Journal of Earth Science*, 49, 1–9.

Rosa, E. A., York, R., & Dietz, T. (2004). Tracking the Antropogenic Drivers of Ecological Impacts. *Ambio*, 33(8), 509–512. 10.1579/0044-7447-33.8.50915666682

Rosenthal, J. K., Crauderueff, R., & Carter, M. (2008). *Urban Heat Island Mitigation Can Improve New York City's Environment*. Sustainable South Bronx.

Run, W. (2023). *Research on safety risk assessment and control of prefabricated building construction based on SEM-FCEM*. CNKI. https://link.cnki.net/doi/10.27784/d.cnki.gahjz.2023.000336

Ruslan, R., Ariffin, K. H. K., Islam, M. A., & Zaidi, N. I. M. (2014). Determinants Students' Selection of Higher Education Institutions in Malaysia. Universiti Malaysia Perlis. *Advances in Environmental Biology, 8*(9), 406-416.

Ruth, M., & Coelho, D. (2007). Understanding and managing the complexity of urban systems under climate change. *Climate Policy*, 7(4), 317–336. 10.1080/14693062.2007.9685659

Saaty, T. L. (1990). *The Analytic Hierarchy Process: Planning, Priority Setting, Resource Allocation*. RWS Publications.

Sabrina Abdul Latif, N., Muhammad Rizwan Majeed, K., Rozzani, N., & Khadijah Saleh, S. (2020). Factors Affecting Housing Prices in Malaysia: A Literature Review. *International Journal of Asian Social Science*, 10(1), 63–67. 10.18488/journal.1.2020.101.63.67

Sackey, A. N. A. (2008). Environmental Impacts of Rapid Urbanization of Ethiopia's Cities in M.P. van Dijk and J. Fransen (eds) *Managing Ethiopian Cities in an Era of Rapid Urbanization.* Eburon Delft.

Saddiqi, M. M., Zhao, W., Cotterill, S., & Dereli, R. K. (2023). Smart management of combined sewer overflows: From an ancient technology to artificial intelligence. *WIREs. Water*, 10(3), e1635. 10.1002/wat2.1635

Saldana, J., & Omasta, M. (2016). *Qualitative Research: Analyzing Life*. Sage.

Saldo, I. J. P., & Walag, A. M. P. (2020). Utilizing problem-based and project-based learning in developing students' communication and collaboration skills in physics. *American Journal of Educational Research*, 8(5), 232–237.

Salem, M., Bose, A., Bashir, B., Basak, D., Roy, S., Chowdhury, I. R., Alsalman, A., & Tsurusaki, N. (2021). Urban Expansion Simulation Based on Various Driving Factors Using a Logistic Regression Model: Delhi as a Case Study. *Sustainability (Basel)*, 13(19), 10805. 10.3390/su131910805

Sallehudin, H. A. S. I. M. I., Razak, R. C., Ismail, M. O. H. A. M. M. A. D., Fadzil, A. M., & Baker, R. O. G. I. S. (2019). Cloud computing implementation in the public sector: Factors and impact. *Asia-Pasific Journal of Information Technology and Multimedia, 7*(2-2), 27-–42.

Sam, S. (2022). *3D BIM Explained: The Foundation of a BIM Journey*. 3D Design Bureau. https://3ddesignbureau.com/blog/3d-bim-explained-the-foundationof-a-bim-journey/

Sanchez, T. W., Shumway, H., Gordner, T., & Lim, T. (2023). The prospects of artificial intelligence in urban planning. *International Journal of Urban Science*, 27(2), 179–194. 10.1080/12265934.2022.2102538

Sangster, N., Bisnath, J., Maharajh, A., & Sammy, A. (2020). Investigation of the Watercourses and Measures to Alleviate Flooding in the Penal/Debe Region. *The International Conference on Emerging Trends in Engineering and Technology (IconETech-2020).* Faculty of Engineering, The UWI. 10.47412/CHIF9112

Sassen, S. (2001). The global city: New York, *London, Tokyo.* Princeton, NJ: Princeton University Press. 10.1515/9781400847488

Saunders, M. N. K., Lewis, P., & Thornhill, A. (2016). Understanding Research Philosophy and Approaches to Theory Development. Research Methods for Business Students. In *Harlow: Pearson Education* (Issue January).

Saunders, M., Lewis, P., & Thornhill, A. (2007). Research Methods for Bussiness Students. In *Pearson* (4th ed.).

Schechter, C. (2011). Switching cognitive gears. *Journal of Educational Administration*, 49(2), 143–165. 10.1108/09578231111116707

Schilling, J. M., & Mallach, A. (2012). *Cities in Transition: A Guide for Practicing Planners.* American Planning Association.

Schilling, J., & Logan, J. (2008). Greening the rust belt: A green infrastructure model for right sizing America's shrinking cities. *Journal of the American Planning Association*, 74(4), 451–466. 10.1080/01944360802354956

ScienceDaily. (2018). *Getting lost: Why older people might lose their way.* ScienceDaily. https://www.sciencedaily.com/releases/2018/03/180315140715.htm

Scobie, M., Gopinathan, U., Kelman, I., & Myhre, S. (2023). The agency of community groups in health and climate change adaptation governance and policy in SIDS: The case of in Toco, Trinidad and Tobago. *Environmental Science & Policy*, 147, 116–125. 10.1016/j.envsci.2023.05.006

Seamon, D., & Sowers, J. (2008). Place and Placelessness, Edward Relph. *Key Texts in Human Geography, London.*, (January), 43–51. 10.4135/9781446213742.n6

Sedgwick, P. (2015). A comparison of parametric and non-parametric statistical tests. *BMJ (Clinical Research Ed.)*, 350(apr17 1), 350. 10.1136/bmj.h205325888112

Seeley, I. H. (1996). *Building Economics.* Macmillan Education UK., 10.1007/978-1-349-13757-2

Sendek, A., Kretz, L., van der Plas, F., Seele-Dilbat, C., Schulz-Zunkel, C., Vieweg, M., Bondar-Kunze, E., Weigelt, A., & Wirth, C. (2021). Topographical features related to flooding frequency promote ecosystem multifunctionality of riparian floodplains. *Ecological Indicators*, 132(108312), 1–11. 10.1016/j.ecolind.2021.108312

Setia, M. S. (2016). Methodology Series Module 3: Cross-Sectional Studies. *Indian Journal of Dermatology*, 61(3), 261–264. 10.4103/0019-5154.18241027293245

Seto, K. C., Güneralp, B., & Hutyra, L. R. (2012). Global forecasts of urban expansion to 2030 and direct impacts on biodiversity and carbon pools. *Proceedings of the National Academy of Sciences of the United States of America*, 109(40), 16083–16088. 10.1073/pnas.121165810922988086

Seyfang, G. (2010). Community action for sustainable housing: Building a low-carbon future. *Energy Policy*, 38(12), 7624–7633. 10.1016/j.enpol.2009.10.027

Shah, A. (2015) *Why Is Biodiversity Important? Who Cares?* Global Issues. https://www.globalissues.org/article/170/why-is-biodiversity-important-who-cares

Shahbaz, M., Shahzad, S. J. H., & Mahalik, M. K. (2018). Is Globalization Detrimental to CO2 Emissions in Japan? New Threshold Analysis. *Environmental Modeling and Assessment*, 23(5), 557–568. 10.1007/s10666-017-9584-0

Shailja Tripathi, D. (2023). *Digital Transformation and Cloud Computing: System Dynamics Modeling Approach*. International Journal For Multidisciplinary Research.

Shakibaei, S., de Jong, G. C., Alpkökin, P., & Rashidi, T. H. (2021). Impact of the COVID-19 pandemic on travel behavior in Istanbul: A panel data analysis. *Sustainable Cities and Society*, 65, 102619. 10.1016/j.scs.2020.10261933251093

Shamshiripour, A., Rahimi, E., Shabanpour, R., & Mohammadian, A. (2020). How is COVID-19 reshaping activity-travel behavior? Evidence from a comprehensive survey in Chicago. *Transportation Research Interdisciplinary Perspectives*, 7, 100216. 10.1016/j.trip.2020.10021634173469

Shang, Y. L. (2023). Research on collaborative optimization management and platform construction of prefabricated construction supply chain enterprises based on BIM technology. *Journal of Shijiazhuang Railway Vocational and Technical College, 22*(3), 1–4+58.

Shao, L. (2022). Research on the visual design of non-heritage cultural and creative products based on sustainable design theory. *Art and Design (Theory)*, 2(4), 87–92. 10.16824/j.cnki.issn10082832.2022.04.020

Shapira, H., Ketchie, A., & Nehe, M. (2017). The integration of Design Thinking and Strategic Sustainable Development. *Journal of Cleaner Production, 140*, 277–287. Q1. 10.1016/j.jclepro.2015.10.092

Sharifi, A., Allam, Z., Bibri, S. E., & Khavarian-Garmsir, A. R. (2024). Smart cities and sustainable development goals (SDGs): A systematic literature review of co-benefits and trade-offs. *Cities, 146*, 104659. 10.1016/j.cities.2023.104659

Sharma, M., Luthra, S., Joshi, S., & Kumar, A. (2022). Implementing challenges of artificial intelligence: Evidence from the public manufacturing sector of an emerging economy. *Government Information Quarterly*, 39(4), 101624. 10.1016/j.giq.2021.101624

Sharma, S. K., Mohapatra, S., Sharma, R. C., Alturjman, S., Altrjman, C., Mostarda, L., & Stephan, T. (2022). Retrofitting Existing Buildings to Improve Energy Performance. *Sustainability (Basel)*, 14(2), 666. 10.3390/su14020666

Sharma, S., Nahid, S., Sharma, M., Sannigrahi, S., Anees, M. M., Sharma, R., Shekhar, R., Basu, A. S., Pilla, F., Basu, B., & Joshi, P. K. (2020). A long-term and comprehensive assessment of urbanization-induced impacts on ecosystem services in the capital city of India. *City and Environment Interactions*, 7, 100047. 10.1016/j.cacint.2020.100047

Sharoff, L. (2019). Creative and innovative online teaching strategies: Facilitation for active participation. *The Journal of Educators Online*, 16(2), n2. 10.9743/JEO.2019.16.2.9

Shen Z. Y., & Sun D. (2022). Research hotspots and trend prospects of high quality development of Vocational education in our country. *Education and profession, 20.* carroll nki. 1004-3985.2022.03.002.10.13615 / j.carol

Shen, J., Duan, H., Zhang, B., Wang, J., Ji, J. S., Wang, J., Pan, L., Wang, X., Zhao, K., Ying, B., Tang, S., Zhang, J., Liang, C., Sun, H., Lv, Y., Li, Y., Li, T., Li, L., Liu, H., & Shi, X. (2020). Prevention and control of COVID-19 in public transportation: Experience from China. *Environmental Pollution*, 266, 115291. Advance online publication. 10.1016/j.envpol.2020.11529132829124

Shenmare, K. (n.d.). Quantitative Research. *Wall Street Mojo.*https://www.wallstreetmojo.com/quantitative-research/

Shen, Q., Liu, Y., & Shen, Q. (2019). Predicted model of ecological footprint based on PLCA. *IOP Conference Series. Materials Science and Engineering*, 688(5), 055035. 10.1088/1757-899X/688/5/055035

Shen, Y. (2022). Evaluation and Selection of Cultural and Creative Design Solutions Based on Grey Correlation Multicriteria Decision Analysis. *Mathematical Problems in Engineering*, 2022, e1755728. 10.1155/2022/1755728

Shen, Y., Xu, M., Lin, Y., Cui, C., Shi, X., & Liu, Y. (2022). Safety risk management of prefabricated building construction based on ontology technology in the BIM environment. *Buildings*, 12(6), 765. 10.3390/buildings12060765

She, Y. (2021). Research on the training system of Engineering cost Professionals in vocational colleges under the background of "Three Education" reform. *Education Observation*, 10(10), 58–60. 10.16070/j.cnki.cn45-1388/g4s.2021.10.017

Shi, J. Y. (2023). Prefabricated construction industrialization worker development path and transformation and upgrading measures and strategies. *Sichuan Architecture*, 43(6), 289–290.

Shrivastav, S., & Joshi, R. (2021). Design Thinking for Long-Term Product Planning. In Chakrabarti, A., Poovaiah, R., Bokil, P., & Kant, V. (Eds.), *Design for Tomorrow—Volume 3* (pp. 387–398). Springer. 10.1007/978-981-16-0084-5_31

Shuster, W. D., Dadio, S., Drohan, P., Losco, R., & Shaffer, J. (2014). Residential demolition and its impact on vacant lot hydrology: Implications for the management of stormwater and sewer system overflows. *Landscape and Urban Planning*, 125, 48–59. 10.1016/j.landurbplan.2014.02.003

Sichuan Provincial Department of Housing and Urban-Rural Development. (2021). *Sichuan province issued the "14th Five-Year Plan" construction industry development plan.* SPD. http://jst.sc.gov.cn/scjst/ghxx/article_list.shtml

Sidin, S., Hussin, S., & Soon, T. (2003). An exploratory study of factors influencing the college choice decision of undergraduate students in Malaysia. *Asia-Pacific Management Review.*, 8, 259–280.

Singh, H. (2021). *5 Tips for Implementing 4D BIM for Construction Scheduling*. https://www.hitechcaddservices.com/news/tips-to-implement-4d-bim-forconstruction-scheduling/

Singh, H., & Gajjar, K. (2022). 5 Benefits of BIM for Contractors During the Pre-construction Stage. *CAD Services: Outsource CAD Drafting, 3D Design & BIM Services.*https://www.hitechcaddservices.com/news/bim-for- contractors-during-pre-construction-stage/

Singh, A., Baalsrud Hauge, J., & Wiktorsson, M. (2021). Simulation-Based Participatory Modelling in Urban and Production Logistics: A Review on Advances and Trends. *Sustainability (Basel)*, 14(17), 17. 10.3390/su14010017

Škerlič, S., Muha, R., & Logožar, K. (2016). A Decision-Making Model for Controlling Logistic Costs. *Tehnicki Vjesnik (Strojarski Fakultet)*, 23(1), 145–156. 10.17559/TV-20141015121023

Smit, B., & Wandel, J. (2006). Adaptation, adaptive capacity, and vulnerability. *Global Environmental Change*, 16(3), 282–292. 10.1016/j.gloenvcha.2006.03.008

Smith, G. A., & Sobel, D. (2014). *Place-and community-based education in schools.* Routledge. 10.4324/9780203858530

Smith, J. A. (2019). *Disaster Response and Recovery*. Wiley.

Smith, J. D. (2019). *Sustainable urban development in the age of climate change: Lessons from Portland.* Routledge.

Smith, J. D., & Johnson, K. (2018). Urban growth modeling of Portland, Oregon: Incorporating land use policy scenarios into cellular automata modeling. *Landscape and Urban Planning*, 175, 103–114.

Soares, N., Martins, C., Goncalves, M., Santos, P., da Silva, L. S., & Costa, J. J. (2019). Laboratory and In-Situ Non-Destructive Methods to Evaluate the Thermal Transmittance and Behavior of Walls, Windows, and Construction Elements with Innovative Materials: A review. *Energy and Building*, 182, 88–110. 10.1016/j.enbuild.2018.10.021

Sohu. (2018). *All classics! 10 well-known prefabricated buildings in china*. Sohu. https://www.sohu.com/a/284605502_120049046

Soini, K., & Birkeland, I. (2014a). Exploring the scientific discourse on cultural sustainability. *Geoforum, 51*, 213–223. Q1. 10.1016/j.geoforum.2013.12.001

Soman, R. K. (2019). Modelling construction scheduling constraints using Shapes Constraint Language (SHACL). In *2019 European Conference on Computing in Construction* (pp. 351-358). IEEE. 10.35490/EC3.2019.170

Song, W. (2022). Building construction design based on particle swarm optimization algorithm. *Journal of Control Science and Engineering*, 2022, 1–8. 10.1155/2022/7139230

Song, Y., & Li, M. (2018). Research on Cultural and Creative Product Development Based on Museum Resources. *IOP Conference Series. Materials Science and Engineering*, 452(2), 022090. 10.1088/1757-899X/452/2/022090

Soon, L. T. S., Hassan, H., & Abidin, N. Z. (2019). Measuring the Effects and Efficiency of 3D modelling in Quantity Surveying Profession. *MCRJ, 2*(7), 58–64. https://expert.taylors.edu.my/file/rems/publication/10 5958_6709_1.

Sopery, L. H. M., & Ayob, M. F. (2023). An Investigation on Sustainable Refurbishment Practice of Building Works in Malaysia. *Journal of Architecture. Planning and Construction Management*, 13(1), 29–39.

Srivastava, R. (2020). Growing Precarity, Circular Migration, and the Lockdown in India. *The Indian Journal of Labour Economics : the Quarterly Journal of the Indian Society of Labour Economics*, 63(0123456789), 79–86. 10.1007/s41027-020-00260-332929315

Staub-French, S., & Fischer, M. A. (2000). *Formalisms and Mechanisms Needed to Maintain Cost Estimates Based on an IFC Product Model.* American Society of Civil Engineers. 10.1061/40513(279)94

Steel, J., Drogemuller, R., & Toth, B. (2012). Model interoperability in building information modelling. *Software & Systems Modeling*, 11(1), 99–109. 10.1007/s10270-010-0178-4

Sterman, J. D. (2000). *Business dynamics: Systems thinking and modeling for a complex world.* Irwin/McGraw-Hill.

Stern, P. C. (2000). Toward a Coherent Theory of Environmentally Significant Behavior. *The Journal of Social Issues*, 56(3), 407–424. 10.1111/0022-4537.00175

Stone, D. H. (1993). Design a questionnaire. *British Medical Journal*, 307(6914), 1264–1266. 10.1136/bmj.307.6914.12648281062

Stonemark, S. (2022). *How Construction Schedules Keep Projects on Time and Budget.* Stonemark Construction Management. https://stonemarkcm.com/blog/how-construction-schedules-keep-projects-on-time-and-budget/

Storch, H., & Downes, N. (2011). A scenario-based approach to assess Ho Chi Minh City's urban development strategies against the impact of climate change. *Cities (London, England)*, 28(6), 517–526. 10.1016/j.cities.2011.07.002

Storrier, A. (2021). *How CAD and BIM Changed the Construction Industry.* Resource Center. https://resources.hexagonppm .com/articles- blogs/how-cad-and-bim-changed-the-construction-industry

Su, A., Xia, M. W., & Tao, X. X. (2024). Research on the Relationship between Education policy and the development of higher vocational Colleges from the perspective of system Theory. *Journal of system science*, 1-8. http://kns.cnki.net/kcms/detail/14.1333.N.20240305.1849.002.html

Subedi, D. (2016). Explanatory Sequential Mixed Method Design as the Third Research Community of Knowledge Claim. *American Journal of Educational Research, 4*(7), 570–577. https://doi.org/10.12691/education-4-7-10

Succar, B. (2009). Building information modelling framework: A research and delivery foundation for industry stakeholders. *Automation in Construction*, 18(3), 357–375. 10.1016/j.autcon.2008.10.003

Succar, B., & Kassem, M. (2015). Macro-BIM adoption: Conceptual structures. *Automation in Construction*, 57, 64–79. 10.1016/j.autcon.2015.04.018

Sugiyama, T., & Thompson, C. W. (2006). Environmental support for outdoor activities and older people's quality of life. *Journal of Housing for the Elderly*, 19(3-4), 167–185. 10.1300/J081v19n03_09

Sukopp, H., Blume, H. P., & Kunick, W. (1979). The soil, flora, and vegetation of Berlin's waste lands. *In Nature in Cities: The Natural Environment in the Design and Development of Urban Green Space*; Wiley. *New Jersy, NY, USA*, 1979, 115–132.

Sulaiman. (2017). Pengaruh Teknologi Komunikasi Terhadap Perubahan Keorganisasian Di Jabatan Penyiaran Malaysia (Effect of Changes in Organizational Communication Technology in the Broadcasting Department of Malaysia). *e-Bangi, 12*(1).

Sullivan, K. T., & Pisoni, R. (2019). BIM and cost estimating: Perceptions, issues, and strategies. *Journal of Management Engineering*, 35(6), 05019008.

Sun, H., Fang, Y., Yin, M., & Shi, F. (2023). Research on the restrictive factors of vigorous promotion of prefabricated buildings in yancheng under the background of "double carbon". *Sustainability (Basel)*, 15(2), 1737. 10.3390/su15021737

Sun, Y., Wang, Z.-J., Deveci, M., & Chen, Z.-S. (2024). Optimal releasing strategy of enterprise software firms facing the competition from cloud providers. *Expert Systems with Applications*, 236, 121264. 10.1016/j.eswa.2023.121264

Sun, Z., Li, X., Fu, W., Li, Y., & Tang, D. (2013). Long-term effects of land use/land cover change on surface runoff in urban areas of Beijing, China. *Journal of Applied Remote Sensing*, 8(1), 084596. 10.1117/1.JRS.8.084596

Su, T. T. (2020). The current situation of prefabricated buildings in china and their application in residential buildings. *Building Materials and Decoration*, (11), 98–99.

Sutrisna, M. (2009). Research Methodology in Doctoral Research: Understanding the Meaning of Conducting Qualitative Research. *Association of Researchers in Construction Management (ARCOM). Doctoral Workshop*, 2(May), 48–57.

Swart, R., Fons, J., Geertsema, W., van Hove, B., Gregor, M., Havranek, M., & Peltonen, L. (2012). *Urban vulnerability indicators: A joint report.* ETC-CCA & ETC-SIA. http://cca.eionet.europa.eu/docs/TP_3-2012,

Syed Zainol Abidin, I. (1995). *Pemeliharaan Warisan Rupa Bandar*. Badan Warisan Malaysia.

Taherdoost, H. (2022). What are Different Research Approaches? Comprehensive Review of Qualitative, Quantitative, and Mixed Method Research, Their Applications, Types, and Limitations. *Journal of Management Science & Engineering Research*, 5(1), 53–63. 10.30564/jmser.v5i1.4538

Tajudeen, A. B. A. (2023). The challenges faced by private higher education in Malaysia from the 1990s to COVID-19. Sejarah: Journal of History Department. *University of Malaya*, 32(1), 166–187. 10.22452/sejarah.vol32no1.10166

Takabi, H., Joshi, J. B., & Ahn, G. J. (2010). Security and privacy challenges in cloud computing environments. *IEEE Security and Privacy*, 8(6), 24–31. 10.1109/MSP.2010.186

Talukder, B., vanLoon, G. W., Hipel, K. W., & Orbinski, J. (2021). COVID-19's implications on agri-food systems and human health in Bangladesh. *Current Research in Environmental Sustainability*, 3, 100033. 10.1016/j.crsust.2021.10003334977606

Tang, M., & Fouad, N. (1997). Understanding Asian Americans' career choices: A social cognitive approach to examine the factors affecting their choices. Paper presented at the Annual Conference of American Psychological Association, Chicago.

Tang, K. H. D. (2019). Climate Change in Malaysia: Trends, Contributors, Impacts, Mitigation and Adaptations. *The Science of the Total Environment*, 650, 1858–1871. 10.1016/j.scitotenv.2018.09.31630290336

Tanko, B. l., & Mbugua, L. (2022). BIM Education in Higher Learning Institutions: A Scientometric Review and the Malaysia Perspective. *International Journal of Built Environment and Sustainability*, *9*(1), 23–37. https://ijbes.utm.my/index.php/ijbes/article/view/864/ 250

Tavares, V., Gregory, J., Kirchain, R., & Freire, F. (2021). What is the potential for prefabricated buildings to decrease costs and contribute to meeting EU environmental targets? *Building and Environment*, 206, 108382. 10.1016/j.buildenv.2021.108382

TeamSystem. (2021). *Advantages of using BIM in Construction.* TeamSystem Construction. https://www.teamsystemconstruction.com/magazine/advantages-of- bim/#:~:text=Pre%20-%20Construction%20Stage&text=-BIM%20is%20helpful%20not%20only,regulations%20regarding%20functi onality%20and%20constructability.

Teng, Y., & Pan, W. (2019). Systematic embodied carbon assessment and reduction of prefabricated high-rise public residential buildings in Hong Kong. *Journal of Cleaner Production*, 238, 117791. 10.1016/j.jclepro.2019.117791

The Centre for Aging. (2011). *Aging: Age-Related Physical Changes.* College of Nursing, University of Cincinnati. https://nursing.uc.edu/content/dam/nursing/docs/CFAWD/Aging %20Series/Part%202%Aging%20 Changes.pdf.

The Geospatial. (2019). What is BIM? Why is it Important? *Medium.* https://medium.com/@thegeospatialnews/what-is- bim-why-is-it-important- a662df34b60b

The Greening of Detroit. (2012). Urban Agriculture and Open Space. The Greening of Detroit. http://detroitagriculture.met/open-space-stories/

The Ministry of Housing and Urban-Rural Development. (2022). *13th Five-Year Plan for the Development of Project Cost Undertakings.* https://www.mohurd.gov.cn/gongkai/fdzdgknr/tzgg/201708/20170804_232859.html

The Peninsular. (2021). *Why do older adults struggle to adapt to new environments?* The Peninsular. https://thepeninsulaqatar.com/article/27/04/2016/Why-do-older-adults-struggle-to-adapt-to-new-environments

The Star. (2011, February 20). University's First Chancellor. *TheStar.*

The Trust for Public Land. (2019). *Our work in Oregon.* TPL. https://www.tpl.org/our-work/oregon

TheCADRoom. (2022, July 28). CAD Vs BIM: What is the Difference? | The CAD Room. *The CAD Room.* https://www.thecadroom.com/difference-between-cad-and- bim/#:~:text=CAD%20has%20rapidly%20been%20overtaken,same%20database%20and%20building%20model

Thongs, G. (2019). Integrating risk perceptions into flood risk management: Trinidad case study. *Natural Hazards*, 98(2), 593–619. 10.1007/s11069-019-03720-2

Tian, G., Zhang, H., Zhou, M., & Li, Z. (2018). AHP, Gray Correlation, and TOPSIS Combined Approach to Green Performance Evaluation of Design Alternatives. *IEEE Transactions on Systems, Man, and Cybernetics. Systems*, 48(7), 1093–1105. 10.1109/TSMC.2016.2640179

Tirachini, A., & Cats, O. (2020). COVID-19 and public transportation: Current assessment, prospects, and research needs. *Journal of Public Transportation*, 22(1), 1–34. 10.5038/2375-0901.22.1.136118518

Tobias, M. (2004). *Elements of Sustainable Architecture.* NY Engineers. https://www.ny-engineers.com/blog/elements-of-sustainable-architecture

Tong, S., Sun, Y., Ranatunga, T., He, J., & Yang, Y. (2012). Predicting plausible impacts of sets of climate and land use change scenarios on water resources. *Applied Geography (Sevenoaks, England)*, 32(2), 477–489. 10.1016/j.apgeog.2011.06.014

Town and Country Planning Organisation. (2014). *Ministry of Urban Development*. Government of India.

Trivedi, A. J., & Mehta, A. (2014). Maslow's Hierarchy of Needs. A Theory of Human Motivation. Psychological. *Psychological Review*, 7(38), 2014. https://kalyan-city.blogspot.com/2010/06/maslow-hierarchy-of-needs-theory-of.html

Troiani, I., Ewing, S., & Periton, D. (2013). *Architecture and Culture: Architecture's Disciplinarity*. Architecture and Culture., 10.2752/175145213X13760412749917

Tsai, P.-H., & Lin, C.-T. (2018). How Should National Museums Create Competitive Advantage Following Changes in the Global Economic Environment? *Sustainability (Basel)*, 10(10), 3749. 10.3390/su10103749

Tu, J.-C., Liu, L.-X., & Cui, Y. (2019). A Study on Consumers' Preferences for the Palace Museum's Cultural and Creative Products from the Perspective of Cultural Sustainability. Sustainability, 11(13). *Sustainability (Basel)*, 13(13), Q3. 10.3390/su11133502

U.S. Census Bureau. (2020). *Census data.* US Census Bureau. https://data.census.gov/

U.S. Department of Housing and Urban Development. (2009). *Revitalizing Foreclosed Properties with Land Banks.* Diane Publishing.

Udika, R. (2010). *Flood Management: An examination of mitigation measures for flooding in Urban areas in Trinidad.* 46th ISOCARP Congress, Nairobi, Kenya. https://www.isocarp.net/data/case_studies/1758.pdf

UIA. (2023). *UNESCO-UIA Charter for Architectural Education.* Union Internationale des Architectes.

Ujwary-Gil, A., Florek-Paszkowska, A., & Kozioł, A. (2023). AHP/ANP Applications in Economic Policy, Business, and Management with the T. L. Saaty Decision Making Hackathon at ICEFM 2022. *International Journal of the Analytic Hierarchy Process*, 14(3). 10.13033/ijahp.v14i3.1071

UN. (2022). *Transforming our World: The 2030 Agenda for Sustainable Development.* UN. https://sustainabledevelopment.un.org/post2015/transformingourworld

UNCHS. (1996). *An Urbanizing World. World Global Report on Human Settlements.* Oxford University Press, New York.

UNDRR. (2020). Disaster Risk Reduction in Malaysia: Status Report 2020. Bangkok, Thailand, United Nations Office for Disaster Risk Reduction (UNDRR), Regional Office for Asia and the Pacific.

UNESCO (Ed.). (2003). *Culture, trade and globalization: Questions and answers.* Unesco.

UNESCO. (2002). *UNESCO universal declaration on cultural diversity: A vision, a conceptual platform, a pool of ideas for implementation, a new paradigm.* UNESCO.

UN-Habitat. (2016). *New Urban Agenda.* UN. https://.org/new-urban-agenda

United Nations. (2018). *World urbanization prospects: The 2018 revision.* UN. https://population.un.org/wup/

United Nations. (2023). (UN) Office for the Coordination of Humanitarian Affairs (UN-OCHA). *Trinidad and Tobago: Floods - Nov 2022.* UN-OCHA. https://reliefweb.int/disaster/fl-2022-000374-tto

UNSDG. (2015). *UN General Assembly, Transforming our world: the 2030 Agenda for Sustainable Development, 21 October 2015, A/RES/70/1.* UN. https://www.refworld.org/docid/57b6e3e44.html

Valentín-Gamazo, D. V., Durango, N. T., Garcia-Herrero, F., Blanco, M. I. M., & de las Moras, M. C. R. (2021). Project-Based Learning with historic buildings: Immersion in real training environments for the Degree in Technical Architecture. *Proceedings of the International Conference of Innovation, Learning and Cooperation 2021, 3129.* IEEE.

van Aalst, M. K., Cannon, T., & Burton, I. (2008). Community level adaptation to climate change: The potential role of participatory community risk assessment. *Global Environmental Change*, 18(1), 165–179. 10.1016/j.gloenvcha.2007.06.002

Van Beek, J. H. G. M., Kirkwood, T. B. L., & Bassingthwaighte, J. B. (2016). Understanding the Physiology of the Ageing Individual: Computational Modelling of Changes in Metabolism and Endurance. *Interface Focus*, 6(2), 20150079. 10.1098/rsfs.2015.007927051508

Van Der Walt, J. L. (2020). Interpretivism-Constructivism as a Research Method in the Humanities and Social Sciences-More to It Than Meets the Eye. *International Journal of Philosophy*, 8(1), 2333–5769. 10.15640/ijpt.v8n1a5

Vasovic, D., Lelovic, S., & Okrajnov-Bajic, R. (2021). Development of new materials for prefabricated concrete elements for building construction in tropical climatic conditions. *IOP Conference Series. Earth and Environmental Science*, 798(1), e012015. 10.1088/1755-1315/798/1/012015

Verhulst, P. F. (1838). Notice sur la loi que la population suit dans son accroissement. *Correspondance Mathématique et Physique*, 10, 113–121.

VIATechnik. (2022). *4D BIM and Scheduling*. VIATechnik. https://www.viatechnik.com/services/construction-administration/4d-bim-scheduling/

Vila-Henninger, L., Dupuy, C., Van Ingelgom, V., Caprioli, M., Teuber, F., Pennetreau, D., Bussi, M., & Le Gall, C. (2022). Abductive Coding: Theory Building and Qualitative (Re)Analysis. In *Sociological Methods and Research.* 10.1177/00491241211067508

Villacañas de Castro, L. S., Moreno-Serrano, L. M., & Giner Real, C. (2022). Museum education, cultural sustainability, and English language teaching in Spain. *Pedagogy, Culture & Society, 30*(2), 201–223. Q1. 10.1080/14681366.2020.1794947

Villarroel, J. D., Antón, Á., Zuazagoitia, D., & Nuño, T. (2017). Young children's environmental judgement and its relationship with their understanding of the concept of living things. *Environmental & Socio-Economic Studies*, 5(1), 1–10. 10.1515/environ-2017-0001

Vinzero. (2019). *What can BIM do after construction is complete?*https://www.a2kstore.com/what-can-bim- do-after-construction-is-complete

Visutsak, P., & Daoudi, M. (2017a). The smart home for the elderly: Perceptions, technologies and psychological accessibilities: The requirements analysis for the elderly in Thailand. *ICAT 2017 - 26th International Conference on Information, Communication and Automation Technologies, Proceedings, 2017-Decem*(October 2017), (pp. 1–6). IEEE. 10.1109/ICAT.2017.8171625

Volpe, S., Sangiorgio, V., Petrella, A., Coppola, A., Notarnicola, M., & Fiorito, F. (2021). Building envelope prefabricated with 3d printing technology. *Sustainability (Basel)*, 13(16), 8923. 10.3390/su13168923

Vries, G. D., & De Vries, G. (2022). Culture and the sustainable development goals: Challenges and opportunities. *Economia Della Cultura*, 4/2022. 10.1446/107171

Wachter, S., Scruggs, G., Voith, R., & Huang, L. (2010). *Redevelopment Authority of the City of Philadelphia Land Use and Policy Study*. Penn Institute for Urban Research, Econsult Corporation.

Wackernagel, M. Schulz, N. B., Deumling, D., Linares, A. C., Jenkins, M., Kapos, V., Monfreda, C., Loh, J., Myers, N., Norgaard, R., & Randers, J. (2002). *Tracking the ecological overshoot of the human economy*. PNAS. www.pnas.org

Wackernagel, Mathis, & Yount, J. D. (2000). *Footprints For Sustainability: The Next Steps Division; WCED-World Commission on Environment and Development*. WWF-World Wide Fund for Nature (originally World Wildlife Fund). http://www.rprogress.org

Wackernagel, M., Lewan, L., & Hansson, C. B. (1999). Evaluating the use of natural capital with the ecological footprint - Applications in Sweden and subregions. *Ambio*, 28(7), 604–612.

Waddell, P. (2002). UrbanSim: Modeling urban development for land use, transportation, and environmental planning. *Journal of the American Planning Association*, 68(3), 297–314. 10.1080/01944360208976274

Wahab, E., Shamsuddin, A., Abdullah, N., & Yi, K. S. (2018). *A Study on the Smart Home Adoption in Malaysia : A Foresight Perspective. 112th The IRES International Conference*, Jeju Island, South Korea.

Wakeland, W., Cholette, S., & Venkat, K. (2012). Food transportation issues and reducing carbon footprint. In *Food Engineering Series* (pp. 211–236). Springer. 10.1007/978-1-4614-1587-9_9

Wang J. et al. (2024). Nearly 20 years in the study of Chinese building energy saving process, hot spot and trend analysis. *Building science,40* (02), 179-188. The DOI: carroll nki. 11-1962 / tu 2024.02.21.10.13614 / j.carol

Wang, P. W. (2019). *Research on the influence of technical and economic policy on technological innovation of prefabricated building* [Master's thesis, Shenyang Jianzhu University]. CNKI. https://cdmd.cnki.com.cn/Article/CDMD-10153-1019255628.htm

Wang, J. Q. (2023). Construction and analysis of knowledge graph of BIM research in railway field. *Journal of Railway Engineering*, 40(11), 87–93.

Wang, M., & Hu, Y. (2021). AHP-TOPSIS-based Attachment Product Design for Children with Autism Spectrum Disorder. *Packaging Engineering*, 42(18), 220–226. 10.19554/j.cnki.1001-3563.2021.18.025

Wang, S., Mursalin, Y., Lin, G., & Lin, C. (2018). Supply chain cost prediction for prefabricated building construction under uncertainty. *Mathematical Problems in Engineering*, 2018, 4580651. 10.1155/2018/4580651

Wang, S., & Sinha, R. (2021). Life cycle assessment of different prefabricated rates for building construction. *Buildings*, 11(11), 552. 10.3390/buildings11110552

Wang, Sh. H., Wei-Chih, W., Kun-Chi, W., & Shih-Yu, Sh. (2015). Applying building information modeling to support fire safety management. *Automation in Construction*, 59, 2015. 10.1016/j.autcon.2015.02.001

Wang, X. M. (2018). Research on Teaching Quality in Chinese Universities: Trajectory, hotspots and Future trend: CiteSpace visualization analysis based on 14 journals of higher Education. *Journal of Education Academic*, (1), 91–103.

Wang, Y. R., Pan, H. P., & Cai, W. Q. (2020). Research on china's prefabricated construction industry policies, current situation and development trends. *Journal of Anhui Jianzhu University*, 28(4), 47–51.

Wang, Y. S., & Fu, Y. Y. (2020). The current situation, hotspots, and trends of international higher education evaluation research in the past decade. *Higher Education Research in Heilongjiang Province*, 38(03), 6–11. 10.19903/j.cnki.cn23-1074/g.2020.03.002

Warden, R., & Woodcock, D. (2005). Historic documentation: A model of project based learning for architectural education. *Landscape and Urban Planning*, 73(2), 110–119. 10.1016/j.landurbplan.2004.11.003

Watkins, J. (2020). Preventing a covid-19 pandemic. In *The BMJ* (Vol. 368). BMJ Publishing Group. 10.1136/bmj.m810

Wegener, M., & Fürst, F. (2011). Land-use transport interaction: State of the art. *SSRN* 1887602.

Wei S. M., & He Z. (2024). Bottleneck Restriction and Practice Direction of Higher Vocational Education Promoting skill Transformation in our country. *China's higher education research,2024* (02), 94-100. DOI: carroll nki. 1004-3667.2024.02.13.10.16298 / j.carol

Weill, J. A., Stigler, M., Deschenes, O., & Springborn, M. R. (2020). Social distancing responses to COVID-19 emergency declarations strongly differentiated by income. *Proceedings of the National Academy of Sciences of the United States of America*, 117(33), 19658–19660. 10.1073/pnas.200941211732727905

Wen, D., & Zhang, Z. (2022). Furniture Design Strategies for Young People Living Alone Based on KANO and TRIZ Theories. *Furniture & Interiors*, 3, 6.

Weng, Q. (2001). A Remote Sensing-GIS Evaluation of Urban Expansion and Its Impact on Surface Temperature in the Zhujiang Delta, China. *International Journal of Remote Sensing*, 22(10), 1999–2014.

Westland, J. (2006). *The Project Management Life Cycle: A Complete Step-by-step Methodology for Initiating Planning Executing and Closing the Project*. Kogan Page Limited.

West, S. E., Owen, A., Axelsson, K., & West, C. D. (2016). Evaluating the Use of a Carbon Footprint Calculator: Communicating Impacts of Consumption at Household Level and Exploring Mitigation Options. *Journal of Industrial Ecology*, 20(3), 396–409. 10.1111/jiec.12372

White, T. J. (2007). Sharing resources: The global distribution of the Ecological Footprint. *Ecological Economics*, 64(2), 402–410. 10.1016/j.ecolecon.2007.07.024

Whitmarsh, L. (2009). Behavioural responses to climate change: Asymmetry of intentions and impacts. *Journal of Environmental Psychology*, 29(1), 13–23. 10.1016/j.jenvp.2008.05.003

Wielechowski, M., Czech, K., & Grz da, Ł. (2020). Decline in mobility: Public transport in Poland in the time of the COVID-19 pandemic. *Economies*, 8(4), 1–24. 10.3390/economies8040078

Wilkinson, S. (2011). *Sustainable Retrofit Potential in Lower Quality Office Stock in the Central Business District*. CIB Management and Innovation in the Sustainable Built Environment.

Witt, E., & Kähkönen, K. (2019). BIM-Enabled Education: a Systematic Literature Review. *10th Nordic Conference on Construction Economics and Organization (Emerald Reach Proceedings Series, Vol. 2)*. Emerald. 10.1108/S2516-285320190000002042

Workie, E., Mackolil, J., Nyika, J., & Ramadas, S. (2020). Deciphering the impact of COVID-19 pandemic on food security, agriculture, and livelihoods: A review of the evidence from developing countries. *Current Research in Environmental Sustainability*, 2, 100014. 10.1016/j.crsust.2020.10001434977605

World Health Organisation. (2020). *Health Statistics and Information Systems*. WHO. https://www.int/healthinfo/survey/aging defnolder/en/. Accessed on October 6 2020.

World Health Organization (WHO). (2023). *Floods*. WHO. https://www.who.int/health-topics/floods#tab=tab_1

Wu, L., & Li, Z. (2020). Study on local market demand and occupational ability composition of engineering cost professionals. *Times Finance*, 161-162.

Wubet, M. T. (2003). *Estimation of Absolute Surface Temperature by Satellite Remote Sensing,* [Unpublished M. Sc Thesis, International Institute for Geoinformation Science and Earth Observation, Netherlands].

Wu, C., Yu, S., Liu, J., & Chen, G. (2020). Development and testing of hybrid precast steel-reinforced concrete column-to-H shape steel beam connections under cyclic loading. *Engineering Structures*, 211, 110460. 10.1016/j.engstruct.2020.110460

Wu, S. Z., Zhen, F., Wu, X., Guo, L., Zhou, X., Zhao, Y., Mao, W., Hu, J., & Liu, Z. (2023). Roundtable discussion: Rethinking and exploring smart city development. *Frontiers of Urban and Rural Planning*, 1(1), 1–10. 10.1007/s44243-022-00003-6

Wu, S., & Liu, Q. (2020). Analysis on the application of BIM and RFID in life cycle management of prefabricated building. *IOP Conference Series. Materials Science and Engineering*, 780(3), e032012. 10.1088/1757-899X/780/3/032012

Xie, L., & Lai, X. (2020). The development of cultural and creative products in red culture memorial museums. *Culture Industry*, 8, 26–27.

Xie, M., & Chen, W. J. (2021). Knowledge graph of Mooc research in China: Hot spots, current situation and trend analysis based on Cite Space. *Journal of Southwest University for Nationalities*, 42(01), 229–235.

Yadav, P., Duckworth, K., & Grewal, P. S. (2012). Habitat structure influences below ground biocontrol services: A comparison between urban gardens and vacant lands. *Landscape and Urban Planning*, 104(2), 238–244. 10.1016/j.landurbplan.2011.10.018

Yang, J., Luo, J. M., & Lai, I. K. W. (2021). Construction of Leisure Consumer Loyalty from Cultural Identity—A Case of Cantonese Opera. *Sustainability, 13*(4), 1980. Q3. 10.3390/su13041980

Yang, Y. (2021). *Application analysis of BIM technology and prefabricated structure construction [Paper presentation].* The Sixth International Conference on Information Management and Technology, Jakarta. 10.1016/j.cie.2021.107316

Yang, B., Dong, M., Wang, C., Liu, B., Wang, Z., & Zhang, B. (2021). IFC-based 4D construction management information model of prefabricated buildings and its application in graph database. *Applied Sciences (Basel, Switzerland)*, 11(16), 7270. 10.3390/app11167270

Yang, B., Fang, T., Luo, X., Liu, B., & Dong, M. (2022). A BIM-based approach to automated prefabricated building construction site layout planning. *KSCE Journal of Civil Engineering*, 26(4), 1535–1552. 10.1007/s12205-021-0746-x

Yang, L., Liu, S., Liu, J., Zhang, Z., Wan, X., Huang, B., Chen, Y., & Zhang, Y. (2020). COVID-19: Immunopathogenesis and Immunotherapeutics. *Signal Transduction and Targeted Therapy*, 5(1), 1–8. 10.1038/s41392-020-00243-232712629

Yang, M. (2022). Risk management of prefabricated building construction based on fuzzy neural network. *Scientific Programming*, 2022, 2420936. 10.1155/2022/2420936

Yang, S.-N., Chang, L.-C., & Chang, F.-J. (2019). AI-based design of urban stormwater detention facilities accounting for carryover storage. *Journal of Hydrology (Amsterdam)*, 575, 1111–1122. 10.1016/j.jhydrol.2019.06.009

Yau, Y. H., & Hasbi, S. J. E. P. (2017). A Comprehensive Case Study of Climate Change Impacts on the Cooling Load in an Air-Conditioned Office Building in Malaysia. *Energy Procedia*, 143, 295–300. 10.1016/j.egypro.2017.12.687

Yazdani, S., Azandehi, S. K., Ghorbani, A., & Shakerian, S. (2018). Explaining the Process of Choosing Clinical Specialties in General Medical Graduates: A Grounded Theory. *Electronic Journal of General Medicine, 15*(6).

Ye, M., Kong, H. B., & Zhang, W. (2018). New Engineering: From Concept to Action. *Higher Engineering Education Research*, 2018(01), 24–31.

Yeo, H. C. (2018, June). Generational Differences Between Employees In Quantity Surveying Firms. *Universiti Teknologi Malaysia.*

Yeomans, J. S., & Kozlova, M. (2023). Extending system dynamics modeling using simulation decomposition to improve the urban planning process. *Frontiers in Sustainable Cities*, 5, 1129316. 10.3389/frsc.2023.1129316

Ye, X., Wang, S., Lu, Z., Song, Y., & Yu, S. (2021). Towards an AI-driven framework for multi-scale urban flood resilience planning and design. *Computational Urban Science*, 1(11), 11. 10.1007/s43762-021-00011-0

Yigitcanlar, T., Corchado, J. M., Mehmood, R., Li, R. Y. M., Mossberger, K., & Desouza, K. (2021). Responsible urban innovation with local government artificial intelligence (AI): A conceptual framework and research agenda. *Journal of Open Innovation*, 7(1), 71. 10.3390/joitmc7010071

Yin, Y. L., & Bai, J. (2015). Exploration and practice of the training mode of applied engineering cost professionals - taking Tianjin University of Technology as an example. *China Engineering Science*, 17(01), 114–119.

Yi, Y. (2019). Cultural and creative product design from the perspective of the relationship between design and culture. *Art Education*, 9, 261–262.

Yohe, G., & Tol, R. (2002). Indicators for social and economic coping capacity: 'Moving towards a working definition of adaptive capacity. *Global Environmental Change*, 12(1), 25–40. 10.1016/S0959-3780(01)00026-7

Yuki, K., Fujiogi, M., & Koutsogiannaki, S. (2020). COVID-19 pathophysiology: A review. In *Clinical Immunology (Vol. 215)*. Academic Press Inc. 10.1016/j.clim.2020.108427

Yunus, R., Handan, R., & Riazi, S. R. M. (2020). Case studies on sustainability factors for industrialised building system (IBS). *International Journal of Sustainabl e Construction Engineering and Technology*, 11(2), 65–71. 10.30880/ijscet.2020.11.02.007

Yu, S., Kong, X., Wang, Q., Yang, Z., & Peng, J. (2023). A new approach of A new approach of Robustness-Resistance-Recovery (3Rs) to assessing flood resilience: A case study in Dongting Lake Basin. *Landscape and Urban Planning*, 230, 104605. 10.1016/j.landurbplan.2022.104605

Yusof, M. F. M. (2017). *Revolusi Perindustrian Keempat / Industri 4.0 / IR 4.0.*

Zaid, N. S. M., Ghazali, M. I., Azmi, N. F., Kamarulzaman, P. N., & Akashah, F. W. (2021). Environmental Sustainability in Malaysian Heritage Building Conservation: Review of Heritage Legistation and Policy. *Malaysian Construction Research Journal (MCRJ)*, 203.

Zalina, Y. (2003, March 10). IPTS, IPTA: Perang Berebut Pelajar Kini Bermula. *Utusan Malaysia*.

Zangirolami-Raimundo, J., Echeimberg, J. de O., & Leone, C. (2018). Research methodology topics: Cross-sectional studies. *Journal of Human Growth and Development*, 28(3), 356–360. 10.7322/jhgd.152198

Zeng, C., Aboagye, E. M., Li, H., & Che, S. (2023). Comments and Recommendations on Sponge City-China's solutions to prevent flooding risks. *Heliyon,9*(e12745):1-11. doi: . 2022.e1274510.1016/j.heliyon

Zeng, L. (2022). *Research on Dynamic Simulation Technology of Urban 3D Landscape Based on Virtual Reality APP Platform*. In 2022 2nd International Conference on Networking, Communications and Information Technology (NetCIT), Manchester, United Kingdom. 10.1109/NetCIT57419.2022.00107

Zhang, J., & Li, D. (2010). Research on 4D virtual construction and dynamic management system based on BIM. In *Proceedings of the International Conference on Computing in Civil and Building Engineering, ICCBE* (pp. 78- 83). IEEE.

Zhang, H. Q. (2016). Study on the Enrollment Ratio of Higher Vocational Majors Based on the Adjustment of Industrial Structure in Shandong Province. *China Vocational and Technical Education*, (35), 14–19.

Zhang, H. Q. (2017). Study on the enrollment ratio of higher vocational majors based on the adjustment of industrial structure in Shandong Province. *China Vocational and Technical Education*, (35), 14–19.

Zhang, H., Kou, G., & Peng, Y. (2019). Soft Consensus Cost Models for Group Decision Making and Economic Interpretations. *European Journal of Operational Research*, 277(3), 964–980. 10.1016/j.ejor.2019.03.009

Zhang, M., & Wang, C. Y. (2020). *Research on the transformation of training mode of Engineering cost professionals under the background of new engineering*. International Public Relations.

Zhang, Y., & Li, J. (2022). Research on Hu bed design based on user research and hierarchical analysis. *Packaging Engineering*, 43(S1), 183–188. 10.19554/j.cnki.1001-3563.2022.S1.031

Zhang, Y., Zhao, L., Zhao, H., & Gao, X. (2021). Urban development trend analysis and spatial simulation based on time series remote sensing data: A case study of Jinan, China. *PLoS One*, 16(10), e0257776. 10.1371/journal.pone.025777634618811

Zhao, D. M., & Li, G. X. (2023). Prefabricated building supply chain risk management based on AHP-DEMATEL. *Journal of Luohe Vocational and Technical College*, 22(6), 32–36.

Zhao, D., Mccoy, A. P., Bulbul, T., Fiori, C., & Nikkhoo, P. (2015). Building Collaborative Construction Skills through BIM-integrated Learning Environment. *International Journal of Construction Education and Research*, 11(2), 97–120. 10.1080/15578771.2014.986251

Zhao, H. H., & Qin, W. (2023). Research on the training of prefabricated construction talents based on the integration of industry and education. *Shanxi Architecture*, 49(17), 196–198. 10.13719/j.cnki.1009-6825.2023.17.052

Zhao, J., & Luo, Y. C. (2019). Some thoughts and preliminary practice on promoting the pilot work of 1+X certificate system. *China Vocational and Technical Education*, (27), 5–8.

Zhao, S., Li, Z., & Li, W. (2005). A modified method of ecological footprint calculation and its application. *Ecological Modelling*, 185(1), 65–75. 10.1016/j.ecolmodel.2004.11.016

Zhao, Y., Cao, C., & Liu, Z. (2022). A framework for prefabricated component hoisting management systems based on digital twin technology. *Buildings*, 12(3), 276. 10.3390/buildings12030276

Zhao, Z. (2022). Analysis of Museum Cultural Creation from the Perspective of Cultural Industry. *Scientific Programming*, 2022, e8138574. 10.1155/2022/8138574

Zhau, S., Da, L., Tang, Z., Fang, H., Song, K., & Fang, J. (2006). Ecological Consequences of Rapid Urban Expansion: Shanghai, China *Frontiers in Ecological Environ. The Ecological Society of America*, 4(7), 341–346.

Zheng, Y. (2020). Design of Museum Cultural and Creative Product Development Based on Educational Function and Connotation: The Case of "Dinosaur Exhibition in China". *Natural Science Museum Research, 5* (2), 47-57+96+99. 10.19628/j.cnki.jnsmr.2020.02.007

Zheng, W., Ye, Y., & Zang, H. (2022). Application of BIM technology in prefabricated buildings based on virtual reality. *Computational Intelligence and Neuroscience*, 2022, 9756255. 10.1155/2022/975625536225548

Zhou, Q. (2014). A Review of Sustainable Urban Drainage Systems Considering the Climate Change and Urbanization Impacts. *Water (Basel)*, 6(4), 976–992. 10.3390/w6040976

Zhou, W., Georgakis, P., Heesom, D., & Feng, X. (2012). Model-based groupware solution for distributed real-time collaborative 4D planning via teamwork. *Journal of Computing in Civil Engineering*, 26(5), 597–611. 10.1061/(ASCE)CP.1943-5487.0000153

Zhuang, X. Z. (2014). Transformation of ordinary undergraduate colleges: Why and What and How to change. *China Vocational and Technical Education*, (21), 84–89.

Zhu, H., Hong, J., Shen, G. Q., Mao, C., Zhang, H., & Li, Z. (2018). The exploration of the life-cycle energy saving potential for using prefabrication in residential buildings in china. *Energy and Building*, 166, 561–565. 10.1016/j.enbuild.2017.12.045

About the Contributors

Ungku Norani Sonet is an architecture and urbanism senior lecturer at Taylor's University. After several years of architectural practise, she opted to pursue postgraduate studies and finished her PhD research. Her primary area of interest in architecture education is the responsive architectural design approach to local context and identity. In addition to the responsive design approach, she is particularly interested in doing research on public participation exercises in development frameworks. Her current research delves deeper to increase public involvement toward a holistic approach of sustainable built environment.

***

Ahmad Faiz Azizi received his bachelor's degree and Ph.D. in Quantity Surveying from Universiti Teknologi Malaysia (UTM) in Skudai, Johor, Malaysia in 2016 and 2021. Currently, he works as a senior lecturer and researcher at UTM, where he is affiliated with the Faculty of Built Environment and Surveying. His research interests lie in the areas of augmented reality, education, building information modelling in construction, and quantity surveying. Currently, Ahmad Faiz Azizi Ahmad Fauzi is a senior lecturer and researcher at UTM, where he is affiliated with the Faculty of Built Environment and Surveying.

Khairool Aizat Ahmad Jamal, is an academician deeply committed to digitalization and sustainability, earned his Bachelor of Architecture (Hons.) from Universiti Teknologi Malaysia and an M.Sc in Integrated Construction Project Management from Universiti Teknologi MARA. Holding registrations as a Graduate Architect (LAM) and Professional Technologist (MBOT), he boasts a diverse background in architecture and construction cultivated through experiences both locally and internationally. As an educator, he imparts valuable insights to students across various disciplines within the built environment. Currently pursuing a Ph.D in Built Environment, his passion for BIM and Digital Architecture propels him to make noteworthy contributions, encompassing presentations, national and international awards, research publications and grants.

Shirley Chin Ai Ling is an experienced Senior Lecturer and Stream Coordinator with a strong background in the field of cost management and professional practice. With a postgraduate degree in Teaching & Learning, as well as a Master's degree in Project Management, she possesses a comprehensive understanding of the subject matter. Sr Shirley has excelled in her teaching and administrative responsibilities at the School of Architecture, Building & Design, where she leads the Bachelor of Quantity Surveying (Honours) program. Most of her subjects taught are related to measurement and management, including subjects such as measurement, construction economics, and quantity surveyor's studio.

Chang Yun Fah received the B.Arts in mathematics, M.Sc. and Ph.D. degrees in applied statistics from University of Malaya, Malaysia, in 1998, 2002, and 2012, respectively. He joined the School of Accounting and Finance, Taylor's University, Malaysia in 2019. He is appointed as Adjunct Professor at the Faculty of Foreign Languages, Dezhou University, China since 2020 and Senior Research Associate at Universiti Tunku Abdul Rahman, Malaysia in 2022. He has published over 70 research articles, book chapters, and books. His research interests include statistics, financial mathematics, image analysis, medical analysis, social statistics, and vocational education. Dr. Chang is the director of Centre for Malaysian Chinese Studies, Panel of Assessor appointed by Malaysia Qualification Agency, and elected member of International Statistics Institute since 2012, 2016, and 2016, respectively. He serves as the member of editorial board, associate editor, guest editor and reviewer for various local and international journals.

Roshnie A. Doon, holds a Ph.D. in Economic Development Policy, MSc. Economics, and BSc. Economics from the University of the West Indies St. Augustine, Trinidad. She has completed a Ph.D. Master Class in International Business from the Henley School of Business at the University of Reading, U.K., and is currently a research affiliate of the Global Labor Organization (GLO), Essen Germany, and an executive committee member of the Caribbean Academy of Sciences Trinidad and Tobago Chapter (CAS-TT). Her academic interests are focused on Applied and Empirical Economics, in the areas of Labour, Gender, and Education Economics. Here she has published on a wide range of areas, which includes Higher Education Instruction, Economic and Social Implications of COVID-19, Migration, Returns to Schooling, Educational Mismatch, STEM Education, Wage Inequality, Gender Wage Gaps, and Economic Development Policy. Apart from being an academic editor, she has published her research in a wide array of journal and book publications, while sharing her research work through guest lectures, and conference presentations domestically, regionally, and internationally in Trinidad, Jamaica, Barbados, Guyana,

About the Contributors

Siti Norlizaiha Harun has PhD in Building Conservation from Universiti Sains Malaysia, Pulau Pinang. She is currently lecturing at the Universiti Teknologi MARA(UiTM) Puncak Alam, Selangor. She was seconded as the Deputy Commissioner of Heritage at the Department of National Heritage Malaysia from 2012 - 2013. She is a registered building conservator under the Department of National Heritage, Malaysia. She has published numerous researches in books and journals; and has wide experience in historic building restoration project.

Huangfu Dongmei is a PhD student of Taylor's University, Malaysia now. In 2016, she was hired in the Investment Management Center of SINOHYDRO BUREAU 14 CO.,LTD. During that period, she participated in the preparation of tender documents for PPP projects of several infrastructure projects. She is National Class1 Cost Engineer and National Class1 Constructor of China. And since 2021, she was hired as a lecturer of Dianchi College of Yunnan University, Kunming, China, teaching for the quantity surveying major. She writes some papers and books on issues of quantity surveying talents and management, PPP, BIM and so on.

Filzani Illia is a Senior Lecturer in Architecture. She is currently working as an academician with more than 5 years of experience in research and practice in the fields of Landscape Architecture, Urban Design and Design Management. She holds a bachelor degree from International Islamic University Malaysia (IIUM), an MSc in Urban Development and Management (UiTM) and a PhD in Specialism of Built Environment from Universiti Teknologi Mara (UiTM) Malaysia. She has also extensive experience in research and journal publications. She was awarded FRGS grant worth RM80,200 for the year of 2014-2016.

Habizah Ilmi has a degree in Quantity Surveying and a master in Integrated Project Management and Construction. She is a member of Royal Institution of Surveyors Malaysia (RISM) and Royal Institution of Chartered Surveyors (RICS) and also a registered Consultant Quantity Surveyor with Board of Quantity Surveyors Malaysia (BQSM). She has 13 years experiences in the construction industry before joining the education industry and has been teaching quantity surveying program for more than 13 years.

Kam Kenn Jhun has a Degree in Building Technology, a Master's Degree in Building Technology, and a Ph.D. in Building Quality and Assessment. He is a Chartered Surveyor from the Royal Institution of Chartered Surveyors (RICS), a registered Professional Technologist with the Malaysian Board of Technologists (MBOT) and a registered Engineering Technologist with the Board of Engineers Malaysia (BEM). He has experience in construction for 5 years and has completed 2 high rises projects from IJM before joining the education industry. He has almost 10 years of experience in academia. He also held a management position such as Associate Dean at his current institution. Most of the subjects that he has been teaching are construction-related subjects such as Construction Technology, Building Services, Research Methodology and Dissertation.

Myzatul Aishah Kamarazaly is the Programme Director for the Bachelor of Quantity Surveying (Hons.) Programme in the School of Architecture, Building & Design at Taylor's University, Lakeside Campus, Malaysia. She brings on board more than 15 years of academic leadership experience. Myzatul received her Doctoral of Philosophy (in Facilities Management) and Master of Philosophy (in Construction) from Massey University, New Zealand, Bachelor of Quantity Surveying (Hons) and Diploma in Quantity Surveying from UiTM, Malaysia. She is an active Associate Member of the Malaysian Association of Facilities Management (MAFM), a member of the Board of Quantity Surveyor Malaysia (BQSM) and Royal Institution of Surveyor Malaysia (RISM). Myzatul has extensive research publications and involvement as International Reviewers and Editorial Board for Journal Publication. Her area of specialization is on Facilities Management, BIM, QS education and profession as well as built-environment related-studies.

Xiaonuo Kang has an undergraduate degree in engineering costing and a master's degree in construction management from the University of Melbourne. She is currently teaching at the School of Civil Engineering, University of Light and Chemical Technology, Zigong in the subject of cost engineering, Sichuan Province, China. Obtained the second-level cost engineer certificate for installation engineering. In addition to teaching work, she is engaged in fire protection engineering site-review,cost management and acceptance work. The author also engaged in prefabricated building site management, and process management of prefabricated components manufacturing etc.

Koh Jing Hao was born in Melaka, Malaysia. He completed his architectural education and graduated with a bachelor's degree at Universiti Sains Malaysia, Penang. He obtained a master's degree in Computer Aided Architectural Design at the University of the West of England, United Kingdom. With the exposure and experience in heritage architectural conservation projects during his study in Penang, Jing Hao developed an interest in heritage architecture which led him to pursue a doctoral study in this domain. He is a lecturer in the School of Architecture, Building & Design at Taylor's University, Malaysia. He has also supervised measured drawing projects in George Town, Melaka, Johor Bahru and Sungai Buloh, involving heritage temples, clan houses, traditional shophouses and traditional dwellings.

Leong Boon Tik is an active researcher who has good connection with industry players. He often listen to practitioners on their challenges to identify research gaps. He published on green and sustainable technologies in construction, he also study the performance of green developers and green contractors. Currently he look into PPVC and BIM in Malaysia context.

Lim Sok Li is currently working as an Associate Professor in Taylor's University, Malaysia. She obtained her PhD in Statistics from Universiti Sains Malaysia. Her main research interest is in Statistical Quality Control, mainly in proposing charts with better sensitivities and lower cost of implementation.

Lim Tze Shwan has a degree in Quantity Surveying and a master degree in Construction Contract Management. She is a member of Royal Institution of Surveyors Malaysia (RISM) and registered Professional Quantity Surveyor with Board of Quantity Surveyors Malaysia (BQSM). She has experiences in the construction industry specializing quantity surveying for 6 years before joining an education industry. Most of the subjects that she had been teaching are quantity surveying subjects such as Measurement, Software Application for Quantity Surveying, Price Analysis, Construction Economic, Industrial Training and Integrated Project.

Nur Nilam Sari is a PhD student focusing on construction safety at Faculty of Built Environment and Surveing in Universiti Teknologi Malaysia. Follow her on this exciting PhD journey as she strive to contribute meaningfully to the improvement of Occupational Safety and Health in the construction industry. Together, let's build a safer and healthier future for construction workers.

Norhazren Izatie Mohd is a senior lecturer at the Faculty of Built Environment and Surveying, Universiti Teknologi Malaysia (UTM). She has expertise in construction safety, information technology in construction, and game development. Her research interests include occupational safety and health in the construction industry, building information modelling, and virtual reality in design. She has published several articles and papers on these areas.

Fara Diva Mustapa is registered a Consultant Quantity Surveyor (CQS) with the Board of Quantity Surveyors, Malaysia (BQSM) and a Fellow member of the Royal Institution of Surveyors Malaysia (RISM) and the Royal Institution of Chartered Surveyors (RICS). She attained her first degree in quantity surveying from Universiti Teknologi Malaysia in 2002, and received the Lawrence Chin Memorial award from the Royal Institution of Surveying Malaysia (then known as Institution of Surveyors) Malaysia (ISM) in 2002 for her outstanding academic performance in quantity surveying studies. She joined the Malaysian construction industry as a quantity surveyor, managing a design and build project worth RM11.5 million. She pursued her Master degree in Construction Economics and Management with Bartlett School of Architecture, University College London (UCL) (UK) in 2004 after receiving a scholarship. She pursued her PhD at Loughborough University, UK, and attained her PhD, specialising in labour economics, in 2014. She is currently a Senior Lecturer at Department of Quantity Surveying, Faculty of Built Environment, Universiti Teknologi Malaysia, where she teaches construction economics-related courses. She is the chairwoman of Royal Institution of Surveyors. Johor branch, session 2021/22 and also an Advisory Panel for Quantity Surveying programme at SEGi University, Kota Damansara and a trainee for BQSM Quantity Surveying Accreditation Council (QSAC). She is also active in research and publications where she won Gold Medal award in IRIISE 2018 and NALI 2018 for her on-going research's outcomes. Currently she have several on-going research grants from MOHE FRGS (worth RM75K) and NAPREC (RM125k) and recently completed her industry grant with TNB on cost benefit analysis for their sub-stations (worth RM150K).

Shahrul Yani Said is an Associate Professor in Heritage Conservation. She started her journey in academics with a BSc Architectural Studies (Hons) and a Bachelor Architecture (Hons) from International Islamic University Malaysia, studying heritage design. She continues her pursuit of heritage conservation in the United Kingdom and received an MA in Conservation Studies (Historic Buildings) in 2008 from the University of York. Her investigation into urban heritage and townscape led to her PhD journey from Oxford Brookes University, where she received her Doctor of Philosophy in 2012. She is an inspiring researcher and produces and reviews many journal articles in the field of heritage conservation. Currently an ICOMOS member, her interest in heritage area nomination and digital documentation of heritage is evident in the number of active research grants received.

Soon Lam Tatt received his Doctor of Philosophy, MSc (Quantity Surveying) and BSc. (Hons) Housing Building and Planning (Quantity Surveying) from Universiti Sains Malaysia. He has more than 15 years of industrial and academic working experience in the field of quantity surveying. He is currently working as a senior lecturer at Taylor's University Malaysia. Prior to that, he held several management positions such as Associate Dean and Stream Coordinator at current institutions. He specializes in Building Information Modelling (BIM) and Real Estate Development. Mr Soon is currently a chartered member of the Royal Institute of Chartered Surveyors and a registered Quantity Surveyor under the Board of Quantity Surveyors Malaysia.

Hamizah Liyana Tajul Ariffin is a professional associated with the construction and academic fields. She is a Programme Coordinator in the Department of Quantity Surveying at Universiti Teknologi Malaysia (UTM) and has been cited in various research works related to construction conflict, occupational safety, and health in the construction industry. Her expertise and main research interests include, but are not limited to, construction contract management, Building Information Modelling, safety in construction and dispute resolution in the construction industry.

Tan Wei Vern is a graduated Bachelor of Quantity Surveying student from Taylor's University.

About the Contributors

Omar Vargas-González is a professor, research assistant and former Head of Computer Systems Department at Tecnologico Nacional de Mexico Campus Ciudad Guzman, professor at Telematic Engineering at Centro Universitario del Sur Universidad de Guadalajara with a master degree in Computer Systems. Has been trained in Innovation and Multidisciplinary Entrepreneurship at Arizona State University (2018) and a Generation of Ecosystems of Innovation, Entrepreneurship and Sustainability for Jalisco course by Harvard University T.H. Chan School of Health. At present conduct research on diverse fields such as Entrepreneurship, Economy, Statistics, Mathematics and Information and Computer Sciences. Has colaborated in the publication of over 30 scientific articles and conducted diverse Innovation and Technological Development projects.

Chiam Winnie is a Bachelor of Quantity Surveying (Honours) graduate student the Taylor's University Malaysia. Her research interests include facility management, BIM in built environment, green building and construction technology.

Siti Norzaini Zainal Abidin is a Senior Lecturer at Taylor's University. Her contribution on the paper 'Strategic Daylight Visualization for the Green Building Process: Integrating Optimization Into An Airport Terminal Design With Skylights' has obtained her the best paper student paper award for ConVR in December 2016 at Regal Kowloon, Hong Kong organised by The Hong Kong University of Science and Technology.

Index

www.ingramcontent.com/pod-product-compliance
Lightning Source LLC
LaVergne TN
LVHW062134300926
849161LV00010B/2527
* 9 7 9 8 3 6 9 3 1 6 5 0 4 *